From Spectroscopic Insights to Structural Wonders: A Theme Issue Dedicated to Professor Jaan Laane

From Spectroscopic Insights to Structural Wonders: A Theme Issue Dedicated to Professor Jaan Laane

Guest Editors

Rui Fausto
Gulce Ogruc Ildiz

MDPI

Basel • Beijing • Wuhan • Barcelona • Belgrade • Novi Sad • Cluj • Manchester

Guest Editors

Rui Fausto
Department of Chemistry
University of Coimbra
Coimbra
Portugal

Gulce Ogruc Ildiz
Department of Physics
Istanbul Kultur University
Istanbul
Türkiye

Editorial Office
MDPI AG
Grosspeteranlage 5
4052 Basel, Switzerland

This is a reprint of the Special Issue, published open access by the journal *Molecules* (ISSN 1420-3049), freely accessible at: https://www.mdpi.com/journal/molecules/special_issues/Professor_Jaan_Laane.

For citation purposes, cite each article independently as indicated on the article page online and as indicated below:

Lastname, A.A.; Lastname, B.B. Article Title. *Journal Name* **Year**, *Volume Number*, Page Range.

ISBN 978-3-7258-4621-4 (Hbk)
ISBN 978-3-7258-4622-1 (PDF)
https://doi.org/10.3390/books978-3-7258-4622-1

Contents

About the Editors

Rui Fausto

Rui Fausto is a professor and the Coordinator of the Computational Chemistry, Spectroscopy, and Thermodynamics research group at the University of Coimbra (Portugal), and a Professor and the ERA-Chair holder of Spectroscopy@IKU: Manipulating and Characterizing Molecular Architectures: From Isolated Molecules to Molecular Crystals at the Istanbul Kultur University (Türkiye). He is the President of the Steering Committee of EUCMOS, as well as the Editor-in-Chief of *Photochemistry and Spectroscopy* (SciLight, Australia), and a member of the editorial boards of several scientific journals. Throughout his career, Rui Fausto has held prominent administrative and scientific roles in academia, including the presidency of the Institute for Interdisciplinary Research, the presidency of the Coimbra Chemistry Research Centre, and the vice-presidency of the Scientific and Directive Boards of the Faculty of Sciences and Technology of the University of Coimbra. He has been awarded the Excellence Prize in Sciences by the Portuguese Science Foundation twice. Rui Fausto has published or edited nearly 50 scientific books and is the author of over 500 research articles, mostly focusing the photochemistry of organic molecules and matrix-isolated reactive intermediates. He is a pioneer in the use of vibrational excitation to induce chemical reactivity in organic molecules under matrix-isolation conditions and has also undertaken extensive and influential work on the photochemistry of reactive intermediates and quantum mechanical tunneling. Besides science, his interests also extend to music, painting, and human rights advocacy.

Gulce Ogruc Ildiz

Gulce Ogruc Ildiz is a Professor of Physics and Vice-rector of Istanbul Kultur University, Türkiye. After graduating from Kültür College, she received her undergraduate degree from Istanbul Technical University in Physics Engineering and her master's degrees from Istanbul Technical University Nuclear Energy Engineering and Istanbul Kültür University Business Administration (MBA) departments. She received her doctorate degrees from Istanbul University Department of Atomic and Molecular Physics and Istanbul Kültür University Department of Civil Engineering Project Management Program. She has been actively involved in academy administration, where she have been acting as the Executive Secretary of Istanbul Kültür University Technology Project Support Unit (TPDB), the coordinator of the Scientific Research Projects (BAP) office, and a member of the Faculty Boards of the IKU Faculties of Engineering, Education and Arts and Sciences. She is the editor of several international journals in the field of spectroscopy and the current coordinator of the ERA-Chair Spectroscopy@IKU: Manipulating and Characterizing Molecular Architectures: From Isolated Molecules to Molecular Crystals. Gulce Ildiz's main scientific interests focus on the investigation of biological phenomena using Raman spectroscopy and chemometrics, matrix isolation, infrared spectroscopy, and computational chemistry.

Preface

This Special Issue of *Molecules* is dedicated to honoring Professor Jaan Laane for his remarkable and enduring contributions to the fields of spectroscopy and molecular structure. Over the course of his illustrious career, Professor Laane established himself as a leading figure whose pioneering research and unwavering dedication have profoundly shaped these disciplines. In keeping with the spirit of this Special Issue, the present Editorial/Preface serves not only as an academic tribute but also as a heartfelt reflection on the individual behind the scientist, illuminating the person whose intellect and character have left an indelible mark on colleagues, students, and the broader scientific community.

The first author of this piece, Lloyd Colegrove, began his scientific journey under Professor Laane's mentorship, joining his research group as an undergraduate in 1982 and earning his Ph.D. in 1989. This formative period, spanning nearly a decade, offered a rare opportunity to work closely with Professor Laane, experiencing first-hand both his scientific brilliance and his personal generosity. Following his academic tenure, Dr. Colegrove devoted nearly 29 years to The Dow Chemical Company, where he held leadership roles in research and manufacturing, culminating in a directorship overseeing advanced applications in data analytics—encompassing both real-time systems and artificial intelligence—as well as spearheading efforts in fundamental problem solving.

The other authors became acquainted with Professor Laane later in their careers, through shared endeavors with the *Journal of Molecular Structure* and the European Congress on Molecular Spectroscopy. Even with differing points of entry, they were quick to recognize his exceptional intellect, profound kindness, and unwavering commitment to advancing scientific knowledge—qualities that soon fostered not only professional respect but genuine friendship.

This Editorial/Preface thus reflects the diverse yet deeply personal connections each author has shared with Professor Laane. While their experiences may differ, the respect, admiration, and gratitude they hold for him are profoundly united.

Lloyd Colegrove: *The Lasting Impact of Jaan's Mentorship*

I first met Professor Jaan Laane during my junior year at Texas A&M University in the "Integrated Laboratory" course—a rigorous, three-afternoon-per-week, four-credit fusion of physical chemistry and inorganic chemistry laboratory work. Only later did I learn that this exceptionally demanding course was, in part, his brainchild. It remains, to this day, the most challenging class I encountered in my undergraduate chemistry studies.

Who was this remarkable individual? As it turns out, Jaan's story was as compelling as his scientific achievements. He was a child refugee of World War II, who, along with his parents, brother, and sister, fled Estonia in the wake of the German retreat and the advancing Russian Army. The family ultimately immigrated to the United States, where Jaan grew up in Polo, Illinois. He pursued his undergraduate studies at the University of Illinois and later earned his Ph.D. under the mentorship of Professor R.C. Lord at the Massachusetts Institute of Technology. After a year on the faculty at Tufts University, Jaan moved to central Texas to begin what would become a distinguished career as a professor and spectroscopist at Texas A&M University. An unlikely "Texan" by origin, this European-bred academic not only embraced his new home but also persuaded his equally accomplished spouse, Dr. Tiiu Virkhaus Laane, to join him in Texas, where together they raised two daughters.

Jaan's research focus was primarily discerning the conformations of small ring molecules through spectroscopic and computational techniques. Although Jaan's interests were not confined to far-infrared and Raman spectroscopy, the bulk of his early research relied on these techniques. Jaan's research also involved two other significant components—the application of computational techniques, empirical or semi-empirical quantum mechanical approaches, and what accidentally became my specialty, inorganic synthetic techniques. In the first half of Jaan's career, when joining his group, one had the choice of studying compounds purchased from the Aldrich Chemical Products catalog or synthesizing compounds that did not exist in nature, or in the Aldrich catalog. I chose this latter path because for some strange reason I like to do hard things. The act of synthesizing molecules that even nature does not bother to make intrigued me. This experience later resonated throughout my professional journey, particularly as I endeavored to foster cultural change within a large chemical corporation—both endeavors demanding perseverance, patience, and, truthfully, more than a little frustration. The scientific training and mindset I developed under Jaan's mentorship during those formative years remained a constant influence throughout my career.

The distinctive nature of Jaan's research group in the 1970s and 1980s—blending spectroscopy, synthesis, and computational methods—offered me a unique perspective that encouraged looking beyond immediate challenges to seek collaboration and broader solutions. The experience of approaching problems from multiple angles, a hallmark of Jaan's mentorship, became a fundamental part of my professional philosophy and greatly influenced my contributions during my later career at The Dow Chemical Company. While spectroscopy was the workhorse of the group, the computational experiences, and molecular dynamics studies we had to perform as part of the analysis opened my eyes to what data science could potentially do. What I ended up making the bulk of my life's work at Dow was based on this very different yet loosely mathematically connected experience. My synthetic experience gave me discipline and perseverance and the realization that the network you build around you will make you a stronger contributor to any endeavor. Oddly, I never relied on my spectroscopic background at Dow. They hired scientists to do that specifically, and I was not one of them.

I believe I may have been the last true "hardcore" synthetic lab rat in Jaan's group, graduating roughly at the midpoint of his distinguished career. After my time, most of the small molecules of interest were either commercially available and his research focus shifted more deeply into advancing new spectroscopic and computational techniques.

Back then, Jaan would present new students with a "bucket list" of dream projects and, true to his philosophy, let each individual's talents, interests, and perseverance shape the direction of their graduate research. Since I was already synthesizing well-documented four- and five-membered silicon-containing ring systems for the group, Jaan casually suggested I try to develop a synthetic pathway to a particular family of five-membered rings containing two silicon atoms. In my dogged pursuit of these challenging molecules, I was blissfully unaware that I could have opted for a more straightforward project when the time came to formalize my research proposal.

In the mid-1980s, there was no substitute for long hours in the library. Weeks were spent combing through journals and dissertations for any clue. The old adage, "six months in the lab can save you an hour in the library", took on a very real meaning for me. For someone who likely had some form of undiagnosed attention deficit disorder, this was sheer torture. Most of my Laane-group colleagues would have avoided this task at all costs—we were spectroscopists, after all, not hardcore synthetic chemists.

After months of frustrating searches, I finally stumbled across an obscure German journal of synthetic inorganic chemistry tucked away in the Evans Library at Texas A&M University. There, a paper—written entirely in German—outlined a route that could get me about 80% of the way to my goal. Though it did not lead directly to the exact compounds I sought, it offered enough of a starting point to chart a synthetic path forward. Learning that my target molecules had never been synthesized or isolated before was at once daunting and exhilarating. I turned to my organic and inorganic chemistry friends for advice, especially about the crucial step of converting a pentane ring into a pentene ring. It was clear this would be a trial-and-error process—the design of experiments was not yet in my vocabulary—and yield losses in the synthetic and isolation steps were constant hurdles. Each additional silicon atom increased the boiling point and complexity of separations, making every step a laborious battle.

It often took three to four months to move from initial synthesis to final isolation of the desired product, and that was once I even figured out what I was doing. From a broader view of my time in Jaan's group, a disproportionate amount of my efforts went into the synthetic work—never mind the coursework, qualifying exams, spectroscopy, and computational analysis that came afterward. The lab became my second home, and I loved every minute of it. In the end, I was proud to have successfully synthesized the world's first isolated samples of several novel molecules and their isotopic variants, studied their molecular dynamics, compared computational predictions to observed vibrational spectra, and completed projects that may have exceeded Jaan's original expectations.

One moment stands out vividly. I recall a Sunday afternoon when I obtained a particularly significant far-infrared spectrum—a textbook signature of the ring-bending mode of 1,3-disilacyclopent-4-ene. I called Jaan from the lab, eager to share the discovery. To my surprise, he drove to campus to see the spectrum in person. His excitement and genuine joy were unforgettable, and it remains one of the most meaningful memories of my scientific career. The foundation that Jaan laid—the trust he placed in his students to tackle difficult problems, the freedom to pursue new paths, and his encouragement at every step—shaped not just my time at Texas A&M but the trajectory of my entire professional life. For that, I remain deeply grateful.

My entire experience with Jaan taught me to be dogged in my pursuit of the knowledge I needed to do the job right, to look at things from more than one perspective, and to question convention. I learned a great deal about computational dynamics and measurement science without having to become an expert computational chemist or inorganic synthetic chemist, and when I walked into my company as a fledgling researcher in R&D, I brought my very broad and well-rounded experience with me that culminated in my many career successes at Dow. I am often mistaken for a data scientist or an engineer. I am neither. I am of the R.C. Lord and Jaan Laane family tree and call myself proudly a chemical physicist by virtue of the output of my spectroscopic, computational, synthetic interests and academic preparation, and I will forever be grateful to Jaan for building that research framework, and then guiding, encouraging, and mentoring me into my career. That I was able to take this experience into my working life to such great effect was a surprise to me, but in the bigger picture, I believe that was Jaan's expectation to begin with.

Gulce O. Ildiz and Rui Fausto: *In the Footsteps of Jaan Laane—Lessons in Science and Humanity*

When we first became involved with the *Journal of Molecular Structure* and the European Congress on Molecular Spectroscopy (EUCMOS), Professor Jaan Laane was already a deeply respected actor in both arenas. As Editor of the journal and a distinguished member of the EUCMOS Steering Committee, he had long been instrumental in shaping the direction and spirit of these scientific communities. From the outset, we had the privilege of observing the thoughtful, meticulous, and visionary way in which he carried out his responsibilities. With his active participation and guidance, the journal and the congress flourished, offering platforms that upheld the highest standards of scientific quality, collaboration, and integrity. Through Jaan's example, we learned much about leadership: how it can be both rigorous and compassionate, ambitious yet inclusive.

Jaan's influence, however, was never confined to administrative leadership. As a scientist, he made seminal contributions that profoundly impacted the fields of spectroscopy and molecular structure. His research laboratory became a hub for the synthesis of new organosilicon and other cyclic compounds, a daunting experimental frontier at the time. Beyond synthesis, Jaan's group fully analyzed the vibrational spectra of these new molecules, delving into the fine structure of molecular vibrations and the subtleties of conformational dynamics. His investigations into the potential energy surfaces for conformationally important vibrations added crucial insights to our understanding of molecular behavior, setting benchmarks that many of us would later build upon. Moreover, Jaan was also at the vanguard of integrating computational methods into experimental spectroscopy, a relatively novel approach at the time. He wrote pioneering computer programs for calculating kinetic and potential energy functions for one-, two-, and three-dimensional vibrational problems, offering tools that dramatically expanded what was possible in the study of molecular motion. His "Laane's Tables," published in 1970, and his programs for modeling periodic potential functions for internal rotations and pseudorotations quickly became essential resources. These works were not just theoretical exercises; they were adopted by dozens of laboratories around the world and became critical instruments in advancing the field. It is no exaggeration to say that many successful studies in molecular spectroscopy owe a debt, whether direct or indirect, to Jaan's computational innovations.

What makes Jaan's career even more remarkable is how he balanced his scientific brilliance with an extraordinary sense of humanity. He has always been an approachable, supportive figure, particularly for young researchers finding their way in a complex and competitive scientific world. It was never enough for him simply to advance his own research; he was committed to lifting up others along the way. His generosity with his time, advice, and encouragement has left a lasting impression on those who had the privilege to work with him or learn from him. In every encounter—whether a formal meeting, a scientific collaboration, or a casual conversation at a conference—he demonstrated a rare combination of wisdom, humility, and kindness.

Another of Jaan's lasting legacies is the atmosphere of collaboration he fostered within the communities he touched. In his view, science was not a zero-sum game but a collective endeavor, where shared knowledge led to greater understanding for all. Under his influence, both the *Journal of Molecular Structure* and EUCMOS maintained cultures of openness, intellectual rigor, and collegial respect. He led by example, showing that it was possible to be exacting in one's standards while also being deeply encouraging of diverse perspectives and approaches. The communities he helped to build continue to benefit from this spirit today.

In reflecting on Professor Jaan Laane's career, it is clear that his impact has been profound and multifaceted. As a researcher, he expanded the frontiers of molecular spectroscopy and computational chemistry. As a leader, he raised the standards and aspirations of the institutions he served. Additionally, as a mentor and colleague, he nurtured generations of scientists, offering not only knowledge but also inspiration and unwavering support. For those of us fortunate enough to work alongside him, to learn from him, and to be influenced by his example, our gratitude runs deep. Jaan's contributions have shaped not only the trajectory of our fields but also the way we approach our own work and relationships. It is an honor to pay tribute to such an extraordinary scientist, leader, and human being.

Brief biographic note: Jaan Laane was born in Estonia on June 20, 1942. When he was two years old, he and his family fled the Soviet regime and were eventually placed in a refugee camp near Augsburg, Germany, in 1945. In 1949 the family immigrated to the USA, settling in the small town of

Polo, Illinois. After graduating from high school, where he was class valedictorian, Laane attended the University of Illinois (Urbana) as an Alfred P. Sloan National Scholar and as a James Scholar. He graduated with Highest Distinction in Chemistry in 1964 and received the Kendall Award as the top senior chemistry major. As an undergraduate, he worked for three years under Professor Theodore L. Brown, carrying out experiments in infrared spectroscopy and organotin synthesis. Following his first of three summers working at Los Alamos Scientific Laboratory (with Llewelyn Jones), he began graduate work at MIT as a National Science Foundation and Woodrow Wilson Fellow. Working in the laboratory of Professor Richard C. Lord, and under the guidance of Professor Dietmar Seyferth for organosilicon synthesis, Laane synthesized silacyclobutane and several other new compounds and studied their vibrational spectra. He wrote an original computer program to accurately calculate the quantum energy levels for ring-puckering vibrations with quartic/quadradic potential functions. He received his Ph.D. in 1967, along with the Kodak Award as the top graduate student in the chemistry department at MIT.

Following a year at Tufts University, Laane moved to Texas A&M University, where he was soon promoted to Associate Professor (1972) and Full Professor (1976) of Chemistry. He then became Professor Emeritus of Chemistry and of Physics and Astronomy. His research there focused on vibrational spectroscopy and the determination of vibrational potential energy surfaces in both ground and excited electronic states. He has been at the forefront of advancing both the fundamental theoretical understanding of molecular vibrations and structures, as well as the experimental methodology in these areas, and has made significant contributions to the understanding of singlet and triplet electronic excited states, the bonding of nitrogen–oxygen compounds, and Raman difference spectroscopy. He has more than 340 publications, including four books and over a dozen book chapters. He was editor of the 2009 book *Frontiers of Molecular Spectroscopy*, a 700-page text with 19 chapters by leading scientists, published by Elsevier, and an updated book in 2019, *Frontiers and Advances of Molecular Spectroscopy*. In addition to this, he has delivered more than 600 presentations, including plenary and invited lectures at chemistry and spectroscopy conferences in the United States and around the world. In 2005, he received the prestigious E. R. Lippincott Award, sponsored jointly by the Optical Society (OSA), Society for Applied Spectroscopy (SAS), and the Coblentz Society. This was presented for "significant contributions to vibrational spectroscopy as judged by his influence on other scientists." He was cited for the "innovative use of vibrational spectroscopy to determine molecular structures and to unravel complex intramolecular dynamics." In 2017, he received the top award from the Society for Applied Spectroscopy and the Honorary Member Award for lifetime contributions to the field of spectroscopy. From 1999 to 2000 he held the Welch Foundation Lectureship, and he has held other lectureships in Japan, Germany, Spain, South Korea, Taiwan, Finland, Belgium, and Estonia, as well as the USA. His research laboratory has synthesized a large number of new organosilicon and other cyclic compounds and fully analyzed their vibrational spectra, including the determination of their potential energy surfaces for conformationally important vibrations. He has been at the forefront of writing computer programs for calculating both the kinetic energy and potential energy functions for one-, two-, and three-dimensional problems. "Laane's Tables", published in 1970, and his programs for calculating periodic potential functions for internal rotations and pseudorotations, have been widely distributed to and utilized by dozens of laboratories around the world. These works have been cited more than 500 times. Laane has supervised the research of more than 40 Ph.D. students and more than 60 undergraduates. Numerous post-docs and visiting professors from around the world have also worked in his laboratories. Professor Laane is a Fellow of the American Physical Society, Society for Applied Spectroscopy, and the American Institute of Chemists. He received the Alexander von Humboldt U.S. Senior Scientist Award in 1979 at the very young age of 37, and a Texas A&M teaching award in 1982. He is an elected member of the Estonian Academy of Sciences and received one of the first honorary D.Sc. degrees from the University of Tartu in 2000. At Texas A&M, he was Chair of the Physical and Nuclear Chemistry Division for many years, and served as Associate Dean of Science and Speaker of the Faculty Senate. He also played the leading role in establishing a Texas A&M campus in Japan while he was Director of the Institute for Pacific Asia in the 1980s. In 1994, he organized and chaired the NATO workshop on "Structures and Conformations of Non-Rigid molecules" in Ulm, Germany. From 2007 to 2009 he served as President of the Alexander von Humboldt Association of America (AvHAA), and in 2008, he organized and chaired the Humboldt Symposium for eminent scientists on "The Universe and the World Arounds Us". Laane served as Editor for the *Journal of Molecular Structure* from 1994 to 2019 and served on the EUCMOS International Advisory Board during this same period. He also has served on the editorial boards of several other journals and on the organizing committees of international conferences.

Jaan has been married for 59 years to Tiiu Laane, Professor Emeritus of German Language and Literature at Texas A&M University and daughter of Estonian composer Leonhard Virkhaus. They met in Cambridge, Massachusetts while Tiiu was earning her Ph.D. at Harvard and Jaan his Ph.D. at MIT. Their daughters Christina Laane (an Ear, Nose, and Throat M.D.) and Lisa Laane (a software engineer) are married to software engineers, Christian Wiedmann and Josh Behnke. Tiiu and Jaan have five grandchildren: Marina Wiedmann (16), Katherine Behnke (15), Marko Wiedmann (14), Mikkel Wiedmann (14), and Erik Behnke (13).

Rui Fausto and Gulce Ogruc Ildiz
Guest Editors

molecules

MDPI

Article

A Pure Rotational Spectroscopic Study of Two Nearly-Equivalent Structures of Hexafluoroacetone Imine, $(CF_3)_2C{=}NH$

Daniel A. Obenchain [1,*], Beppo Hartwig [1], Daniel J. Frohman [2], G. S. Grubbs II [3], B. E. Long [4], Wallace C. Pringle [2], Stewart E. Novick [2,†] and S. A. Cooke [5,*]

1 Institute of Physical Chemistry, University of Göttingen, Tammannstrasse 6, 37077 Göttingen, Germany; bhartwi@gwdg.de
2 Department of Chemistry, Wesleyan University, 52 Lawn Avenue, Middletown, CT 06459, USA; dfrohman@wesleyan.edu (D.J.F.); wpringle@wesleyan.edu (W.C.P.)
3 Department of Chemistry, Missouri University of Science and Technology, 104 Schrenk Hall, 400 W. 11th St, Rolla, MO 65409, USA; grubbsg@mst.edu
4 Department of Chemistry and Biochemistry, James Madison University, Harrisonburg, VA 22807, USA; long2be@jmu.edu
5 School of Natural and Social Sciences, Purchase College SUNY, 735 Anderson Hill Rd, Purchase, NY 10577, USA
* Correspondence: daniel.obenchain@uni-goettingen.de (D.A.O.); stephen.cooke@purchase.edu (S.A.C.)
† Deceased, October 2023.

Abstract: Rotational spectra for hexafluoroacetone imine, the singly substituted ^{13}C isotopologues, and the ^{15}N isotopologue, have been recorded using both cavity and chirped pulse Fourier transform microwave spectrometers. The spectra observed present as being doubled with separations between each pair of transitions being on the order of a few tens of kilohertz which is consistent with a large amplitude motion producing two torsional substates. The observed splitting is most likely due to the combined motions of the CF3 groups, for which the calculated barrier is small. However, no transitions between states could be observed and, similarly, no Coriolis coupling parameters were required to achieve a satisfactory fit for the transition frequencies. Hence, and somewhat curiously, the two states have been fit independently of each other such that the two states may simply be considered near-equivalent conformers. The structural properties of hexafluoroacetone imine are compared with two isoelectronic molecules hexafluoroisobutene and hexafluoroacetone. Rotational constants, quartic centrifugal distortion constants, and the ^{14}N nuclear electric quadrupole coupling tensor have been determined and are presented together with supporting quantum chemical calculations.

Keywords: hexafluoroacetone imine; 1,1,1,3,3,3-hexafluoro-2-propanimine; rotational spectroscopy; large amplitude motions

1. Introduction

Hexafluoroacetone imine may be considered a member of a grouping of six molecules shown in Figure 1. This grouping consists of two isoelectronic series, the first being isobutene, acetone imine, and lastly acetone, and the second being the hexafluorinated analogs of the first series. Except for hexafluoroacetone imine, all these molecules have had their rotational spectra recorded at high resolution. These spectra are reported, or most recently discussed, in the following references [1–5]. Isobutene and acetone are well-known examples of rotational spectra in which the effects of the internal rotation of two equivalent methyl tops are manifested. At high resolution, rotational transitions from these types of molecules will appear as quartets owing to transitions occurring from within the *AA*, *EE*, *AE*, and *EA* torsional sub-levels.

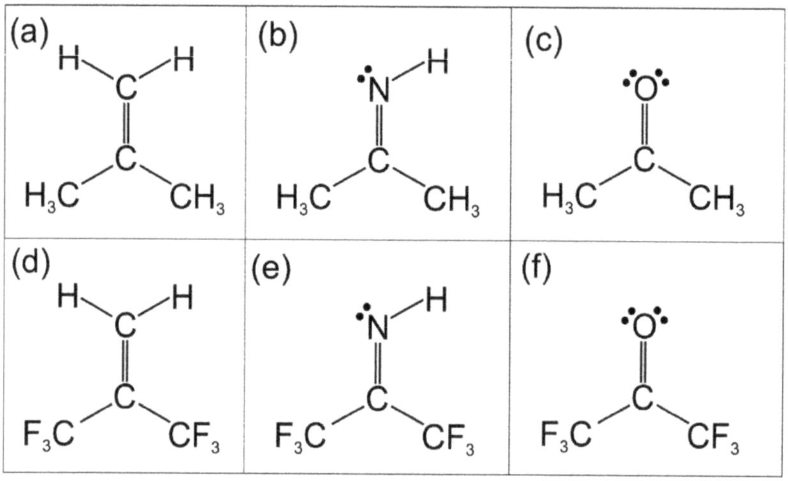

Figure 1. Two series of isoelectronic molecules. (**a–c**) is isobutene, acetone imine, and acetone. (**d–f**) are the hexafluorinated analogs.

These spectra have been interpreted to yield effective CH_3 internal rotation tunneling barriers of 761(1) cm^{-1} for isobutene [1] and 251.4(26) cm^{-1} for acetone [3]. One would correctly assume that the tunneling barrier to CH_3 internal rotation in acetone imine would be intermediate between isobutene and acetone. However, the spectra are more complex owing to the two CH_3 rotors being non-equivalent. In that case, the effects of the two non-equivalent CH_3 rotors result in five torsional substrates, one with A_1 symmetry and four doubly degenerate states with E symmetry. Hence, rotational transitions from within these substates present to the observer as quintets. An analysis of these spectra by Zou et al. [2] has resulted in internal rotation barrier heights of 531.956(64) cm^{-1}, attributed to the CH_3 furthest from the imine hydrogen, and 465.013(26) cm^{-1}, attributed to the CH_3 nearest to the imine hydrogen. Taken as a series, then we observe that the CH_3 barriers to internal rotation increase in the order of increasing hydrogenation of the apex atom, i.e., in the order O, N-H, H-C-H. From a purely classical mechanics standpoint, this is rational in the sense that the apex hydrogen(s) clearly impede the CH_3 internal rotations.

We then consider the hexafluorinated analogs. The rotational spectrum of hexafuoroacetone has been recorded at high resolution with spectral line widths on the order of 7 kHz at full width, half maximum height [5]. However, none of the transitions observed for hexafluoroacetone were observed as multiplets, or broadened as they were for acetone. Now, it is generally true that, the higher the barrier heights to internal motion the closer together in energy will be the torsional substates and, hence, the rotational transitions from within those different torsional substates will appear closer together in frequency. Given that the barrier heights to CF_3 internal rotation are significantly higher than those for CH_3 internal rotation, together with the very small internal rotational constant for a CF_3 group, $F \approx 10$ GHz (compared to that for a CH_3 group, $F \approx 160$ GHz), then effects due to CF_3 internal rotation are most often unresolvable. A helpful discussion on this topic is presented in reference [6]. Accordingly, the rotational spectrum of hexafluoroacetone presents like that of an "ordinary" semirigid rotor, with no evidence of any internal motions.

Given that isobutene has a higher barrier to CH_3 internal rotation compared to that of acetone, it may be assumed that the barrier to CF_3 internal rotation in hexaflouroisobutene would be higher than that in hexafluoroacetone and, therefore, would have a rotational spectrum with no observable evidence of internal motion. However, the recorded rotational spectrum of hexafluoroisobutene presented as being doubled with spacings between the rotational transitions being on the order of tens of megahertz. The spectral analyses revealed that the bis-trifluoromethyl groups of hexafluoroisobutene are staggered in the equilibrium configuration, and that a novel, out-of-phase rotation through an F-CCC-F planar configuration with a low barrier (<100 cm^{-1}), leads to the observed doubled rotational spectra [4]. This bistrifluoromethy effect has subsequently been explored in related molecules [7,8].

In this article, we present the measurement and analyses of rotational spectra for the remaining member of the molecules in Figure 1, hexafluoroacetone imine. We ask the question of how well the spectra are present, like that of hexafluoroacetone or that of hexafluoroisobutene. And further, how do we interpret the spectra?

2. Results

2.1. Spectral Analysis

The ground-state rotational spectrum of hexafluoroacetone imine, and all of the singly substituted ^{13}C and ^{15}N isotopologues, exhibits both a-type and b-type transitions. The spectra consist of "doubled" transitions separated by a few tens of kilohertz, we refer to a transition as either belonging to State I or State II. An example set of transitions is shown in Figure 2.

For both State I and State II, transition quantum number assignments and measured frequencies are available in the Supplementary Data. For each state, spectroscopic constants were fitted to the observed frequencies using the SPCAT/SPFIT software [9]. The Hamiltonian used was of the simple form $H = H_R + H_Q$ constructed in the coupled symmetric rotor basis set $\mathbf{I} + \mathbf{J} = \mathbf{F}$. The Watson A reduction in the I^r representation was chosen for the semirigid rotor Hamiltonian H_R [10], but only terms up to the fourth power in angular momentum were required. The second operator, H_Q, represents the well-known interaction energy of the nuclear electric quadrupole moment with the electric field gradient at the nitrogen nucleus. The term H_Q is not needed for the ^{15}N isotopologue as the nuclear spin, $I = 1/2$, and it therefore does not possess a quadrupole moment. In all cases, the χ_{aa} and χ_{cc} were used in the fits with χ_{bb} being determined from the requirement that $\chi_{aa} + \chi_{bb} + \chi_{cc} = 0$. The determined spectroscopic constants for the two states of the parent isotopologue are presented in Table 1. The spectroscopic constants of the minor isotopologues are presented in Table 2 (State I), and Table 3 (State II). Many attempts were made to (i) locate transitions that span States I and II, and (ii) to use Coriolis constants in order to link the states. However, only those constants shown in Tables 1–3 were required to fit the transition data sets.

2.2. Theoretical Calculations

The MP2/6-311G++(2d,2p) [11–14] equilibrium geometry of hexafluoroisobutene is shown in Figure 3 and this structure produces the rotational constants shown in Table 1.

Calculated centrifugal distortion constants are also shown in Table 1. It is easily seen that the agreement is very good, with only small discrepancies appearing for some of the centrifugal distortion constants. This is unsurprising given the likely internal motions.

As discussed in the introduction, curiosity surrounds the possibility of observing the hexafluoroacetone imine spectra effects due to a similar internal motion to that observed for hexafluoroisobutene. The relaxed 2D potential energy scan for hexafluoroisobutene is shown in Figure 4 (see Figure 3 for atom labeling). For reference, the staggered nature of the F-CCC-F atoms in the equilibrium structure is shown on the right-hand side of Figure 3.

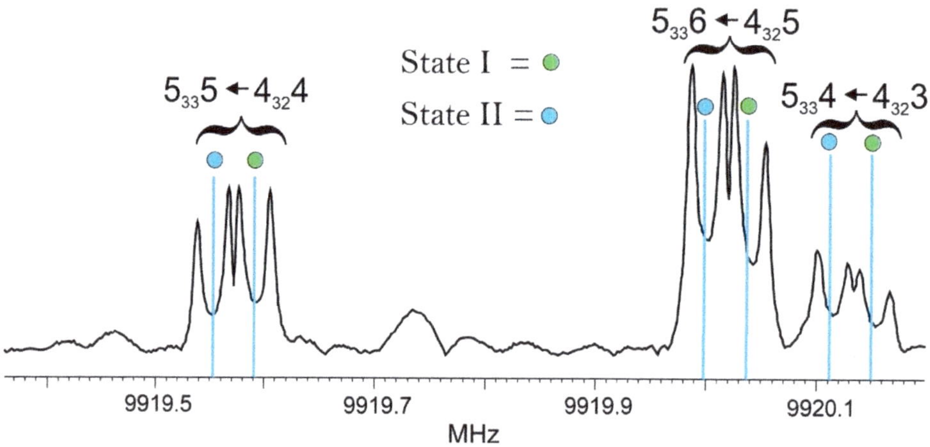

Figure 2. A section of the spectra recorded for hexafluoroacetone imine using a cavity Fourier transform microwave spectrometer. The quantum number assignments given follow $J'_{K-1K+1}F \leftarrow J''_{K-1K+1}F$. The portion of spectra shown is the Fourier transform of 75 free-induction decays averaged together. Transitions appear as Doppler doublets and the average of the two Doppler components is taken as the line center. The vertical lines indicate the predicted frequencies for these transitions using the rotational constants given in Table 1.

Table 1. Calculated and experimentally determined effective ground-state rotational constants, centrifugal distortion constants, and nitrogen nuclear electric quadrupole coupling constants for hexafluoroacetone imine.

	MP2/6-311G++(2d,2p)	State I	State II
A (MHz)	2158.610	2170.17085(35)	2170.16744(35)
B (MHz)	1045.446	1043.25444(14)	1043.24681(14)
C (MHz)	937.933	936.37030(13)	936.37059(13)
Δ_J (kHz)	0.054	0.0576(10)	0.0525(10)
Δ_{JK} (kHz)	0.15	0.2047(62)	0.2219(62)
Δ_K (kHz)	−0.11	−0.363(26)	−0.457(26)
δ_J (kHz)	0.0061	0.00734(44)	0.00471(43)
δ_K (kHz)	−0.35	−0.475(41)	−0.563(41)
χ_{aa} (MHz)	-	−5.0732(33)	
χ_{bb} (MHz)	-	2.9924(40)	
χ_{cc} (MHz)	-	2.0809(40)	
RMS (kHz) [1]	-	2.7	
N	-	176	177

[1] Microwave root mean square = $\sqrt{\sum \left[(obs - calc)^2\right] / N}$ where N is the number of transitions.

Table 2. Experimentally determined effective ground-state rotational constants, centrifugal distortion constants, and nitrogen nuclear electric quadrupole coupling constants for the parent and isotopologues for State I of hexafluoroacetone imine.

	Parent	$^{13}C_1$	$^{13}C_2$	$^{13}C_3$	^{15}N
A (MHz)	2170.17085(35)	2170.2055(79)	2170.1985(44)	2165.3028(24)	2134.1858(28)
B (MHz)	1043.25444(14)	1039.72015(93)	1039.65915(41)	1043.33695(20)	1043.25063(14)
C (MHz)	936.37030(13)	933.51356(21)	933.47028(12)	935.530070(95)	929.59788(10)
Δ_J (kHz)	0.0576(10)	[0.0576]	[0.0576]	[0.0576]	[0.0576]
Δ_{JK} (kHz)	0.2047(62)	[0.2047]	[0.2047]	[0.2047]	[0.2047]
Δ_K (kHz)	−0.363(26)	[−0.363]	[−0.363]	[−0.363]	[−0.363]
δ_J (kHz)	0.00734(44)	[0.00734]	[0.00734]	[0.00734]	[0.00734]
δ_K (kHz)	−0.475(41)	[−0.475]	[−0.475]	[−0.475]	[−0.475]
χ_{aa} (MHz)	−5.0732(33)	−5.052(31)	−5.109(24)	−5.070(21)	-
χ_{bb} (MHz)	2.9924(55)	2.97(16)	2.974(89)	2.961(70)	-
χ_{cc} (MHz)	2.0809(40)	2.08(16)	2.135(86)	2.109(67)	-
RMS (kHz) [1]	2.7	2.8	4.3	1.9	2.9
N	176	28	39	27	23

[1] Microwave root mean square = $\sqrt{\sum\left[(obs-calc)^2\right]/N}$ where N is the number of transitions.

Table 3. Experimentally determined effective ground-state rotational constants, centrifugal distortion constants, and nitrogen nuclear electric quadrupole coupling constants for the parent and isotopologues for State II of hexafluoroacetone imine.

	Parent	$^{13}C_1$	$^{13}C_2$	$^{13}C_3$	^{15}N
A (MHz)	2170.16744(35)	2170.2155(79)	2170.2607(41)	2165.2925(25)	2134.1836(21)
B (MHz)	1043.24681(14)	1039.71300(93)	1039.65821(44)	1043.32924(20)	1043.24306(13)
C (MHz)	936.37059(13)	933.51364(21)	933.46923(11)	935.530474(96)	929.59823(10)
Δ_J (kHz)	0.0525(10)	[0.0525]	[0.05764]	[0.05764]	[0.05764]
Δ_{JK} (kHz)	0.2219(62)	[0.2219]	[0.2046]	[0.2046]	[0.2046]
Δ_K (kHz)	−0.457(26)	[−0.457]	[−0.364]	[−0.364]	[−0.364]
δ_J (kHz)	0.00471(43)	[0.00471]	[0.00738]	[0.00738]	[0.00738]
δ_K (kHz)	−0.563(41)	[−0.563]	[−0.473]	[−0.473]	[−0.473]
χ_{aa} (MHz)	−5.0732(33)	−5.052(31)	−5.109(24)	−5.070(21)	-
χ_{bb} (MHz)	2.9924(55)	2.97(16)	2.974(89)	2.961(70)	-
χ_{cc} (MHz)	2.0809(40)	2.08(16)	2.135(86)	2.109(67)	-
RMS (kHz) [1]	2.7	3.7	4.3	1.9	2.9
N	177	28	39	27	23

[1] Microwave root mean square = $\sqrt{\sum\left[(obs-calc)^2\right]/N}$ where N is the number of transitions.

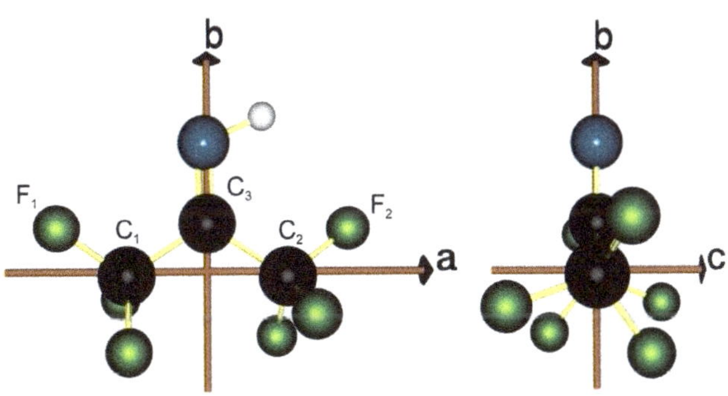

Figure 3. Hexafluoroacetone imine calculated at the MP2/6-311G++(2d,2p) level in the *ab*- and *bc*-planes of its principal axes system.

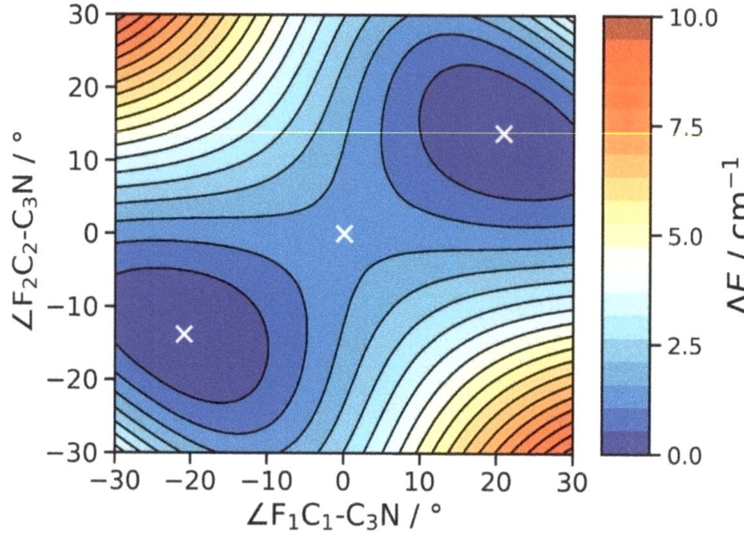

Figure 4. A 2-D potential energy scan for hexafluoroacetone imine. Stationary points are shown with a white cross. Please refer to Figure 3 for the atomic number scheme, and to the text for further discussion.

The zero-degree/zero-degree center of the scan corresponds to a planar F-CCC-F configuration. A one-dimensional, diagonal slice (lower left minimum through to upper right minimum) of the potential energy scan for hexafluoroacetone imine shows a double minimum with a barrier height of approximately 116 cm^{-1}. The barrier heights at other levels of theory are shown in Table 4, and also in the Supplementary Tables, and are generally consistent. This is, approximately, double that of the analogous barrier height in hexafluoroisobutene, and using alternative quantum mechanical methods Shahi and Arunan [8] have shown the barrier quadruples in magnitude. For reference, the Shahi and Arunan calculations have been reproduced and are shown in the Supplementary Data. The increased barrier height for this out-of-phase rotation of CF_3 groups through a planar F-CCC-F local maximum is consistent with the rotational transitions within each torsional substrate being closer together in frequency than for those observed in hexafluoroisobutene.

Table 4. Overview of the electronic ($\Delta E_{el}{}^{\ddagger}$) and zero-point corrected ($\Delta E_0{}^{\ddagger}$) barrier for the inversion. The singular imaginary mode ($\tilde{v}_{imag}$) is also given associated with the transition state. All methods used (i) Grimme's dispersion correction was used throughout in conjunction with Becke-Johnson damping, D3(BJ), and (ii) the aug-cc-pVTZ basis sets were used.

Method	$\Delta E_{el}{}^{\ddagger}/cm^{-1}$	$\Delta E_0{}^{\ddagger}/cm^{-1}$	$\tilde{v}_{imag}/i\ cm^{-1}$
B3LYP	118	117	32.0
PBE0	112	112	31.3
CAM-B3LYP	112	112	31.7
B2PLYP	127	126	33.0
DSD-PBEP86-D3	139	138	34.2

The results of a nudged-elastic-band scan are shown in the Supplementary Data for hexafluoroacetone, hexafluoroacetone imine, and hexafluoroisobutene. The angles used here are the average of the two dihedral angles defining the rotation of the CF3 groups. The scans illustrate that both the width as well as the barrier height significantly increase in going from hexafluoroisobutene to hexafluoroacetone imine to hexafluoroacetone. From the scans, you can also see that in hexafluoroacetone and hexafluoroacetone imine the lone pair side dihedral angles are actually very similar.

For the purpose of deriving semi-experimental equilibrium structures r_e^{SE}, we conducted vibrational perturbation theory of second-order calculations (VPT2) [15] to obtain vibrational corrections for the rotational constants. In particular, we used Gaussian's generalized VPT2 (GVPT2) implementation [16,17]. Note that the VPT2 calculation was conducted separately from the geometry optimization.

2.3. Structure

From the calculations detailed above, a very accurate structure was determined that reproduced the observed rotational constants very well. The best method used was the PBE0-D3(BJ)/aug-cc-pVTZ level of the theory and key structural features are presented in Table 5. The computed rotational constants are shown in Table 6.

Table 5. Key structural parameters for hexafluoroacetone imine determined from the zero-point corrected PBE0-D3(BJ)/aug-cc-pVTZ structure. See Figure 3 for the atom labeling scheme.

Parameter	Value
r (C_1-C_3)/Å	1.534
r (C_2-C_3)/Å	1.536
r (C_3-N)/Å	1.984
$\angle$ (C_1-C_3-C_2)/degrees	116.4
D(F_1-C_1-C_3-N)/degrees	−20.5

Table 6. The PBE0-D3(BJ)/aug-cc-pVTZ equilibrium X_e, and zero-point corrected X_0 rotational constants for hexafluoroacetone imine. All values are in MHz.

	A_e	A_0	B_e	B_0	C_e	C_0
Parent	2180.96	2168.545	1045.109	1038.330	938.792	933.093
^{13}C=N	2175.995	2163.619	1045.109	1038.444	937.871	932.272
^{15}N	2145.054	2132.83	1045.102	1038.319	932.072	926.398
^{13}C, N lone pair side	2180.916	2168.619	1041.447	1034.739	935.829	930.183
^{13}C, NH side	2180.895	2168.594	1041.510	1034.811	935.875	930.243

Attempts were made to perform r_0, r_e^{SE} (semi-experimental [18]) and $r_m^{(2)}$ [19] fits using the determined experimental rotational constants and, where appropriate, with quantum mechanically calculated vibrational parameters. For a comprehensive review of these approaches, please see reference [20]. For the C=N bond, the values determined were satisfactory, but in all the attempts, the two unique C-C bonds are not in agreement with the calculated structure, having large uncertainties in the bonds, and non-converging fits. The r_m^2 structure converges with both c_a and d_a, but again the bonds have non-physical values. It is most likely that these failures have to do with the effective nature of the rotational constants used together with, possibly, poorly captured large amplitude motions in the quantum mechanical calculations.

Given the availability of the rotational constants for all the singly substituted ^{13}C- and ^{15}N-isotopologues Kraitchman analysis [21] we can obtain the substitution principal atomic coordinates of each of the substituted atoms, with the origin of these coordinates being the center of mass of the parent isotopologue. The results are shown in Table 7. As expected, it is found that both State I and II share nearly-equivalent structures which agree well with the quantum mechanical calculations.

Table 7. Kraitchman substitution coordinates, in Angstroms, for the heavy atoms in hexafluoroacetone imine. The *c*-coordinates are either imaginary * or are zero to two significant figures.

	State I		State II		PBE0-D3(BJ)/aug-cc-pVTZ	
	a	*b*	*a*	*b*	*a*	*b*
C_1	−1.2867(12)	0 *	−1.2876(12)	0 *	−1.29	−0.08
C_2	1.2968(11)	0 *	1.2980(12)	0 *	1.30	0.07
C_3	0 *	0.7244(21)	0 *	0.7229(21)	0	0.73
N	0.060(25)	1.9910(7)	0.073(20)	1.9904(7)	0.06	1.98

* These coordinates are imaginary consistent with the atom being close, or on, a principal axis. We have taken them to indicate a coordinate of zero.

Regarding the structure, it is interesting to compare the second moment in the direction of the *c*-principal axes, $P_{cc} = \sum_i m_i c_i^2$, for the three molecules: hexafluoroisobutene, hexafluoroacetone imine, and hexafluoroacetone. All these structures have two CF_3- groups which are staggered with respect to each other, and so the only contributions to P_{cc} will be from the six out-of-plane fluorine atoms. The P_{cc} values may be determined from the experimental rotational constants and are found to be 89.20 amu Å^2, 88.79 amu Å^2, and 89.15 amu Å^2, respectively. For two CF_3 groups, Bohn [22] has shown that the anticipated value of P_{cc} should be approximately 90 amu-Å^2. This is in good agreement with the present data set and is indicative of the staggered/helical tendencies of many perfluorinated molecules [23].

2.4. Nitrogen Nuclear Electric Quadrupole Coupling Tensor

The ^{14}N-nitrogen nuclear quadrupole coupling tensor components for hexafluoroacetone imine are compared to those for CH_2NH [24] and CF_2NH [25,26] in Table 8. Direct comparison is difficult as each molecule has its own principal axes system. However, in all three cases, the *c*-principal axes are perpendicular to the C=N-H plane and therefore the χ_{cc} values are comparable. For χ_{cc}, we find that the magnitudes change in the order CH_2NH > $(CF_3)_2CNH$ > CF_2NH. This trend may be rationalized by an appeal to the electronegativities of H, F, and CF3, and it is found that the ordering of electronegativities on the Allred-Pauling scale [27,28] is F (3.98) > CF3 (2.99) > H (2.20), the opposite trend to that observed for χ_{cc}. Nuclear quadrupole coupling tensors are often related to ionicity [29] where the quadrupole coupling tensor for an ion will be very close to zero owing to the spherical symmetry of either empty (or full) *p*-orbitals. So, one may rationalize that χ_{cc} will decrease in magnitude with an increase in the electronegativity of the attached groups which is consistent with experimental observations.

Table 8. The values of the ^{14}N-nuclear quadrupole coupling tensor components, χ, for hexafluoroacetone imine, methanimine (CH_2NH) [24] and difluomethanimine (CF_2NH) [25,26].

	CH_2NH	CF_2NH	$(CF_3)_2CNH$
χ_{aa}/MHz	−0.9131(16)	1.029(20)	−5.0732(33)
χ_{bb}/MHz	−2.6688(14)	−2.56(17)	2.9924(40)
χ_{cc}/MHz	3.5819(21)	1.531(22)	2.0809(40)

3. Discussion

The pure rotational spectrum of hexafluoroacetone imine presents as consisting of two nearly-equivalent conformers. However, it is undoubtedly the case that these two sets of measured rotational transitions arise from two torsional substates, State I and State II, resulting from a relatively low barrier between two equivalent F-CCC-F staggered configurations. The torsional substates most likely arise from an out-of-phase internal rotation of the two CF_3 groups which is governed by an unusual potential energy function with six minima, three high barriers (V1) and three low barriers (V2) as observed in the related molecule hexafluoroisobutene [4]. The substrates arise via tunneling through the low barrier. An example of the potential energy function is given in Figure 5. For the hexafluorinated species, the global minimum corresponds to a staggered F_1-CCC-F_2 configuration exemplified in Figure 3.

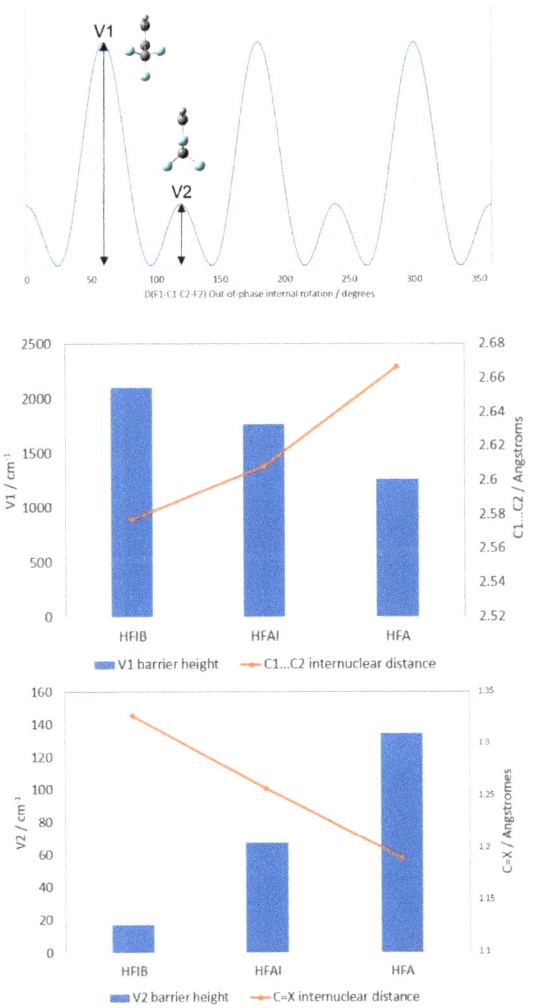

Figure 5. Top: an example, i.e., not to any scale but just showing form, of a relevant potential energy plot with the x-axis being the F1-C1-C2-F2 dihedral angle. Middle: Bar chart showing how, for the series HFIB, HFAI, HFA the barrier V1 decreases as the distance between the C1 and C2 atoms increases. Bottom: Bar chart showing how the V2 barrier height increases as the C=X distance decreases, X=C, N, and O. HFIB is hexafluoroisobutene, HFAI is hexafluoroacetone imine, and HFA is hexafluoroacetone. See Figure 3 for the atom labeling scheme.

Careful calculations by Shahi and Arunan [8] have revealed that for the isoelectronic sequence of molecules hexafluoroisobutene (HFIB), hexafluoroacetone imine (HFAI), and hexafluoacetone (HFA) the barrier height V1 varies as HFIB > HFAI > HFA, whereas the V2 barrier height varies as HFA > HFAI > HFIB. In all three cases, the V1 barrier height is too high for tunneling effects to manifest in the observed spectra consistent with the absence of groupings of rotational transitions from within four or five torsional substates. The V2 tunneling barrier height is more nuanced. With the molecule unable to tunnel through V1, we are left with a tunneling motion through a much lower barrier separating two equivalent F-CCC-F staggered configurations, more akin to a double minimum potential often observed in ring-puckering problems. The barrier V2 is lowest for hexafluoroisobutene consistent with rotational transitions from the two torsional states being separated by tens of MHz. Whereas the barrier V2 is highest for hexafluoroacetone consistent with no observed "doublets" in the rotational spectrum. Hexafluoroacetone imine is intermediate, again consistent with the observation of paired rotational transitions separated by tens of kHz.

The above trends in both V1 and V2 may be rationalized through an appeal to the molecular geometries. In regards to the trend in V1, we note that all three molecules possess a C_1-C_3-C_2 structural component, see Figure 3. It is found that the distance $C_1 \ldots C_2$ decreases in the order HFIB > HFAI > HFA. A large value of $C_1 \ldots C_2$ means that the two CF_3-groups are further away from one another, and thus easier to rotate, compared to a small value of $C_1 \ldots C_2$ which corresponds to the CF_3-groups being closer together. The correspondence between V1 and $C_1 \ldots C_2$ is shown in Figure 5. In regards to V2 in which the top of the smaller barrier corresponds to a planar F_1-CCC-F_2 (see Figure 5), it is useful to compare the C=X distance where X = CH_2, NH, and O. This C=X distance decreases in the order HFIB > HFAI > HFA. The larger C=X distance in hexafluoroisobutene corresponds to a small V2, i.e., it is easier for F-CCC-F to pass through a planar configuration, compared to the longer C=X distance in hexafluoroacetone where the planar F-CCC-F configuration is more greatly interfered with by the oxygen lone pairs. The correspondence between V2 and C=X is also shown in Figure 5.

The above arguments are supported and augmented by calculations that reveal non-covalent interactions. Following the work of Johnson et al. [30] non-covalent interactions based on the electron density and its derivatives have been located for hexafluoroisobutene, hexafluoroacetone imine, and hexafluoroacetone. The results are presented in Figure 6. In this figure, green indicates weak attractive interactions whereas the orange and reddish colors indicate repulsive interactions. To obtain these plots, the electron density at the B3LYP level was analyzed using the Multiwfn [31,32] program and then visualized using VMD [33].

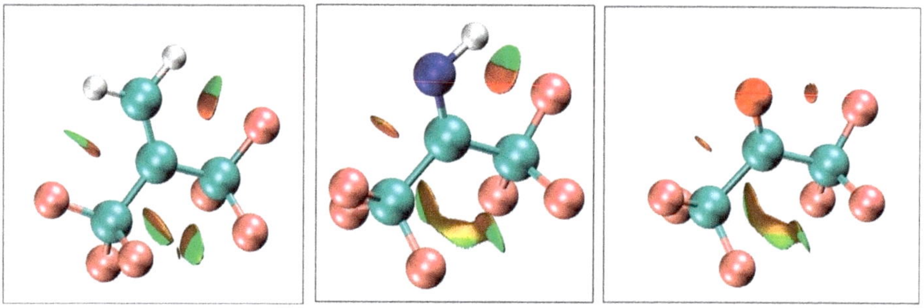

Figure 6. From left to right regions of most attractive (green) and most repulsive (red) interactions for hexafluoroisobutene, hexafluoroacetone imine, and hexafluoroacetone. Colors between red and green indicate areas of intermediate attraction/repulsion.

Within the figure, it is observed that in going from hexafluoroacetone-to-hexafluoroacetone imine to hexafluorisobutene the attractive interactions increase between the CF_3 groups and O, NH and CH_2, which in turn explains why the rotation of the CF_3 groups becomes easier in that series. It is also observed that interactions with the lone pairs are purely repulsive whereas interactions with the NH and CH_2 are partially attractive.

4. Materials and Methods

4.1. Experimental Methods

Hexafluoroacetone imine (95%, b.p. 16.5°) was purchased from Synquest Labs and used without further purification. A gas tank was pressurized to approx. 0.25 bar of hexafluoroacetone imine and then diluted with argon to a final pressure of approximately 6 bar. This solution of gases was pulsed through a solenoid valve into vacuum chambers held at approximately 10^{-4} bar causing a rotational cooling to approximately $T^{rot} \approx 3$ K. Rotational spectra were recorded between 5.5 GHz and 21 GHz on two types of spectrometers first a chirp-pulsed Fourier transform spectrometer, and then a Balle-Flygare cavity Fourier transform spectrometer. Both spectrometers have been explained in detail elsewhere [34–38]. Common to both instruments was the use of microsecond pulses of microwave radiation to bring about bulk rotational coherence of the cold target molecules. A free-induction decay, FID, was collected as a function of time as the bulk coherence is lost. FIDs are averaged to reduce random noise, and were then fast Fourier transformed to produce a frequency domain spectrum. The key difference between the spectrometers is that the chirp-pulsed spectrometer utilizes a microwave pulse of radiation containing a sweep, i.e., chirp, of frequencies from, say, 8–18 GHz. This chirp was then greatly amplified to powers of, in the present case, 5 W and then broadcast onto the molecules through a horn antenna. The FIDs collected were amplified and consisted of 800,000 points which was fast Fourier transformed on a suitably broad-banded oscilloscope. Line widths with this method were typically on the order of 80 kHz and line centers possess uncertainties of approximately 10 kHz. In the cavity experiment, monochromatic pulses of microwave radiation are amplified using a Fabry-Perot resonator which resulted in very high interrogation path lengths. FIDs were collected in the same way as above; however, the FIDs consist of approximately 1600 data points and, after mixing down, were fast Fourier transformed on a narrow band PC oscilloscope card. Line widths with this method were approximately 10 kHz and line center uncertainties were approximately 2 kHz.

4.2. Quantum Chemical Calculations

Initially, geometry optimizations and vibrational frequencies were calculated at the MP2/6-311G++(2d,2p) [11–14] level of theory using the Gaussian 16 rev. B01 software [39]. This method was relatively fast and sufficient to allow initial assignments of experimental spectroscopic transitions.

Higher-level quantum mechanical calculations were also pursued. We utilized the commonly used B3LYP [40–43], PBE0 [44,45] and CAM-B3LYP [46] hybrid density functionals as well as the B2PLYP [47,48] and DSD-PBEP86 [49,50] double-hybrid functionals. The CAM-B3LYP and DSD-PBEP86 functionals may be more accurately referred to as a range-separated hybrid functional and spin component scaled double-hybrid functional, respectively. To compensate for the inaccurate description of dispersion interactions within density functionals, Grimme's D3 dispersion correction was used throughout in conjunction with Becke-Johnson damping (D3(BJ)) [51–53]. As a basis set, we utilized Dunning's augmented triple zeta basis set aug-cc-pVTZ [54,55].

Geometry optimizations were carried out using the *VeryTight* optimization criterion and the very dense *SuperFine* integration grid. For the transition state search, the Hessian matrix was recomputed exactly at every third step. Subsequently, analytic harmonic frequency calculations were conducted to confirm the presence of a minimum of (first order) transition states. In addition, we carried out 2D relaxed surface scans with the B3LYP, PBE0 and CAM-B3LYP functionals to better understand the potential energy surface. To this end, we varied the $F_1C_1C_3N$ and $F_2C_2C_3N$ dihedral angles from 30° to −30° in steps of 2° (see Figure 3 for atom labeling). To reduce computational cost during the scan, the optimization criterion was lowered to *Tight*.

The zero-degree/zero-degree center of the scan corresponds to a planar F-CCC-F configuration. A one-dimensional, diagonal slice (lower left minimum through to upper right minimum) of the potential energy scan for hexafluoroacetone imine shows a double minimum.

For deriving semi-experimental equilibrium structures r_e^{SE}, we conducted vibrational perturbation theory of second-order calculations (VPT2) [15] to obtain vibrational corrections for the rotational constants. In particular, we used Gaussian's generalized VPT2 (GVPT2) implementation [16,17]. Note that the VPT2 calculation was conducted separately from the geometry optimization.

Lastly, Nudged-Elastic-Band [56] scans for the fluorinated variants using ORCA 6.0.1 [57,58] and 60 images at the B3LYP-D3(BJ)/aVTZ level were performed to further investigate the barrier heights.

Supplementary Materials: The following supporting information can be downloaded at https://www.mdpi.com/article/10.3390/molecules30092051/s1, Table S1: Transition frequencies and quantum number assignments for hexafluoroacetone imine. Table S2: Overview of the electronic ($\Delta E^{\ddagger}_{el}$) and zero-point corrected ($\Delta E^{\ddagger}_{0}$) barrier for the inversion. The singular imaginary mode ($\widetilde{v}_{imag}$) is also given associated with the transition state. Figure S1: 2-D potential energy scan for the six molecules of interest, acetone (top left), acetone imine (middle left), isobutene (bottom left), and, in each case, the hexafluorinated versions are shown to their right. Stationary points for the hexafluorinated species are shown with a white cross. See text for methods used. Figure S2: NEB scans for the fluorinated variants using ORCA 6.0.1 and 60 images and B3LYP-D3(BJ)/aVTZ (using Gaussian's definition of B3LYP). The angle here is the average of the two dihedral angles defining the rotation of the CF3 groups.

Author Contributions: Conceptualization, S.A.C.; methodology, D.A.O., S.A.C. and B.H.; validation, D.A.O. and S.A.C.; formal analysis, D.A.O., B.H., W.C.P. and S.A.C.; investigation, D.A.O., D.J.F., G.S.G.II and B.E.L.; resources, S.E.N.; writing—original draft preparation, S.A.C.; writing—review and editing, D.A.O., B.H., D.J.F., G.S.G.II, B.E.L., W.C.P. and S.A.C.; visualization, B.H. and S.A.C.; supervision, D.A.O., S.E.N. and S.A.C.; project administration, D.A.O., S.E.N. and S.A.C.; funding acquisition, D.A.O., S.E.N. and S.A.C. All authors have read and agreed to the published version of the manuscript.

Funding: This research was funded by the National Science Foundation, grant number CHE-1011214. This work was also funded by the Deutsche Forschungsgemeinschaft (DFG, German Research Foundation)—389479699/GRK2455. This work was funded by the Deutsche Forschungsgemeinschaft (DFG, German Research Foundation)—OB 535/1-1.

Institutional Review Board Statement: Not applicable.

Informed Consent Statement: Not applicable.

Data Availability Statement: All data presented in this study are available on request from the corresponding authors.

Acknowledgments: S. A. C. acknowledges support from the Taina Chao Fellowship.

Conflicts of Interest: The authors declare no conflicts of interest.

References

1. Gutowsky, H.S.; Germann, T.C. Internal Rotation and Structure of Isobutylene. *J. Mol. Spectrosc.* **1991**, *147*, 91–99. [CrossRef]
2. Zou, L.; Guillemin, J.-C.; Belloche, A.; Jørgensen, J.K.; Margulès, L.; Motiyenko, R.A.; Groner, P. Millimeter-Wave Spectrum of 2-Propanimine. *Mon. Not. R. Astron. Soc.* **2023**, *520*, 4089–4102. [CrossRef]
3. Groner, P. Experimental Two-Dimensional Torsional Potential Function for the Methyl Internal Rotors in Acetone. *J. Mol. Struct.* **2000**, *550–551*, 473–479. [CrossRef]
4. Grubbs, G.S.; Novick, S.E.; Pringle, W.C.; Laane, J.; Ocola, E.J.; Cooke, S.A. Bis-Trifluoromethyl Effect: Doubled Transitions in the Rotational Spectra of Hexafluoroisobutene, $(CF_3)_2C{=}CH_2$. *J. Phys. Chem. A* **2012**, *116*, 8169–8175. [CrossRef] [PubMed]
5. Grabow, J.U.; Heineking, N.; Stahl, W. The Microwave Spectrum and Dipole Moment of Hexafluoropropanone. ZNA-1991-46a-0229. *Z. Naturforsch* **1991**, *46a*, 229–232. [CrossRef]
6. Tolles, W.M.; Handelman, E.T.; Gwinn, W.D. Microwave Spectrum and Barrier to Internal Rotation in Trifluoronitromethane. *J. Chem. Phys.* **1965**, *43*, 3019–3024. [CrossRef]
7. Cooke, S.A.; Minei, A.J. Are the CF_3 Groups in 2,2-Bis(Trifluoromethyl)Oxirane Eclipsed or Staggered? Insights from Rotational Spectroscopy and Quantum Chemical Calculations. *Chem. Phys. Lett.* **2012**, *535*, 35–39. [CrossRef]
8. Shahi, A.; Arunan, E. Microwave Spectrum of Hexafluoroisopropanol and Torsional Behavior of Molecules with a CF_3–C–CF_3 Group. *J. Phys. Chem. A* **2015**, *119*, 5650–5657. [CrossRef]
9. Pickett, H.M. The Fitting and Prediction of Vibration-Rotation Spectra with Spin Interactions. *J. Mol. Spectrosc.* **1991**, *148*, 371–377. [CrossRef]
10. Watson, J.K.G. Aspects of Quartic and Sextic Centrifugal Effects on Rotational Energy Levels. In *Vibrational Spectra and Structure*; Durig, J.R., Ed.; Elsevier Scientific Publishing Company: Amsterdam, The Netherlands, 1977; Volume 6.
11. Møller, C.; Plesset, M.S. Note on an Approximation Treatment for Many-Electron Systems. *Phys. Rev.* **1934**, *46*, 618–622. [CrossRef]
12. Ditchfield, R.; Hehre, W.J.; Pople, J.A. Self-Consistent Molecular-Orbital Methods. IX. An Extended Gaussian-Type Basis for Molecular-Orbital Studies of Organic Molecules. *J. Chem. Phys.* **1971**, *54*, 724–728. [CrossRef]
13. Krishnan, R.; Binkley, J.S.; Seeger, R.; Pople, J.A. Self-Consistent Molecular Orbital Methods. XX. A Basis Set for Correlated Wave Functions. *J. Chem. Phys.* **1980**, *72*, 650–654. [CrossRef]
14. Clark, T.; Chandrasekhar, J.; Spitznagel, G.W.; Schleyer, P.V.R. Efficient Diffuse Function-augmented Basis Sets for Anion Calculations. III. The 3-21+G Basis Set for First-row Elements, Li–F. *J. Comput. Chem.* **1983**, *4*, 294–301. [CrossRef]
15. Nielsen, H.H. The Vibration-Rotation Energies of Molecules. *Rev. Mod. Phys.* **1951**, *23*, 90–136. [CrossRef]
16. Barone, V. Anharmonic Vibrational Properties by a Fully Automated Second-Order Perturbative Approach. *J. Chem. Phys.* **2005**, *122*, 14108. [CrossRef]
17. Barone, V.; Bloino, J.; Guido, C.A.; Lipparini, F. A Fully Automated Implementation of VPT2 Infrared Intensities. *Chem. Phys. Lett.* **2010**, *496*, 157–161. [CrossRef]
18. Clabo, D.A.; Allen, W.D.; Remington, R.B.; Yamaguchi, Y.; Schaefer, H.F. A Systematic Study of Molecular Vibrational Anharmonicity and Vibration—Rotation Interaction by Self-Consistent-Field Higher-Derivative Methods. Asymmetric Top Molecules. *Chem. Phys.* **1988**, *123*, 187–239. [CrossRef]
19. Watson, J.K.G.; Roytburg, A.; Ulrich, W. Least-Squares Mass-Dependence Molecular Structures. *J. Mol. Spectrosc.* **1999**, *196*, 102–119. [CrossRef]
20. Demaison, J. Experimental, Semi-Experimental and Ab Initio Equilibrium Structures. *Mol. Phys.* **2007**, *105*, 3109–3138. [CrossRef]
21. Kraitchman, J. Determination of Molecular Structure from Microwave Spectroscopic Data. *Am. J. Phys.* **1953**, *21*, 17–24. [CrossRef]
22. Bohn, R.K.; Montgomery, J.A.; Michels, H.H.; Fournier, J.A. Second Moments and Rotational Spectroscopy. *J. Mol. Spectrosc.* **2016**, *325*, 42–49. [CrossRef]
23. Bailey, W.C.; Bohn, R.K.; Dewberry, C.T.; Grubbs, G.S., II; Cooke, S.A. The Structure and Helicity of Perfluorooctanonitrile, CF_3-$(CF_2)_6$-CN. *J. Mol. Spectrosc.* **2011**, *270*, 61–65. [CrossRef]
24. Krause, H.; Sutter, D.H. The Molecular Zeeman Effect of Imines. I. Methanimine, Its Molecular g-Tensor, Its Magnetic Susceptibility Anisotropies, Its Molecular Electric Quadrupole Moment, Its Electric Field Gradient at the Nitrogen Nucleus, and Its Nitrogen Spin-Rotation Coupling. *Z. Naturforschung A* **1989**, *44*, 1063–1078. [CrossRef]
25. Groner, P.; Nanaie, H.; Durig, J.R.; DesMarteau, D.D. The Microwave Spectrum of Difluoromethanimine, CF2=NH. *J. Chem. Phys.* **1988**, *89*, 3983–3985. [CrossRef]
26. Möller, K.; Winnewisser, M.; Pawelke, G.; Bürger, H. The Microwave and Millimeter Wave Spectrum of Difluoromethanimine, F2C NH: Rotational and Nuclear Hyperfine Structure Analysis. *J. Mol. Struct.* **1988**, *190*, 343–355. [CrossRef]
27. Allred, A.L. Electronegativity Values from Thermochemical Data. *J. Inorg. Nucl. Chem.* **1961**, *17*, 215–221. [CrossRef]
28. True, J.E.; Thomas, T.D.; Winter, R.W.; Gard, G.L. Electronegativities from Core-Ionization Energies: Electronegativities of SF_5 and CF_3. *Inorg. Chem.* **2003**, *42*, 4437–4441. [CrossRef]

29. Gordy, W.; Cook, R.L. *Microwave Molecular Spectra*, 3rd ed.; John Wiley & Sons: Hoboken, NJ, USA, 1984.
30. Johnson, E.R.; Keinan, S.; Mori-Sánchez, P.; Contreras-García, J.; Cohen, A.J.; Yang, W. Revealing Noncovalent Interactions. *J. Am. Chem. Soc.* **2010**, *132*, 6498–6506. [CrossRef] [PubMed]
31. Lu, T.; Chen, F. Multiwfn: A Multifunctional Wavefunction Analyzer. *J. Comput. Chem.* **2012**, *33*, 580–592. [CrossRef]
32. Lu, T. A Comprehensive Electron Wavefunction Analysis Toolbox for Chemists, Multiwfn. *J. Chem. Phys.* **2024**, *161*, 082503. [CrossRef]
33. Humphrey, W.; Dalke, A.; Schulten, K. VMD: Visual Molecular Dynamics. *J. Mol. Graph.* **1996**, *14*, 33–38. [CrossRef] [PubMed]
34. Grubbs, G.S., II; Dewberry, C.T.; Etchison, K.C.; Kerr, K.E.; Cooke, S.A. A Search Accelerated Correct Intensity Fourier Transform Microwave Spectrometer with Pulsed Laser Ablation Source. *Rev. Sci. Instrum.* **2007**, *78*, 096106. [CrossRef] [PubMed]
35. Grubbs, G.S.; Powoski, R.A.; Jojola, D.; Cooke, S.A. Some Geometric and Electronic Structural Effects of Perfluorinating Propionyl Chloride. *J. Phys. Chem. A* **2010**, *114*, 8009–8015. [CrossRef]
36. Balle, T.J.; Flygare, W.H. Fabry–Perot Cavity Pulsed Fourier Transform Microwave Spectrometer with a Pulsed Nozzle Particle Source. *Rev. Sci. Instrum.* **1981**, *52*, 33–45. [CrossRef]
37. Walker, A.R.H.; Chen, W.; Novick, S.E.; Bean, B.D.; Marshall, M.D. Determination of the Structure of HBr OCS. *J. Chem. Phys.* **1995**, *102*, 7298–7305. [CrossRef]
38. Grubbs, G.S.; Obenchain, D.A.; Pickett, H.M.; Novick, S.E. H2—AgCl: A Spectroscopic Study of a Dihydrogen Complex. *J. Chem. Phys.* **2014**, *141*, 114306. [CrossRef]
39. Frisch, M.J.; Trucks, G.W.; Schlegel, H.B.; Scuseria, G.E.; Robb, M.A.; Cheeseman, J.R.; Scalmani, G.; Barone, V.; Petersson, G.A.; Nakatsuji, H.; et al. *Gaussian 16*; Revision B.01; Gaussian, Inc.: Wallingford, CT, USA, 2016.
40. Becke, A.D. Density-functional Thermochemistry. III. The Role of Exact Exchange. *J. Chem. Phys.* **1993**, *98*, 5648–5652. [CrossRef]
41. Lee, C.; Yang, W.; Parr, R.G. Development of the Colle-Salvetti Correlation-Energy Formula into a Functional of the Electron Density. *Phys. Rev. B* **1988**, *37*, 785–789. [CrossRef]
42. Vosko, S.H.; Wilk, L.; Nusair, M. Accurate Spin-Dependent Electron Liquid Correlation Energies for Local Spin Density Calculations: A Critical Analysis. *Can. J. Phys.* **1980**, *58*, 1200–1211. [CrossRef]
43. Stephens, P.J.; Devlin, F.J.; Chabalowski, C.F.; Frisch, M.J. Ab Initio Calculation of Vibrational Absorption and Circular Dichroism Spectra Using Density Functional Force Fields. *J. Phys. Chem.* **1994**, *98*, 11623–11627. [CrossRef]
44. Ernzerhof, M.; Scuseria, G.E. Assessment of the Perdew–Burke–Ernzerhof Exchange-Correlation Functional. *J. Chem. Phys.* **1999**, *110*, 5029–5036. [CrossRef]
45. Adamo, C.; Barone, V. Toward Reliable Density Functional Methods without Adjustable Parameters: The PBE0 Model. *J. Chem. Phys.* **1999**, *110*, 6158–6170. [CrossRef]
46. Yanai, T.; Tew, D.P.; Handy, N.C. A New Hybrid Exchange–Correlation Functional Using the Coulomb-Attenuating Method (CAM-B3LYP). *Chem. Phys. Lett.* **2004**, *393*, 51–57. [CrossRef]
47. Grimme, S. Semiempirical Hybrid Density Functional with Perturbative Second-Order Correlation. *J. Chem. Phys.* **2006**, *124*, 034108. [CrossRef] [PubMed]
48. Schwabe, T.; Grimme, S. Double-Hybrid Density Functionals with Long-Range Dispersion Corrections: Higher Accuracy and Extended Applicability. *Phys. Chem. Chem. Phys.* **2007**, *9*, 3397–3406. [CrossRef]
49. Kozuch, S.; Martin, J.M.L. DSD-PBEP86: In Search of the Best Double-Hybrid DFT with Spin-Component Scaled MP2 and Dispersion Corrections. *Phys. Chem. Chem. Phys.* **2011**, *13*, 20104–20107. [CrossRef] [PubMed]
50. Kozuch, S.; Martin, J.M.L. Spin-component-scaled Double Hybrids: An Extensive Search for the Best Fifth-rung Functionals Blending DFT and Perturbation Theory. *J. Comput. Chem.* **2013**, *34*, 2327–2344. [CrossRef]
51. Grimme, S.; Antony, J.; Ehrlich, S.; Krieg, H. A Consistent and Accurate Ab Initio Parametrization of Density Functional Dispersion Correction (DFT-D) for the 94 Elements H-Pu. *J. Chem. Phys.* **2010**, *132*, 154104–154119. [CrossRef]
52. Smith, D.G.A.; Burns, L.A.; Patkowski, K.; Sherrill, C.D. Revised Damping Parameters for the D3 Dispersion Correction to Density Functional Theory. *J. Phys. Chem. Lett.* **2016**, *7*, 2197–2203. [CrossRef]
53. Grimme, S.; Ehrlich, S.; Goerigk, L. Effect of the Damping Function in Dispersion Corrected Density Functional Theory. *J. Comput. Chem.* **2011**, *32*, 1456–1465. [CrossRef]
54. Dunning, T.H. Gaussian Basis Sets for Use in Correlated Molecular Calculations. I. The Atoms Boron through Neon and Hydrogen. *J. Chem. Phys.* **1989**, *90*, 1007–1023. [CrossRef]
55. Kendall, R.A.; Dunning, T.H.; Harrison, R.J. Electron Affinities of the First-Row Atoms Revisited. Systematic Basis Sets and Wave Functions. *J. Chem. Phys.* **1992**, *96*, 6796–6806. [CrossRef]
56. Henkelman, G.; Uberuaga, B.P.; Jónsson, H. A Climbing Image Nudged Elastic Band Method for Finding Saddle Points and Minimum Energy Paths. *J. Chem. Phys.* **2000**, *113*, 9901–9904. [CrossRef]

57. Neese, F.; Wennmohs, F.; Becker, U.; Riplinger, C. The ORCA Quantum Chemistry Program Package. *J. Chem. Phys.* **2020**, *152*, 224108. [CrossRef]
58. Neese, F. Software Update: The ORCA Program System—Version 5.0. *WIREs Comput. Mol. Sci.* **2022**, *12*, e1606. [CrossRef]

molecules

MDPI

Article

Two Coupled Low-Barrier Large Amplitude Motions in 3,5-Dimethylanisole Studied by Microwave Spectroscopy

Safa Khemissi [1,*], Lynn Ferres [2] and Ha Vinh Lam Nguyen [1,3,*]

1 Univ Paris Est Créteil and Université Paris Cité, CNRS, LISA, F-94010 Créteil, France
2 Institute of Physical Chemistry, RWTH Aachen University, Landoltweg 2, D-52074 Aachen, Germany; lynn.ferres@rwth-aachen.de
3 Institut Universitaire de France (IUF), F-75231 Paris, France
* Correspondence: safa.khemissi@lisa.ipsl.fr (S.K.); lam.nguyen@lisa.ipsl.fr (H.V.L.N.)

Abstract: The microwave spectrum of 3,5-dimethylanisole was recorded using a pulsed molecular jet Fourier transform microwave spectrometer, covering the frequency range from 2.0 to 26.5 GHz. Splittings from internal rotations of the *syn-m* and *anti-m*-methyl groups were observed, analyzed, and modeled using the *XIAM* and the *ntop* programs for a data set including 622 rotational lines. The torsional barriers of the *syn-m* and *anti-m*-methyl groups were determined to be 58.62367(53) cm^{-1} and 36.28449(69) cm^{-1}, respectively. The low barriers to internal rotation of both methyl groups posed significant challenges for spectral analysis and modeling. The successful assignment was achieved using combination difference loops and separately fitting the five torsional components. Comparing the torsional barriers observed in various toluene derivatives with methyl groups at *meta*-positions supports the assumption that electrostatic effects contribute more significantly than steric effects in the low-barrier cases of aromatic molecules.

Keywords: microwave spectroscopy; large amplitude motions; internal rotation; low barriers

1. Introduction

Rotational spectroscopic investigations into aromatic molecules containing methyl internal rotors are relatively scarce, and studies on those with low torsional barriers are even more limited [1]. This is most likely due to the complexity involved in analyzing and modeling the fine splittings caused by these Large Amplitude Motions (LAMs) [2]. For molecules with one methyl group, all rotational lines split into A-E doublets. The extent of these splittings depends significantly on the methyl torsional barrier. In general, the lower the barrier, the larger the splittings, and the more challenging the spectral analysis and modeling.

Low torsional barriers (say < 100 cm^{-1}) often arise from a steric-hindrance-free environment around the methyl rotors, as observed in several monomethyl-substituted aromatic molecules such as *p*-toluic acid (7.9 cm^{-1}) [3], *m*-tolunitril (14.2 cm^{-1}) [4], *m*-nitrotoluene (6.8 cm^{-1}) [5], *p*-tolualdehyde (28.1 cm^{-1}) [6], 2-methylthiazole (34.1 cm^{-1}) [7], and 3-methylphenylacetylene (11.4 cm^{-1}) [8]. The variation in barrier heights is often attributed to electronic effects on the π-electron conjugation system caused by other substituents on the aromatic rings. However, a comprehensive understanding of the steric and electronic influences on these barriers remains elusive due to the limited number of studies available.

The complexity increases when dealing with molecules containing two methyl groups, where only a few studies have been reported. In the case of two equivalent methyl groups, the rotational lines split into quartets; with two inequivalent methyl groups, they split into quintets [2]. Most two-top investigations in the literature have focused on molecules with two high barriers, as they are easier to analyze. Molecules with one intermediate barrier and one low barrier, such as *syn*-2,5-dimethylbenzaldehyde [9], 2,5-dimethylfluorobenzene [10], 2,5-dimethylanisole [11], 2,4-dimethylthiazole [12], and 2,3-dimethylanisole [13], are more challenging but have also been studied. Only two toluene derivatives with two low torsional barriers have been investigated: xylene (4.5 cm^{-1}) [14] and 3,5-dimethylbenzaldehyde (*syn-m* 53 cm^{-1} and *anti-m* 25.3 cm^{-1}) [9]. For these cases, the large splittings between the torsional species significantly complicate the spectral analysis, making accurate modeling of the spectra particularly difficult.

Several program codes can handle two methyl internal rotors. The *XIAM* program [15] is the most widely used in the microwave spectroscopic community. *XIAM* is user-friendly and offers a good balance between calculation accuracy and speed, thanks to efficient matrix factorization and basis transformations. The fitting results are generally satisfactory when the barrier(s) hindering the methyl internal rotation(s) is intermediate (~200–600 cm^{-1}) or high (>600 cm^{-1}) [16–21]. However, *XIAM* often struggles with low barrier cases, failing to achieve standard deviations within the measurement accuracy due to the limited number of available higher-order parameters. Other programs, such as *BELGI-C_s-2Tops* [22], *PAM-C_{2v}-2tops* [23], and *ERHAM* [24], can address this issue more effectively by fitting higher-order parameters [10,25,26]. Another program, *ntop,* also handles low barriers of multiple methyl rotors efficiently and has been successfully applied to two two-top molecules: 2,4-dimethylanisole [27] and 4-methylacetophenone [28,29].

In the present study, we report the microwave spectrum of 3,5-dimethylanisole (35DMA, illustrated in Figure 1), which contains two inequivalent methyl groups in the *meta* positions, both undergoing internal rotation hindered by low barriers. We initially applied the *XIAM* program for spectral assignment and fitting, but it failed to achieve a fit with standard deviation within the measurement accuracy. This goal was reached by a subsequent fit with the *ntop* code. We then compared the low torsional barrier values determined for 35DMA with those of the 2,3-, 2,4-, and 3,4-isomers, as well as other toluene derivatives with *meta*-methyl groups, to gain a better understanding of the electronic effects on the methyl torsional barriers.

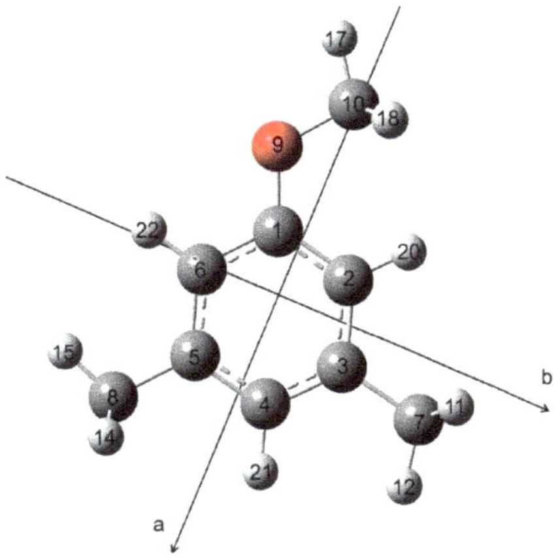

Figure 1. Molecular structure of 35DMA optimized at the B3LYP-D3BJ/6-311++G(d,p) level of theory in the principal axes of inertia. The oxygen atom is red, the carbon atoms are grey, and the hydrogen atoms are white.

2. Quantum Chemical Calculations

2.1. Conformational Analysis

All quantum chemical calculations were performed using the *GAUSSIAN 16* software package [30]. To identify the possible conformers of 35DMA, a conformational analysis was conducted by varying the dihedral angle $\beta = \angle(C_2, C_1, O_9, C_{10})$ in 10° increments, corresponding to a rotation about the C_1–O_9 bond (see Figure 1 for atom labeling), while all other geometry parameters were optimized at the B3LYP-D3BJ/6-311++G(d,p) level of theory [31–35], which was used in previous studies on other dimethylanisoles. The potential energy curve is shown in Figure 2. This level of theory predicted two equivalent minima at β = 0° and β = 180°, corresponding to two identical molecular structures of 35DMA.

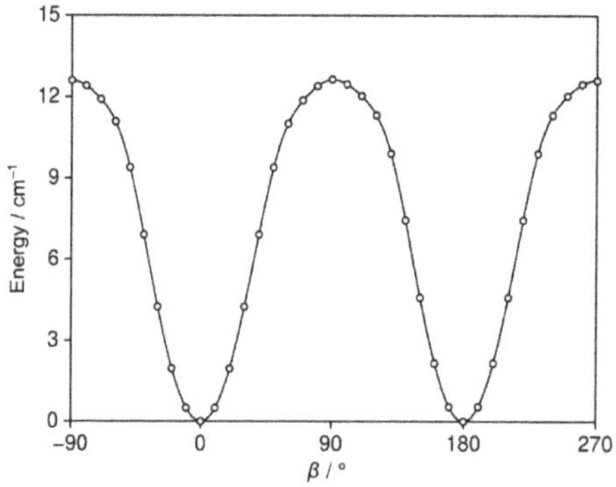

Figure 2. The potential energy curve of 35DMA obtained at the B3LYP-D3BJ/6-311++G(d,p) level of theory by varying the dihedral angle $\beta = \angle(C_2, C_1, O_9, C_{10})$ in 10° steps, corresponding to the rotation of the methoxy group. The energies are given relative to the lowest value of −425.5592664 Hartree.

2.2. Basis Set Variation

Geometry optimizations were carried out at the B3LYP-D3BJ/6-311++G(d,p) [31–35], MP2/6-311++G(d,p) [35,36], and MP2/6-31G(d,p) [35,36] levels of theory. These levels were chosen for their ability to produce reliable rotational constants that closely match experimental values for various six-membered aromatic molecules with methyl groups, such as the five isomers of the dimethylfluorobenzene family [10,37–40] and the 2,3- [13], 2,4- [27], 3,4-dimethylanisole [41] isomers. Geometry optimizations were followed by harmonic frequency calculations to verify that the obtained structure is a real minimum and anharmonic frequency calculations to access ground state rotational constants and centrifugal distortion constants. The main contribution of the difference between the ground state rotational constants and the equilibrium rotational constants is the vibration–rotation interaction, which is, to a first approximation, one-half the sum of the α values from each of the fundamental vibrations. The structure optimized at the B3LYP-D3BJ/6-311++G(d,p) level is illustrated in Figure 1. The calculated equilibrium rotational constants, vibrational ground state rotational constants, centrifugal distortion constants, and dipole moment components are presented in Table 1. The atomic Cartesian coordinates are given in Table S1 in the Supplementary Materials.

For benchmarking purposes, and to offer alternative levels beyond MP2/6-31G(d,p), MP2/6-311++G(d,p), and B3LYP-D3BJ/6-311++G(d,p), geometry optimizations were performed using various methods and basis sets. Calculations employed density functional theory (DFT) methods including B3LYP-D3 [31–33], B3LYP-D3BJ [31–34], CAM-B3LYP-D3BJ [42], M06-2X [43], ωB97X-D [44], MN15 [45], and PBE0 [46]. Additionally, the ab initio MP2 [36] and CCSD [47] methods were used. These methods were combined with a range of Pople [35] and Dunning [48] basis sets, except for the coupled-cluster method, which was combined exclusively with the cc-pVDZ basis set. The predicted rotational constants are listed in Table S2 of the Supplementary Materials.

Table 1. Equilibrium rotational constants (A_e, B_e, C_e), vibrational ground state rotational constants (A_0, B_0, C_0), centrifugal distortion constants in the symmetrically reduced Hamiltonian obtained from anharmonic frequency calculations, dipole moment components, and V_3 potential for the *syn-m-* and *anti-m*-methyl groups at the 3- and 5-positions, respectively, calculated at the MP2/6-31G(d,p), MP2/6-311++G(d,p), and B3LYP-D3BJ/6-311++G(d,p) levels of theory.

Par.	Unit	MP2/6-31G(d,p)	MP2/6-311++G(d,p)	B3LYP-D3BJ
A_e	MHz	1736.4	1728.6	1740.3
B_e	MHz	1095.5	1094.0	1092.6
C_e	MHz	680.3	678.6	679.7
A_0	MHz	1724.2	1717.8	1726.9
B_0	MHz	1085.8	1084.5	1084.0
C_0	MHz	674.9	673.3	674.6
D_J	kHz	0.073437	0.074241	0.074614
D_{JK}	kHz	−0.104737	−0.105306	−0.097750
D_K	kHz	0.037086	0.036934	0.029037
d_1	kHz	−0.016684	−0.016884	−0.016691
d_2	kHz	0.001187	0.001205	0.002070
$\mid\mu_a\mid$	D	0.5	0.5	0.5
$\mid\mu_b\mid$	D	1.1	1.2	1.2
$\mid\mu_c\mid$	D	0.1	0.0	0.0
$V_{3,syn\text{-}m}$	cm^{-1}	55.2	69.2	67.2
$V_{3,anti\text{-}m}$	cm^{-1}	36.8	34.6	44.0

2.3. Methyl Internal Rotations

The molecule 35DMA features a methoxy group and two inequivalent methyl groups undergoing internal rotation, located at the *syn-meta* and *anti-meta* positions. Previous studies have estimated the barrier hindering the methoxy methyl internal rotation in anisole to be approximately 1200 cm^{-1} [49]. Prior research by Ferres et al. on 2,3- [13], 2,4- [27], and 3,4-dimethylanisole [41] has demonstrated that no splittings resulting from the methoxy methyl torsion are observed in their microwave spectra. For 35DMA, the barrier to internal rotation was calculated at the MP2/6-31G(d,p), MP2/6-311++G(d,p), and B3LYP-D3BJ/6-311++G(d,p) levels of theory by varying the dihedral angle $\varphi = \angle(C_1, O_9, C_{10}, H_{17})$ in 10° increments. Due to the three-fold C_3 symmetry of the methyl group, a rotation of 120° was sufficient. The calculated V_3 potential terms were 1081.8 cm^{-1}, 981.0 cm^{-1}, and 1070.5 cm^{-1}, respectively. These results indicate that no resolvable splittings arising from the internal rotation of the methoxy methyl group are expected to be observed in the microwave spectrum.

Low barriers hindering the internal rotations of the two inequivalent methyl groups at the 3- and 5-positions (*syn-m* and *anti-m*, respectively) were anticipated due to their steric-free environment. To calculate the barrier heights, the respective dihedral angles $\alpha_1 = \angle(C_2, C_3, C_7, H_{12})$ and $\alpha_2 = \angle(C_4, C_5, C_8, H_{15})$ were also varied in 10° increments, while optimizing all other parameters. The resulting potential energies were parameterized using a 1D-Fourier expansion, with the corresponding coefficients provided in Table S3 of the Supplementary Materials. The potential energy curves are shown in Figures 3 and 4. At the B3LYP-D3BJ/6-311++G(d,p) level of theory, calculations yielded typical three-fold potentials for both the *syn-m* and *anti-m*-methyl groups, with V_3 terms of 67.2 cm^{-1} and 44.0 cm^{-1}, respectively. Small V_6 terms of about 4.8 cm^{-1} and 7.2 cm^{-1} were obtained. At both the MP2/6-311++G(d,p) and MP2/6-31G(d,p) levels, asymmetric potential energy curves were observed, with significant V_6 contributions. For the *syn-m*-methyl group, the V_3/V_6 ratios were calculated as 69.2 cm^{-1}/29.4 cm^{-1} and 55.2 cm^{-1}/26.2 cm^{-1}, respectively. For the *anti-m*-methyl group, the respective V_3/V_6 ratios are 34.6 cm^{-1}/26.2 cm^{-1} and 36.8 cm^{-1}/22.0 cm^{-1}. The asymmetry observed in the potential energy curves at the MP2 levels suggests a significant interaction between the internal rotors, as further evidenced by the potential energy surfaces, as described in Section 2.4.

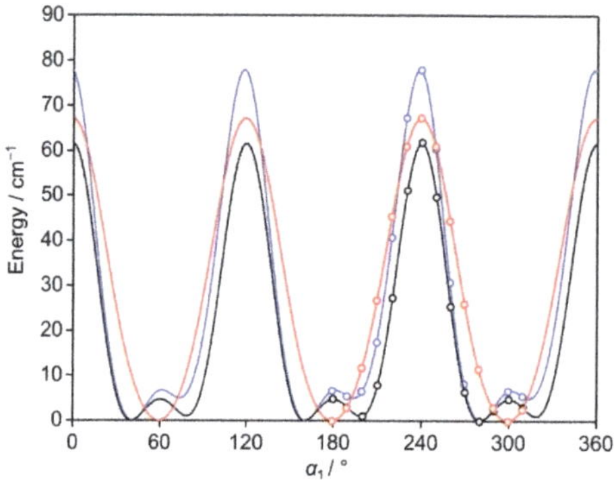

Figure 3. The potential energy curves of 35DMA calculated at the B3LYP-D3BJ/6-311++G(d,p) (red curve), MP2/6-311++G(d,p) (blue curve), and MP2/6-31G(d,p) (black curve) levels of theory by varying the dihedral angle $\alpha_1 = \angle(C_2, C_3, C_7, H_{12})$ in 10° increments, corresponding to the rotation of the 3-methyl group (*syn-m*) about the C_3–C_7 bond.

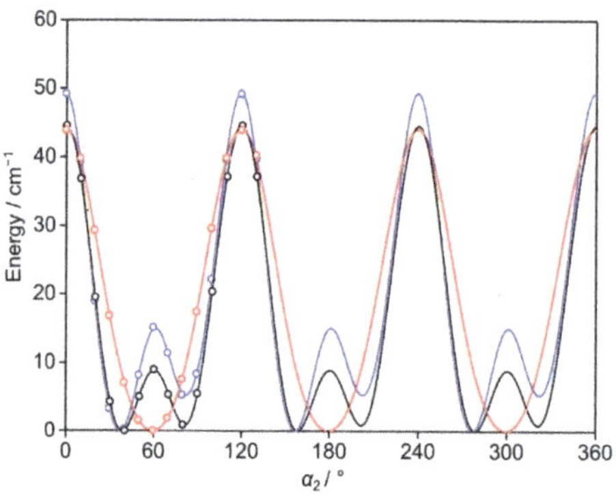

Figure 4. The potential energy curves of 35DMA calculated at the B3LYP-D3BJ/6-311++G(d,p) (red curve), MP2/6-311++G(d,p) (blue curve), and MP2/6-31G(d,p) (black curve) levels of theory by varying the dihedral angle $\alpha_2 = \angle(C_4, C_5, C_8, H_{15})$ in 10° increments, corresponding to the rotation of the 5-methyl group (*anti-m*) about the C_5–C_8 bond.

2.4. Potential Energy Surfaces

The coupling between the methyl internal rotations in aromatic molecules through π conjugation, particularly when hindered by low barriers, can complicate the spectral assignment and modeling. To gain insight into this coupling, we calculated two-dimensional potential energy surfaces (2D-PES) based on the dihedral angles α_1 and α_2 at the B3LYP-D3BJ/6-311++G(d,p), MP2/6-311++G(d,p), and MP2/6-31G(d,p) levels of theory. Both α_1 and α_2 were varied in 10° increments, while all other geometry parameters were optimized. The resulting potential energy points were parameterized using a 2D-Fourier expansion, with the coefficients provided in Table S4 of the Supplementary Materials. The 2D-PES plots are shown in Figures 5 and 6. The 2D-PES calculated at the B3LYP-D3BJ/6-311++G(d,p) level exhibits minima with an oblate, stretched shape rather than a circular shape, indicating a coupling between the two tops. This coupling can also be recognized by the significant V_{cc} term $\cos(3\alpha_1)\cos(3\alpha_2)$, as shown in Table S4 of the Supplementary Materials. The 2D-PES calculated at the MP2/6-311++G(d,p) and MP2/6-31G(d,p) levels revealed the presence of asymmetric double minima, consistent with the 1D potential curves illustrated in Figures 3 and 4 and confirmed the substantial coupling of the two rotors (note also the V_{cc} terms in Table S4).

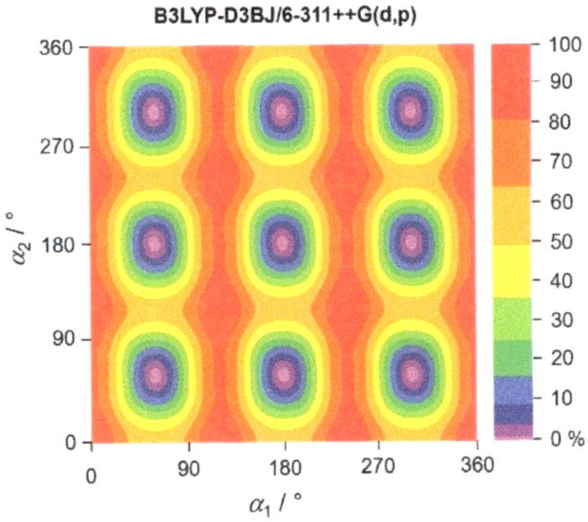

Figure 5. Two-dimensional potential energy surface of 35DMA as a function of the dihedral angles $\alpha_1 = \angle(C_2, C_3, C_7, H_{12})$ and $\alpha_2 = \angle(C_4, C_5, C_8, H_{15})$, calculated at the B3LYP-D3BJ/6-311++G(d,p) level of theory. The dihedral angles were varied in 10° increments, while all other geometry parameters were optimized. The numbers in the color code indicate the energy (in percent) relative to the energetic minimum (0%) and maximum (100%), with $E_{min} = -425.559266$ Hartree and $E_{max} = -425.558897$ Hartree. Note that a non-linear scale is used.

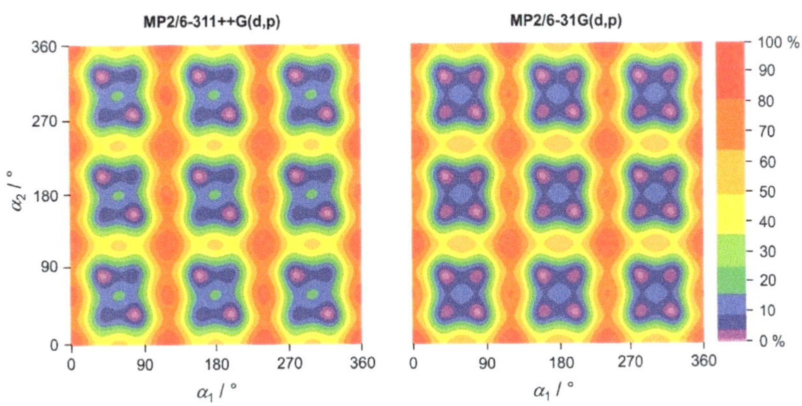

Figure 6. Two-dimensional potential energy surfaces of 35DMA as a function of the dihedral angles α_1 and α_2, calculated at the MP2/6-311++G(d,p) and MP2/6-31G(d,p) levels of theory. The dihedral angles were varied in 10° increments, while all other geometry parameters were optimized. The numbers in the color code indicate the energy (in percent) relative to the energetic minimum (0%) and maximum (100%), with $E_{min} = -424.244142$ Hartree and $E_{max} = -424.243628$ Hartree for the MP2/6-311++G(d,p) level, and $E_{min} = -424.083418$ Hartree and $E_{max} = -424.082994$ Hartree for the MP2/6-31G(d,p) level.

3. Microwave Spectroscopy

3.1. Measurements

The microwave spectra were recorded using a pulsed molecular jet Fourier transform microwave spectrometer with a coaxially-oriented beam-resonator arrangement (COBRA) [50] in the frequency range of from 2.0 to 26.5 GHz. The substance, with a purity of 99%, was purchased from Alfa Aesar, Karlsruhe, Germany. A few drops of the sample were placed on a 5 cm piece of pipe cleaner inserted into a stainless steel tube mounted upstream of the nozzle. Helium was passed over the sample at a backing pressure of 2 bar, and the resulting Helium-35DMA mixture was expanded into the cavity. In the frequency range from 9.8 to 13.3 GHz, a survey scan was recorded by overlapping spectra with 50 co-added free induction decays (FIDs) for each spectrum at a step width of 0.25 MHz. All signals observed in the survey scan were remeasured at higher resolution, where each line appeared as doublets due to the Doppler effect caused by the COBRA setup. The measurement accuracy is estimated to be approximately 4 kHz. A scan portion and a typical high-resolution spectrum showing the (00) and (11) torsional species of the $7_{07} \leftarrow 6_{06}$ and $7_{17} \leftarrow 6_{06}$ rotational transitions, respectively, are shown in Figure 7.

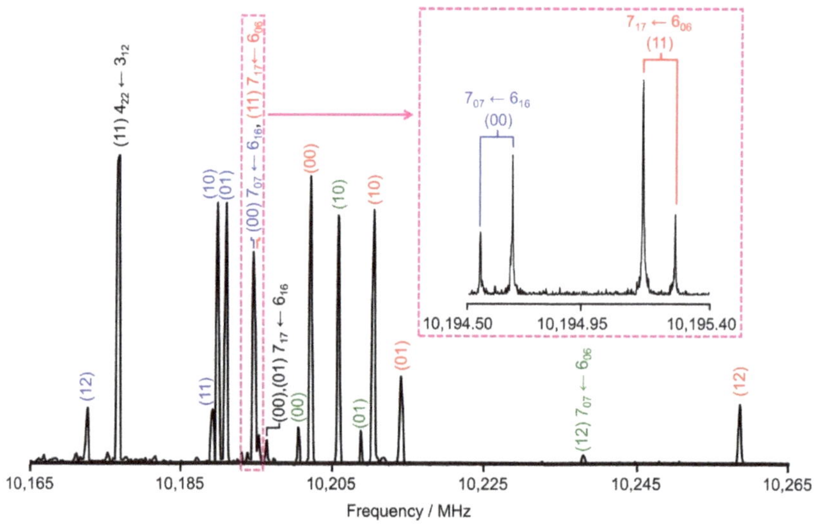

Figure 7. A portion of the survey scan from 10,165 MHz and 10,265 MHz recorded by overlapping spectra with 0.25 MHz step width and 50 co-added FIDs per each single measurement. The absolute intensity is presented in arbitrary units. All lines are labeled with their torsional species (00), (10), (01), (11), and (12) as well as their corresponding rotational quantum numbers $J'_{K'_aK'_c} \leftarrow J_{K_aK_c}$. Inset: a measurement at high resolution showing the (00) and (11) torsional species of the $7_{07} \leftarrow 6_{16}$ and $7_{17} \leftarrow 6_{06}$ rotational transitions, respectively. Doppler doublets are marked by brackets.

3.2. Spectral Assignments

As a first step, the assignment was initiated by treating the molecule as a rigid-rotor with no internal rotation effects, focusing only on the (00) torsional species. Using the rotational constants and dipole moment components calculated at the B3LYP-D3BJ/6-311++G(d,p) level of theory (see Table 1), we predicted the theoretical microwave spectrum using the *XIAM* program [15]. This level of theory was chosen because it has predicted rotational constants that closely match experimental values in previous studies on other dimethylanisole isomers [13,27,41], and was also specifically recommended to guide the spectral assignment of similar molecular systems, such as the dimethylfluorobenzene family [10,37–40]. Given the dipole moment components in Table 1, the spectrum is expected to display strong b-type transitions, weaker a-type transitions, and an absence of c-type transitions. We concentrated first on identifying intense b-type transitions with low J values (J_{max} = 5) and low K_a values (K_a = 0, 1), followed by a-type transitions. This assignment process was straightforward. As a next step, we searched for transitions with higher J and K_a values, completing the rigid-rotor assignment.

Due to the high barrier to internal rotation of the methoxy methyl group, the spectrum is expected to exhibit torsional splittings arising from the internal rotations of only the two inequivalent *syn-m* and *anti-m*-methyl groups. Each rotational transition splits into five torsional components (00), (10), (01), (11), and (12) [41]. The low barriers to internal rotation of both methyl groups result in large spectral splittings, making the assignment process challenging. To analyze these complex LAMs, we first focused on the (10) species associated with the *syn-m*-methyl torsion, given its higher predicted barrier (see Table 1). We calculated a theoretical one-top spectrum using the rotational constants obtained from the rigid-rotor fit as well as the $V_{3,1}$ potential term and the angle $\angle(i_1,a)$ from the B3LYP-D3BJ/6-311++G(d,p) calculations. Again, we targeted first the intense b-type transitions with low J and K_a values, especially the $(J+1)_{0(J+1)} \leftarrow J_{1J}$ branch, as this branch has the smallest splittings of all b-type lines. We then moved on to assign a-type transitions, aiming to find closed combination difference loops (Ritz cycles) [51] to verify the accuracy of the assigned lines (see Figure 8). Subsequently, we fitted the loop-checked (10) lines separately using the *SFLAMS* program [28], incorporating odd power parameters in the $\mathbf{H_{op}}$ term of the Hamiltonian, as shown in Equation (1):

$$\mathbf{H_{op}} = (q + q_J\mathbf{P}^2 + q_K\mathbf{P_z^2})\mathbf{P_z} + (r + r_J\mathbf{P}^2)\mathbf{P_x} + \frac{1}{2}r_K\left\{\mathbf{P_z^2}, \mathbf{P_x}\right\}. \quad (1)$$

With predictions from the separate fit, further higher J and K_a transitions were identified, extending the loops as well as the *XIAM* and *SFLAMS* fits.

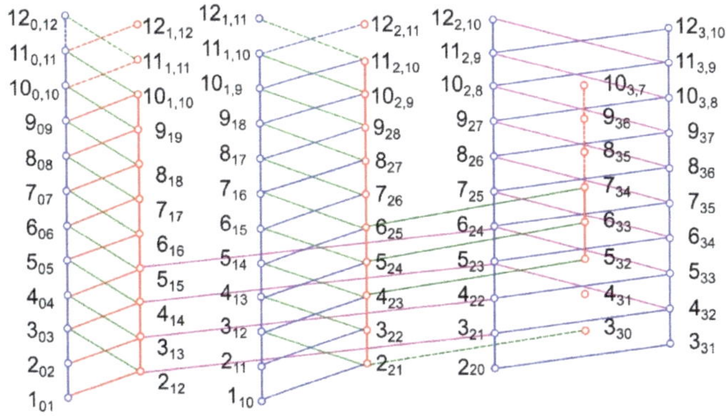

Figure 8. Schematic illustration of almost all observed rotational transitions for the (10), (01), (11), and (12) torsional components of 35DMA. The solid lines connecting the circles represent transitions verified by combination difference loops, which sum to a measurement accuracy of 4 kHz. Some weak transitions with high J and K_a could not be measured, and some loops remain open. Transitions involved in such loops are indicated as dashed lines.

After assigning the (10) torsional species, we repeated the same procedure to assign the (01) species associated with the internal rotation of the *anti-m*-methyl group, including checking with Ritz cycles and separate fits. Then, we combined the two *XIAM* one-top fits to a two-top fit to predict and search for the interaction (11) and (12) components. After numerous trials, we successfully assigned a sufficient number of lines which were verified using combination difference loops and separate fits with *SFLAMS*. Predictions from these fits subsequently guided further assignments. Given the C_s symmetry of the molecular frame, *c*-type transitions were not expected in the microwave spectrum of 35DMA, as the *c*-component of the dipole moment is zero. However, some *c*- and *x*-type forbidden transitions were observed for the (10), (01), (11), and (12) torsional species. For those perturbation-allowed transitions, the K_a and K_c quantum numbers have lost their meaning and only indicate the energy order. These transitions are highlighted in the frequency list in Table S5 of the Supplementary Materials. The molecular parameters of the five separate fits are given in Table 2.

Table 2. Molecular parameters of the (00), (10), (01), (11), and (12) separate fits obtained with the program *SFLAMS*.

Par. [a]	Unit	Fit (00)	Fit (10)	Fit (01)	Fit (11)	Fit (12)
A	MHz	1744.77418(29)	1744.00056(64)	1735.59250(52)	1734.7419(13)	1735.1469(14)
B	MHz	1101.19864(18)	1096.03548(24)	1097.88093(18)	1093.31277(46)	1092.68724(31)
C	MHz	680.562851(68)	680.563975(88)	680.531514(74)	680.55101(12)	680.53608(16)
D_J	kHz	0.0286(14)	0.0203(30)	0.0288(24)	0.0278(76)	0.0326(35)
D_{JK}	kHz	0.1740(66)	0.170(26)	0.330(22)	0.285(60)	0.153(27)
d_1	kHz	−0.01947(87)	−0.0124(14)	−0.0235(11)	−0.0204(39)	−0.0181(19)
d_2	kHz	−0.00821(35)	−0.00525(53)	−0.01183(43)	−0.0092(16)	−0.00443(74)
q	MHz		1817.842(14)	1104.1998(14)	1237.8649(25)	881.224(36)
r	MHz		563.3230(16)	630.8680(12)	70.043(10)	1127.8147(20)
q_J	kHz		−14.844(99)	−44.003(69)	−55.39(13)	
q_K	kHz		13.52(25)	−21.05(13)	−7.81(17)	
r_J	kHz		−23.138(55)	−15.644(43)	8.18(42)	−34.871(83)
r_K	kHz		−23.90(31)	−120.59(22)	−94.1(22)	−137.82(54)
N [b]		141	148	144	100	89
rms [c]	kHz	3.6	4.8	3.3	4.8	3.0

[a] All parameters refer to the principal axis system. Waston's S reduction in I^r representation was used. [b] Number of lines. [c] Root-mean-square deviation of the fit.

4. Results of the Global Fits and Discussion

In total, 622 torsional lines were assigned. The data set was fitted globally with the *XIAM* program [15], yielding a standard deviation of 347.5 kHz, while the measurement accuracy is 4 kHz. To improve the quality of the global fit, we applied the *ntop* program [27] to the same data set, successfully decreasing the rms deviation to 5.3 kHz by incorporating a larger number of higher-order parameters. The molecular parameters obtained using the *XIAM* and *ntop* codes are presented in Table 3. A list of all fitted frequencies along with their residuals is available in Table S5 in the Supplementary Materials. Due to the strong correlation between the V_3 terms and the internal rotation constants F_0, we fixed F_0 to 160 GHz, a value often found for methyl groups, in both the *XIAM* and *ntop* fits. Note that if we fit F_0 in *XIAM* and include additional high-order parameters (V_J, V_K, and V_-, also referred to as D_{c3J}, D_{c3K}, and D_{c3-}) [28,52], the standard deviation can be decreased to 115.9 kHz, but the F_0 values are unusual for methyl groups (152.4 GHz for one and 164.4 GHz for the other, while typical values range between 158 GHz and 160 GHz) and the V_3 terms are less precisely determined. The molecular parameters obtained with this fit are given in Table S6 in the Supplementary Materials.

Table 3. Molecular parameters of 35DMA in the principal axis system obtained using the *XIAM* and *ntop* programs. Rotor 1 refers to the *syn-m* and rotor 2 to the *anti-m*-methyl group.

Par. [a]	Unit	Fit *XIAM*	Fit *ntop*	Calc. [b]
A	MHz	1737.126(11)	1742.502(62)	1740.3
B	MHz	1095.3062(52)	1094.343(15)	1092.6
C	MHz	680.5538(17)	680.688(11)	679.7
D_J	kHz		0.01311(38)	0.074614
D_{JK}	kHz		0.1113(29)	−0.097750
d_1	kHz		−0.00679(22)	−0.016691
d_2	kHz		−0.003364(96)	0.002070
V_{cc}	cm^{-1}	−7.7494(50)	−9.803586(86)	−7.8
V_{ss}	cm^{-1}		0.78428(51)	0.6
$V_{3,1}$	cm^{-1}	60.1207(56)	58.62367(53)	67.2
$V_{3,2}$		37.5936(64)	36.28449(69)	44.0
$D_{\pi^2J,1}$	MHz	4.41(71)		
$D_{\pi^2K,1}$	MHz	−0.060(11)		
$D_{\pi^2-,1}$	MHz	0.0261(22)		
$D_{\pi^2J,2}$	MHz		0.02403(75)	
$D_{\pi^2K,2}$	MHz	0.3687(78)	−0.5658(79)	
$D_{mK,2}$	MHz		0.4228(64)	
$d_{m,2}$	MHz		0.0636(16)	
$V_{J,1}$	MHz	−44.7(72)	−0.46049(65)	
$V_{K,1}$	MHz		2.0337(70)	
$V_{-,1}$	MHz		−0.31988(69)	
$V_{J,2}$	MHz		1.076(14)	
$V_{K,2}$	MHz		−10.36(11)	
$V_{-,2}$	MHz		1.155(15)	
$Q_{zJ,1}$	kHz		−3.360(26)	
$Q_{zK,1}$	kHz		8.806(88)	
$Q_{z-,1}$	kHz		4.014(79)	
$\angle(i_1,a)$	°	78.0849(22)	78.15036(51)	78.6
$\angle(i_1,b)$	°	11.9151(22)	11.84964(51)	11.4
$\angle(i_1,c)$	°	90.0 [c]	90.0 [c]	90.0
$\angle(i_2,a)$	°	41.7799(22)	41.9409(13)	41.5
$\angle(i_2,b)$	°	131.7799(22)	131.9409(13)	131.5
$\angle(i_2,c)$	°	90.0 [c]	90.0 [c]	90.0
N [d]		622	622	
rms [e]	kHz	347.5	5.3	

[a] All molecular parameters refer to the principal axis system. Watson's S reduction in I^r representation was used. [b] Calculated at the B3LYP-D3BJ/6-311++G(d,p) level of theory. The rotational constants refer to the equilibrium structure. The centrifugal distortion constants are obtained from anharmonic frequency calculations. [c] Fixed due to symmetry. [d] Number of lines. [e] Root-means-square deviation of the fit.

The experimental rotational constants were compared with theoretical values. Since the values obtained from *XIAM* and *ntop* differ slightly, we use the experimental values from *XIAM*. This choice is due to the fact that fewer parameters are fitted, resulting in lower correlation and a clearer physical meaning of the rotational constants. All method and basis set combinations mentioned in Section 2.2 produced satisfactory results (see Table S2 in the Supplementary Materials), yielding equilibrium rotational constants in good agreement with the experimental values. An excellent agreement was found between the experimental values and those calculated at the MP2/6-31G(d,p), ωB97X-D/cc-pVDZ, MN15/cc-pVDZ, and MN15/aug-cc-pVDZ levels of theory, with deviations under 0.04%. With these results, we continue to recommend the MP2/6-31G(d,p) level for assignment guidance of small rigid molecules containing a benzene ring. From a theoretical perspective, the experimental rotational constants $B_{0,\text{exp.}}$ should be compared to the zero-point vibrationally corrected rotational constants $B_{0,calc.}$ rather than to the equilibrium rotational constants $B_{e,calc.}$. However, performing anharmonic frequency calculations at a sufficiently high level of theory is computationally demanding for an experimental laboratory, and is not necessary, particularly when the primary goal is to obtain predicted values to aid in spectral assignment. For this purpose, we rely on benchmarking studies of equilibrium rotational constants $B_{e,calc.}$ to identify cost-efficient levels of theory that yield a small $B_{0,exp.}$-$B_{e,calc.}$ difference due to error compensation. Our experience suggests that, if a given level of theory provides a small $B_{0,exp.}$-$B_{e,calc.}$ difference for one molecule, it often does so for structurally similar molecules, which significantly facilitates initial spectral assignments in new projects.

The torsional barriers of the *syn-m* and *anti-m*-methyl groups in 35DMA were determined to be 60.1207(56) cm^{-1} and 37.5936(64) cm^{-1}, respectively, using the *XIAM* program. Similar values of 58.62367(53) cm^{-1} and 36.28449(69) cm^{-1} were obtained with *ntop*. Compared with the predicted values shown in Table 1, the agreement is reasonably good. When comparing the torsional barriers found for the two *m*-methyl groups in 35DMA (molecule **8** in Figure 9) with other anisole derivatives, we observed that 35DMA is a combination of *anti-m* (**2a**) [53] and *syn-m*-methylanisole (**2b**) [53]. Specifically, the torsional barrier of 58.6 cm^{-1} for the *syn-m*-methyl group in 35DMA (**8**) closely matches the value of *syn-m*-methylanisole (55.8 cm^{-1}) (**2b**). Similarly, the barrier of 36.3 cm^{-1} for the *anti-m*-methyl group is nearly identical to the 36.6 cm^{-1} barrier observed in *anti-m*-methylanisole (**2a**). In contrast, a slightly higher value of 65.7 cm^{-1} was obtained for the *m*-methyl group in 2,5-dimethylanisole (**7**) [11]. The difference might arise from coupling between the two rotors, facilitated by electrostatic interactions transmitted through π-electron delocalization in the benzene ring. This is supported by the significant V_{cc} term derived from both the experimental fits (see Table 3) and the 2D-PES calculations (Table S4). Electrostatic effects also explain the slight differences between barrier values for steric-free methyl groups at the *meta*-position and those at the *para*-position in *p*-methylanisole (**3**) (49.6 cm^{-1}) [26] and 2,4-dimethylanisole (**6**) (47.6 cm^{-1}) [27]. Much higher barriers, ranging from 441.1 cm^{-1} to 451.7 cm^{-1}, are observed for methyl groups at the *ortho*-position in *o*-methylanisole (**1**) [54], 2,4-dimethylanisole (**6**) [27], and 2,5-dimethylanisole (**7**) [11] due to steric hindrance. Steric effects also explain the intermediate barriers for the *meta* and *para*-methyl groups in *anti*-3,4-dimethylanisole (**5a**) (499.6 cm^{-1} and 533.5 cm^{-1}, respectively) [41] and *syn*-3,4-dimethylanisole (**5b**) (430.0 cm^{-1} and 467.9 cm^{-1}, respectively) [41], as well as the *m*-methyl group in 2,3-dimethylanisole (**4**) (518.7 cm^{-1}) [13]. An exception to the steric hindrance trend is the remarkably low barrier of 26.9 cm^{-1} for the *o*-methyl group in 2,3-dimethylanisole (**4**), which is squeezed between a methyl group and a methoxy group.

The molecule 3,5-dimethylbenzaldehyde (**9**) [9] is structurally similar to 35DMA (**8**), with two methyl groups in identical positions. Though similar, the barrier heights of the respective methyl groups of 3,5-dimethylbenzaldehyde (**9**) are lower than those of 35DMA (**8**). The discrepancy probably arises from the different electrostatic contributions of the methoxy and aldehyde groups to the π-electron system. This observation is further supported while comparing the barriers for 3,5-dimethylbenzaldehyde (**9**) with those for *anti-m*-methylbenzaldehyde (**10a**) [55] and *syn-m*-methylbenzaldehyde (**10b**) [55]. Not like in the case of 35DMA, the barriers are very different. Specifically, *anti-m*-methylbenzaldehyde (**10a**) has an extremely low barrier of 4.6 cm^{-1}, whereas 3,5-dimethylbenzaldehyde (**9**) shows a higher barrier of 25.3 cm^{-1}. Similarly, *syn-m*-methylbenzaldehyde (**10b**) has a barrier of 35.9 cm^{-1}, which is notably lower than the value of 53.0 cm^{-1} for 3,5-dimethylbenzaldehyde (**9**). These differences highlight the significant influence of electronic effects, functional groups, and the coupling between the two rotors on their torsional barriers. Given the limited number of molecules with two methyl groups undergoing internal rotations hindered by low barriers available in the literature, a comprehensive understanding of these effects remains challenging. Additional benchmarking studies on such molecules are needed to confirm and refine the assumptions regarding electrostatic and steric contributions.

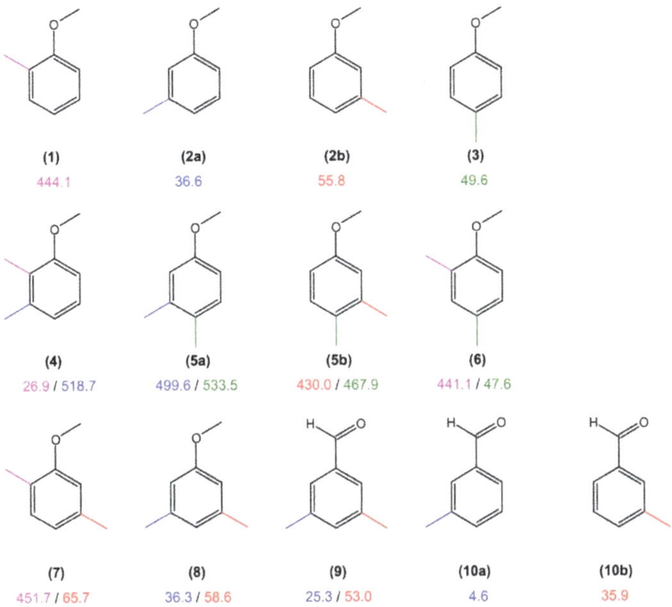

Figure 9. Comparison of the internal rotation barriers of the *syn-m* and *anti-m*-methyl groups in 35DMA with the values of other toluene derivatives. (**1**) *o*-methylanisole [54], (**2a**) *anti-m*-methylanisole [53], (**2b**) *syn-m*-methylanisole [53], (**3**) *p*-methylanisole [26], (**4**) 2,3-dimethylanisole [13], (**5a**) *anti*-3,4-dimethylanisole [41], (**5b**) *syn*-3,4-dimethylanisole [41], (**6**) 2,4-dimethylanisole [27], (**7**) 2,5-dimethylanisole [11], (**8**) 3,5-dimethylanisole (this work), (**9**) 3,5-dimethylbenzaldehyde [9], (**10a**) *anti-m*-methylbenzaldehyde [55], and (**10b**) *syn-m*-methylbenzaldehyde [55].

5. Conclusions

The microwave spectrum of 35DMA was recorded using a pulsed molecular jet Fourier transform microwave spectrometer operating in the frequency range from 2.0 to 26.5 GHz. The high barrier to internal rotation of the methoxy methyl group, exceeding 1000 cm^{-1}, resulted in unresolvable torsional splittings in the spectrum. In contrast, the low barrier heights of less than 60 cm^{-1} for the internal rotations of the 3- (*syn-m*) and the 5-methyl (*anti-m*) groups caused large torsional splittings, with each rotational transition appearing as a quintet. This made the spectral assignment particularly challenging. The initial fits of the five torsional species were performed using the *XIAM* program, but the final fit failed to achieve a standard deviation close to the measurement accuracy of 4 kHz. To verify the assigned frequencies, combination difference loops (Ritz cycles) were employed during the assignment process. Subsequently, separate fits for each torsional species were performed using the *SFLAMS* program, achieving standard deviations of approximately 4 kHz, thereby confirming the correctness of all assigned frequencies. A global fit was then carried out using the *ntop* program, which reduced the deviation from 347.5 kHz (*XIAM*) to 5.3 kHz. The barriers to internal rotation of the *syn-m* and *anti-m*-methyl groups were determined to be 58.62367(53) cm^{-1} and 36.28449(69) cm^{-1}, respectively. Comparisons with other toluene derivatives containing *meta*-methyl groups suggest that electrostatic effects significantly contribute to the barrier values observed in aromatic molecules.

Supplementary Materials: The following Supplementary Materials can be downloaded at: https://www.mdpi.com/article/10.3390/molecules30061195/s1, Table S1: Nuclear coordinates of 35DMA in the principal axis system calculated at the MP2/6-31G(d,p), MP2/6-311++G(d,p), and B3LYP-D3BJ/6-311++G(d,p) levels of theory; Table S2: Rotation constants of 35DMA (in MHz) calculated at different levels of theory; Table S3: Coefficients of Fourier expansion for the one-dimensional potential energy curves of 35DMA given in Figures 3 and 4 calculated at the B3LYP-D3BJ/6-311++G(d,p), MP2/6-311++G(d,p), and MP2/6-31G(d,p) levels of theory; Table S4: Coefficients of the two-dimensional Fourier expansions of the potential energy surfaces of 35DMA given in Figures 5 and 6 calculated at the B3LYP-D3BJ/6-311++G(d,p), MP2/6-311++G(d,p), and MP2/6-31G(d,p) levels of theory; Table S5: Fitted frequencies ($\nu_{obs.}$) of 35DMA. The $\nu_{obs.} - \nu_{calc.}$ residuals are obtained with *XIAM* and *ntop* programs given in Table 3. Table S6: Molecular parameters of 35DMA in the principal axis system obtained using the *XIAM* program by fitting the rotational constants $F_{0,1}$ and $F_{0,2}$ of the internal rotors.

Author Contributions: Conceptualization, L.F. and H.V.L.N.; methodology, S.K., L.F. and H.V.L.N.; software, H.V.L.N.; validation, H.V.L.N.; formal analysis, S.K., L.F. and H.V.L.N.; software, H.V.L.N.; investigation, S.K., L.F. and H.V.L.N.; software, H.V.L.N.; resources, H.V.L.N.; data curation, L.F.; writing—original draft preparation, S.K.; writing—review and editing, S.K., L.F. and H.V.L.N.; software, H.V.L.N.; visualization, S.K., L.F. and H.V.L.N.; software, H.V.L.N.; supervision, H.V.L.N.; project administration, H.V.L.N.; funding acquisition, H.V.L.N. All authors have read and agreed to the published version of the manuscript.

Funding: This work was supported by the European Union (ERC, 101040480-LACRIDO). The views and opinions expressed are, however, those of the authors only and do not necessarily reflect those of the European Union. Neither the European Union nor the granting authority can be held responsible for them.

Institutional Review Board Statement: Not applicable.

Informed Consent Statement: Not applicable.

Data Availability Statement: Data are contained within the article and Supplementary Materials.

Acknowledgments: Part of the simulations were performed with computing resources granted by the RWTH Aachen University under the project rwth0506.

Conflicts of Interest: The authors declare no conflicts of interest.

References

1. Nguyen, H.V.L.; Caminati, W.; Grabow, J.-U. The LAM of the Rings: Large Amplitude Motions in Aromatic Molecules Studied by Microwave Spectroscopy. *Molecules* **2022**, *27*, 3948. [CrossRef] [PubMed]
2. Nguyen, H.V.L.; Kleiner, I. Understanding (Coupled) Large Amplitude Motions: The Interplay of Microwave Spectroscopy, Spectral Modeling, and Quantum Chemistry. *Phys. Sci. Rev.* **2022**, *7*, 679–726. [CrossRef]
3. Schnitzler, E.G.; Seifert, N.A.; Kusuma, I.; Jäger, W. Rotational Spectroscopy of *p*-Toluic Acid and Its 1:1 Complex with Water. *J. Phys. Chem. A* **2017**, *121*, 8625–8631. [CrossRef]
4. Bruhn, T.; Mäder, H. The Microwave Spectrum of *m*-Tolunitrile: Methyl Internal Rotation and ^{14}N Nuclear Quadrupole Coupling. *J. Mol. Spectrosc.* **2000**, *200*, 151–161. [CrossRef] [PubMed]
5. Roucou, A.; Kleiner, I.; Goubet, M.; Bteich, S.; Mouret, G.; Bocquet, R.; Hindle, F.; Meerts, W.L.; Cuisset, A. Towards the Detection of Explosive Taggants: Microwave and Millimeter-Wave Gas-Phase Spectroscopies of 3-Nitrotoluene. *ChemPhysChem* **2018**, *19*, 1056–1067. [CrossRef]
6. Saal, H.; Grabow, J.-U.; Walker, A.H.; Hougen, J.; Kleiner, I.; Caminati, W. Microwave Study of Internal Rotation in *para*-ToluAldehyde: Local versus Global Symmetry Effects at the Methyl-Rotor Site. *J. Mol. Spectrosc.* **2018**, *351*, 55–61. [CrossRef]
7. Nguyen, T.; Van, V.; Gutlé, C.; Stahl, W.; Schwell, M.; Kleiner, I.; Nguyen, H.V.L. The Microwave Spectrum of 2-Methylthiazole: ^{14}N Nuclear Quadrupole Coupling and Methyl Internal Rotation. *J. Chem. Phys.* **2020**, *152*, 134306. [CrossRef]
8. Obenchain, D.A.; Pinacho, P.; Zinn, S.; Schnell, M. The Low-Barrier Methyl Internal Rotation in the Rotational Spectrum of 3-Methylphenylacetylene. *J. Mol. Struct.* **2020**, *1213*, 128109. [CrossRef]

9. Tudorie, M.; Kleiner, I.; Jahn, M.; Grabow, J.-U.; Goubet, M.; Pirali, O. Coupled Large Amplitude Motions: A Case Study of the Dimethylbenzaldehyde Isomers. *J. Phys. Chem. A* **2013**, *117*, 13636–13647. [CrossRef]
10. Sun, H.; Khemissi, S.; Kleiner, I.; Nguyen, H.V.L. Low Barriers to Internal Rotation in the Microwave Spectrum of 2,5-Dimethylfluorobenzene. *J. Chem. Phys.* **2024**, *160*, 094302. [CrossRef]
11. Sun, H.; Ferres, L.; Kleiner, I.; Nguyen, H.V.L. Coupled Methyl Internal Rotations with Intermediate and Low Torsional Barriers in 2,5-Dimethylanisole Investigated by Microwave Spectroscopy. *J. Chem. Phys.* **2024**, *161*, 224304. [CrossRef] [PubMed]
12. Khemissi, S.; Van, V.; Schwell, M.; Kleiner, I.; Nguyen, H.V.L. Low Barrier Methyl Internal Rotations and ^{14}N Quadrupole Coupling in the Microwave Spectrum of 2,4-Dimethylthiazole. *J. Phys. Chem. A* **2023**, *127*, 5779–5789. [CrossRef] [PubMed]
13. Ferres, L.; Truong, K.-N.; Stahl, W.; Nguyen, H.V.L. Interplay Between Microwave Spectroscopy and X-ray Diffraction: The Molecular Structure and Large Amplitude Motions of 2,3-Dimethylanisole. *ChemPhysChem* **2018**, *19*, 1781–1788. [CrossRef]
14. Thomsen, C.; Dreizler, H. The Microwave Spectra of *m*-Xylene and *m*-Xylene-d10. Determination of the Low Methyl Internal Rotation Barrier. *Z. Naturforsch. A* **2001**, *56*, 635–640. [CrossRef]
15. Hartwig, H.; Dreizler, H. The Microwave Spectrum of *trans*-2,3-Dimethyloxirane in Torsional Excited States. *Z. Naturforsch. A* **1996**, *51*, 923–932. [CrossRef]
16. Welzel, A.; Hellweg, A.; Merke, I.; Stahl, W. Structural and Torsional Properties of *o*-Cresol and *o*-Cresol-OD as Obtained from Microwave Spectroscopy and ab Initio Calculations. *J. Mol. Spectrosc.* **2002**, *215*, 58–65. [CrossRef]
17. Schmitz, D.; Shubert, V.A.; Giuliano, B.M.; Schnell, M. The Broadband Microwave Spectra of the Monoterpenoids Thymol and Carvacrol: Conformational Landscape and Internal Dynamics. *J. Chem. Phys.* **2014**, *141*, 034304. [CrossRef]
18. Écija, P.; Evangelisti, L.; Vallejo, M.; Basterretxea, F.J.; Lesarri, A.; Castanño, F.; Caminati, W.; Cocinero, E.J. Conformational Flexibility of Mephenesin. *J. Phys. Chem. B* **2014**, *118*, 5357–5364. [CrossRef]
19. Schnitzler, E.G.; Zenchyzen, B.L.M.; Jäger, W. High-Resolution Fourier-Transform Microwave Spectroscopy of Methyl-and Dimethylnapthalenes. *Astrophys. J.* **2015**, *805*, 141. [CrossRef]
20. Schnitzler, E.G.; Zenchyzen, B.L.M.; Jäger, W. Rotational Spectroscopy of the Atmospheric Photo-Oxidation Product *o*-Toluic Acid and its Monohydrate. *Phys. Chem. Chem. Phys.* **2016**, *18*, 448–457. [CrossRef]
21. Gurusinghe, R.M.; Tubergen, M.J. Probing the Electronic Environment of Methylindoles using Internal Rotation and ^{14}N Nuclear Quadrupole Coupling. *J. Phys. Chem. A* **2016**, *120*, 3491–3496. [CrossRef] [PubMed]
22. Tudorie, M.; Kleiner, I.; Hougen, J.T.; Melandri, S.; Sutikdja, L.W.; Stahl, W. A Fitting Program for Molecules with Two Inequivalent Methyl Tops and a Plane of Symmetry at Equilibrium: Application to New Microwave and Millimeter-Wave Measurements of Methyl Acetate. *J. Mol. Spectrosc.* **2011**, *269*, 211–225. [CrossRef]
23. Ilyushin, V.V.; Hougen, J. A Fitting Program for Molecules with Two Equivalent Methyl Tops and C_{2v} Point-Group Symmetry at Equilibrium: Application to Existing Microwave, Millimeter, and Sub-millimeter Wave Measurements of Acetone. *J. Mol. Spectrosc.* **2013**, *289*, 41–49. [CrossRef]
24. Groner, P. Effective Rotational Hamiltonian for Molecules with Two Periodic Large-Amplitude Motions. *J. Chem. Phys.* **1997**, *107*, 4483–4498. [CrossRef]
25. Nguyen, H.V.L.; Stahl, W. The Microwave Spectrum of Isopropenyl Acetate—An Asymmetric Molecule with Two Internal Rotors. *J. Mol. Spectrosc.* **2010**, *264*, 120–124. [CrossRef]
26. Ferres, L.; Stahl, W.; Kleiner, I.; Nguyen, H.V.L. The Effect of Internal Rotation in *p*-Methyl Anisole Studied by Microwave Spectroscopy. *J. Mol. Spectrosc.* **2018**, *343*, 44–49. [CrossRef]
27. Ferres, L.; Stahl, W.; Nguyen, H.V.L. Low Torsional Barrier Challenges in the Microwave Spectrum of 2,4-Dimethylanisole. *J. Chem. Phys.* **2019**, *151*, 104310. [CrossRef]
28. Herbers, S.; Fritz, S.M.; Mishra, P.; Nguyen, H.V.L.; Zwier, T.S. Local and Global Approaches to Treat the Torsional Barriers of 4-Methylacetophenone Using Microwave Spectroscopy. *J. Chem. Phys.* **2020**, *152*, 074301. [CrossRef]
29. Herbers, S.; Zingsheim, O.; Nguyen, H.V.L.; Bonah, L.; Heyne, B.; Wehres, N.; Schlemmer, S. Internal Rotation Arena: Program Performances on the Low Barrier Problem of 4-Methylacetophenone. *J. Chem. Phys.* **2021**, *155*, 224302. [CrossRef]
30. Frisch, M.J.; Trucks, G.W.; Schlegel, H.B.; Scuseria, G.E.; Robb, M.A.; Cheeseman, J.R.; Scalmani, G.; Barone, V.; Petersson, G.A.; Nakatsuji, H.; et al. *Gaussian16, Revision B.01*; Gaussian Inc.: Wallingford, CT, USA, 2016.
31. Becke, A.D. Density-Functional Thermochemistry. III. The Role of Exact Exchange. *J. Chem. Phys.* **1993**, *98*, 5648–5652. [CrossRef]
32. Lee, C.; Yang, W.; Parr, R.G. Development of the Colle-Salvetti Correlation-Energy Formula into a Functional of the Electron Density. *Phys. Rev. B* **1988**, *37*, 785–789. [CrossRef] [PubMed]
33. Grimme, S.; Antony, J.; Ehrlich, S.; Krieg, H.A. Consistent and Accurate Ab Initio Parametrization of Density Functional Dispersion Correction (DFT-D) for the 94 Elements H-Pu. *J. Chem. Phys.* **2010**, *132*, 154104. [CrossRef] [PubMed]
34. Grimme, S.; Ehrlich, S.; Goerigk, L. Effect of the Damping Function in Dispersion Corrected Density Functional Theory. *J. Comput. Chem.* **2011**, *32*, 1456–1465. [CrossRef] [PubMed]
35. Frisch, M.J.; Pople, J.A.; Binkley, J.S. Self-Consistent Molecular Orbital Methods 25. Supplementary Functions for Gaussian Basis Sets. *J. Chem. Phys.* **1984**, *80*, 3265–3269. [CrossRef]

36. Møller, C.; Plesset, M.S. Note on an Approximation Treatment for Many-Electron Systems. *Phys. Rev.* **1934**, *46*, 618–622. [CrossRef
37. Khemissi, S.; Schwell, M.; Kleiner, I.; Nguyen, H.V.L. Approaching the Free Rotor Limit: Extremely Low Methyl Torsional Barrie Observed in the Microwave Spectrum of 2,4-Dimethylfluorobenzene. *Phys. Chem. Chem. Phys.* **2024**, *26*, 402–411. [CrossRef]
38. Mélan, J.; Khemissi, S.; Nguyen, H.V.L. Steric Effects on Two Inequivalent Methyl Internal Rotations of 3,4-Dimethylfluorobenzene *Spectrochim. Acta A* **2021**, *253*, 119564. [CrossRef] [PubMed]
39. Khemissi, S.; Pérez Salvador, A.; Nguyen, H.V.L. Large Amplitude Motions in 2,3-Dimethylfluorobenzene: Steric Effects Failing to Interpret Hindered Methyl Torsion. *J. Phys. Chem. A* **2021**, *125*, 8542–8548. [CrossRef]
40. Khemissi, S.; Nguyen, H.V.L. Two Equivalent Internal Rotations in the Microwave Spectrum of 2,6-Dimethylfluorobenzene *ChemPhysChem* **2020**, *21*, 1682–1687. [CrossRef]
41. Ferres, L.; Cheung, J.; Stahl, W.; Nguyen, H.V.L. Conformational Effect on the Large Amplitude Motions of 3,4-Dimethylanisole Explored by Microwave Spectroscopy. *J. Phys. Chem. A* **2019**, *123*, 3497–3503. [CrossRef]
42. Yanai, T.; Tew, D.P.; Handy, N.C. A New Hybrid Exchange–Correlation Functional using the Coulomb-Attenuating Method (CAM-B3LYP). *Chem. Phys. Lett.* **2004**, *393*, 51. [CrossRef]
43. Zhao, Y.; Truhlar, D.G. The M06 Suite of Density Functionals for Main Group Thermochemistry, Thermochemical Kinetics Noncovalent Interactions, Excited States, and Transition Elements: Two New Functionals and Systematic Testing of Four M06-Class Functionals and 12 Other Functionals. *Theor. Chem. Acc.* **2008**, *120*, 215–241. [CrossRef]
44. Chai, J.-D.; Head-Gordon, M. Long-Range Corrected Hybrid Density Functionals with Damped Atom-Atom Dispersion Corrections. *Phys. Chem. Chem. Phys.* **2008**, *10*, 6615–6620. [CrossRef] [PubMed]
45. Yu, H.S.; He, X.; Li, S.L.; Truhlar, D.G. MN15: A Kohn–Sham Global-Hybrid Exchange–Correlation Density Functional with Broad Accuracy for Multi-Reference and Single-Reference Systems and Noncovalent Interactions. *Chem. Sci.* **2016**, *7*, 5032–5051 [CrossRef]
46. Adamo, C.; Barone, V. Toward Reliable Density Functional Methods without Adjustable Parameters: The PBE0 Model. *J. Chem Phys.* **1999**, *110*, 6158–6170. [CrossRef]
47. Bartlett, R.J.; Musiał, M. Coupled-Cluster Theory in Quantum Chemistry. *Rev. Mod. Phys.* **2007**, *79*, 291–352. [CrossRef]
48. Dunning, T.H. Gaussian Basis Sets for Use in Correlated Molecular Calculations. I. The Atoms Boron through Neon and Hydrogen *J. Chem. Phys.* **1989**, *90*, 1007–1023. [CrossRef]
49. Reinhold, B.; Finneran, I.A.; Shipman, S.T. Room Temperature Chirped-Pulse Fourier Transform Microwave Spectroscopy of Anisole. *J. Mol. Spectrosc.* **2011**, *270*, 89. [CrossRef]
50. Grabow, J.-U.; Stahl, W.; Dreizler, H.A. Multioctave Coaxially Oriented Beam-resonator Arrangement Fourier-Transform Microwave Spectrometer. *Rev. Sci. Instrum.* **1996**, *67*, 4072–4084. [CrossRef]
51. Zhao, Y.; Nguyen, H.V.L.; Stahl, W.; Hougen, J.T. Unusual Internal Rotation Coupling in the Microwave Spectrum of Pinacolone *J. Mol. Spectrosc.* **2015**, *318*, 91–100. [CrossRef]
52. Herbers, S.; Nguyen, H.V.L. Next Level Achievement of The *XIAM* Code in Modeling the Microwave Spectrum of *m*-Methylanisole *J. Mol. Spectrosc.* **2020**, *370*, 111289. [CrossRef]
53. Ferres, L.; Stahl, W.; Nguyen, H.V.L. Conformational Effects on the Torsional Barriers in *m*-Methylanisole Studied by Microwave Spectroscopy. *J. Chem. Phys.* **2018**, *148*, 124304. [CrossRef] [PubMed]
54. Ferres, L.; Mouhib, H.; Stahl, W.; Nguyen, H.V.L. Methyl Internal Rotation in the Microwave Spectrum of *o*-Methyl Anisole. *ChemPhysChem* **2017**, *18*, 1860–1865. [CrossRef] [PubMed]
55. Shirar, A.J.; Wilcox, D.S.; Hotopp, K.M.; Storck, G.L.; Kleiner, I.; Dian, B.C. Impact of Molecular Conformation on Barriers to Internal Methyl Rotation: The Rotational Spectrum of *m*-Methylbenzaldehyde. *J. Phys. Chem. A* **2010**, *114*, 12187–12194. [CrossRef]

molecules

MDPI

Article

Scaling of Rotational Constants

Denis S. Tikhonov [1,*,†,‡], Colin J. Sueyoshi [2,§], Wenhao Sun [1], Fan Xie [1], Maria Khon [1,3], Eva Gougoula [1], Jiayi Li [1,3], Freya Berggötz [1,4], Himanshi Singh [1,3], Christina M. Tonauer [1] and Melanie Schnell [1,3,*]

1 Deutsches Elektronen-Synchrotron DESY, Notkestr. 85 , 22607 Hamburg, Germany
2 Department of Chemistry, Amherst College, Amherst, MA 01002-5000, USA
3 Institute of Physical Chemistry, Christian-Albrechts-Universität zu Kiel, 24118 Kiel, Germany
4 Department of Physics, Universität Hamburg, 22607 Hamburg, Germany
* Correspondence: denis.tikhonov@desy.de (D.S.T.); melanie.schnell@desy.de (M.S.)
† Current address: Center for Free-Electron Laser Science CFEL, Deutsches Elektronen-Synchrotron DESY, Notkestr. 85, 22607 Hamburg, Germany.
‡ Free Moscow University; https://freemoscow.university/.
§ Work was done during an internship at Deutsches Elektronen-Synchrotron DESY.

Abstract: This manuscript introduces the concept of scaling factors for rotational constants. These factors are designed to bring computed equilibrium rotational constants closer to experimentally fitted ground-state-averaged rotational constants. The parameterization of the scaling factors was performed for several levels of theory, namely DF-Dn/def2-mVP (DF = B3LYP, PBE0, $n = 3(\mathrm{BJ}), 4$, $m = \mathrm{S}, \mathrm{TZ}$), PBEh-3c, and r^{2}SCAN-3c. The obtained scaling factors systematically improved the consistency between the theoretical and experimental rotational constants.

Keywords: rotational constants; scaling factors; density functional theory

1. Introduction

Rotational spectroscopy, consisting of microwave (MW) [1,2], millimeter wave (MMW) [3], and terahertz (THz) [4–6] spectroscopies, is a powerful high-resolution experimental technique that provides unprecedented structural sensitivity for the different structural (including conformational) and constitutional isomerisms of molecules [7–9], as well as for isotopic substitutions and various large-amplitude vibrational motions [10], such as proton transfer [11], internal rotation [12], inversion [13], and even movements of an entire molecule across another molecule's surface [9]. This kind of spectroscopic technique can be used in various applications, from monitoring the chemical composition of mixtures [14] and reactions [3,15] to detecting atmospheric [16,17] and interstellar molecular species [8,18].

In the zeroth approximation, the rotational spectrum of a molecule is given by the rigid rotor model, which in the case of non-linear molecules is parametrized using three rotational constants, which are denoted as A, B, and C or, equivalently, as B_a, B_b, and B_c, respectively [19–21]. These constants, usually expressed in MHz, are related to the moments of inertia I_α along the given α-th principal axis of the molecule through the following expression [19,20]:

$$B_\alpha = \frac{\hbar}{4\pi I_\alpha}, \quad (1)$$

where $\alpha = a, b, c$ is the given principal axis, $\hbar = 1.05 \times 10^{-34}$ J·s is the reduced Planck constant, and the moment of inertia is given as

$$I_\alpha = \sum_{n=1}^{N} m_n \cdot r_{\alpha n}^2, \quad (2)$$

where n enumerates the atoms in the molecule, N is the overall number of atoms in the molecule, m_n is the mass of the n-th atom, and $r_{\alpha n}$ is the distance of the n-th atom to the α-th principal axis. In the experiment, the so-called vibrationally averaged rotational constants are obtained, usually in the ground vibrational state, and they are commonly denoted as A_0, B_0, and C_0 [21].

The initial structural assignment of a spectroscopically observed species is usually conducted based on quantum–chemical (QC) calculations. The candidate structures are optimized at a chosen level of theory, usually with a dispersion-corrected density functional theory (DFT) calculation, and then the theoretical rotational constants are compared with the experimentally determined ones [22]. However, such a comparison does not always yield an unambiguous assignment of the molecular structures, and the reason for this is two-fold. First of all, the optimized structure corresponds to the so-called equilibrium geometry with the corresponding equilibrium rotational constants A_e, B_e, and C_e, in which all vibrational effects are absent [21,23]. Therefore, the experimental A_0, B_0, and C_0 values and theoretical A_e, B_e, and C_e are not equal due to the vibrational anharmonic shifts, which usually expand the molecular size, thus increasing the moments of inertia (Equation (2)) and consequently decreasing the corresponding rotational constant (Equation (1)) [24]. In other words, it is generally expected that $B_{0,\alpha} \leq B_{e,\alpha}$. The second reason is the quality of the QC approximation, which can distort the equilibrium structure due to the complicated underestimation and/or overestimation of various intra- and intermolecular chemical bonds and non-covalent interactions. This systematic error does not have a preferred shift of the rotational constant values with respect to their experimental counterparts and thus can be of any type [21].

In this work, we propose to systematically improve the inconsistency between experimental ground-state-averaged rotational constants (A_0, B_0, C_0) and their theoretical equilibrium counterparts (A_e, B_e, C_e) by applying tabulated scaling factors. Such an approach, where the band shifts due to anharmonic effects and QC approximation failure, has been demonstrated to be fruitful in the case of vibrational (e.g., infrared) spectroscopy [25–32]. Therefore, it is interesting to investigate whether a similar systematic improvement for the lower-frequency spectral range can be achieved as well. First, we will introduce the procedure of the scaling, the fitting model, and the training dataset; then, we will provide the scaling factors, and in the end, we will give an application example using a few recently studied systems with the usage of the PBE0-D3(BJ)/def2-TZVP level of theory.

2. Scaling Procedure

We propose to perform the scaling of the theoretical equilibrium constants obtained from QC calculations (A_e^{QC}, B_e^{QC}, C_e^{QC}) with a single global scaling factor, s, for a given QC approximation, such that the adjusted constants A_s, B_s, and C_s are defined as

$$\begin{cases} A_s = s \cdot A_e^{QC}\,, \\ B_s = s \cdot B_e^{QC}\,, \\ C_s = s \cdot C_e^{QC}\,. \end{cases} \tag{3}$$

Since the rotational constants are inversely proportional to the moments of inertia (see Equations (1) and (2)), such a scaling procedure is effectively equivalent to the global scaling of the atomic coordinates:

$$\mathbf{r}_{s,n} = \frac{\mathbf{r}_{e,n}}{\sqrt{s}}\,, \tag{4}$$

where $\mathbf{r}_{e,n}$ and $\mathbf{r}_{s,n}$ are the equilibrium and scaled positions of the n-th atom in the molecule (see Equation (2)).

The scale factors are supposed to work as follows. The user optimizes the molecular geometry at a given level of theory to produce the equilibrium rotational constants (A_e^{QC}, B_e^{QC}, C_e^{QC}). Then, a single tabulated scale factor, s, is taken. These rotational constants are multiplied with this factor, as given in Equation (3), to produce scaled rotational constants (A_s, B_s, C_s), which by design should better resemble the experimentally measured values.

Then, the comparison of the experimental and scaled rotational constants can be made, e.g., to identify particular conformers in the gas phase (see refs. [7,9,15,22,33]).

To determine and tabulate the scaling factors s for given QC approximations, we need a benchmark dataset. In this case, we chose the set of molecules used for obtaining the scaling factors for harmonic frequencies from ref. [34], which were computed using ORCA 5 software [35,36]. From the set of 441 neutral singlet molecules, 174 non-linear molecules with up to 17 atoms with experimentally available A_0, B_0, and C_0 values were selected. Only the relatively high-quality QC approximations were considered here, namely DF-Dn/def2-mVP (DF $=$ B3LYP, PBE0; $n = 3(\mathrm{BJ}), 4$; $m = \mathrm{S}, \mathrm{TZ}$) [37–43], PBEh-3c [44], and r^{2}SCAN-3c [45] levels of theory, while the calculations with the 6-31G basis set were left out from the discussion here. This was performed since rotational spectroscopy, as a high-resolution technique, generally requires more high-quality data for comparison than rotationally unresolved infrared spectroscopy. In addition, MW spectroscopy, as the lowest frequency gas-phase spectroscopic technique to this date, is practically limited by the size of the systems that can be brought into the gas phase in sufficient amounts and by the spectral resolution, as for large systems, the spectra will become dense and uninterpretable, despite the experimental resolution being of the order of $\Delta\nu/\nu \sim 10^{-6}$–$10^{-7}$ in the standard arrangement [46].

As the values of the rotational constants $A \geq B \geq C$ can differ by orders of magnitude, the metrics that use the absolute deviations between the experimental and theoretical rotational constants are essentially useless, as they will mostly fit the A-rotational constants for small-size molecules. Therefore, an advantageous approach is switching to relative values fitting, which was introduced in ref. [34]. In its application to rotational constants, the least-squares problem can be written in the following way:

$$\mathrm{rRSMD}^2(s) = \\ = \sum_{k=1}^{N_{\mathrm{mol}}} \left(\left(s \cdot \frac{A_{\mathrm{e},k}^{\mathrm{QC}}}{A_{0,k}^{\mathrm{exp}}} - 1 \right)^2 + \left(s \cdot \frac{B_{\mathrm{e},k}^{\mathrm{QC}}}{B_{0,k}^{\mathrm{exp}}} - 1 \right)^2 + \left(s \cdot \frac{C_{\mathrm{e},k}^{\mathrm{QC}}}{C_{0,k}^{\mathrm{exp}}} - 1 \right)^2 \right) \to \min, \quad (5)$$

where rRMSD denotes the relative root mean square deviation (rRMSD) of the rotational constants, k enumerates the molecules in the dataset, and N_{mol} is the total number of molecules in the dataset. The optimal scaling factors are thus given via the following equation [34]:

$$s_{\mathrm{opt}} = \underset{s}{\mathrm{argmin}} \left(\mathrm{rRSMD}^2(s) \right) = \frac{\sum_{k=1}^{N_{\mathrm{mol}}} \left(\frac{A_{\mathrm{e},k}^{\mathrm{QC}}}{A_{0,k}^{\mathrm{exp}}} + \frac{B_{\mathrm{e},k}^{\mathrm{QC}}}{B_{0,k}^{\mathrm{exp}}} + \frac{C_{\mathrm{e},k}^{\mathrm{QC}}}{C_{0,k}^{\mathrm{exp}}} \right)}{\sum_{k=1}^{N_{\mathrm{mol}}} \left(\left(\frac{A_{\mathrm{e},k}^{\mathrm{QC}}}{A_{0,k}^{\mathrm{exp}}} \right)^2 + \left(\frac{B_{\mathrm{e},k}^{\mathrm{QC}}}{B_{0,k}^{\mathrm{exp}}} \right)^2 + \left(\frac{C_{\mathrm{e},k}^{\mathrm{QC}}}{C_{0,k}^{\mathrm{exp}}} \right)^2 \right)}, \quad (6)$$

where argmin denotes the minimal value of s, which minimizes the corresponding function value. The fitting uncertainty of this value is given by the following equation:

$$\sigma_{\mathrm{opt}} = \frac{\mathrm{rRSMD}(s_{\mathrm{opt}})}{\sqrt{\sum_{k=1}^{N_{\mathrm{mol}}} \left(\left(\frac{A_{\mathrm{e},k}^{\mathrm{QC}}}{A_{0,k}^{\mathrm{exp}}} \right)^2 + \left(\frac{B_{\mathrm{e},k}^{\mathrm{QC}}}{B_{0,k}^{\mathrm{exp}}} \right)^2 + \left(\frac{C_{\mathrm{e},k}^{\mathrm{QC}}}{C_{0,k}^{\mathrm{exp}}} \right)^2 \right)}}, \quad (7)$$

where $\mathrm{rRSMD}(s_{\mathrm{opt}})$ is the value of the rRMSD (Equation (5)), with the optimal scaling factor given in Equation (6). Note that we do not use weighting with the standard deviations of the experimental fits here because the theoretical calculations have much larger systematic uncertainties that we cannot account for.

3. Resulting Scaling Factors

The resulting scaling factors for various levels of theory, as well as the rRMSD values (Equation (5)) for the unscaled and optimally scaled theoretical rotational constants, are given in Table 1. Several trends can be observed from these results. First, the scaling improved the match between the theory and experiment in most cases, except for B3LYP-Dn/def2-TZVP (n = 3(BJ), 4), which we will discuss later. We also see that the increase in the basis set quality from def2-SVP to def2-TZVP improved the agreement between the experiment and theory in both scaled and unscaled cases. Applying either the D3(BJ) or D4 dispersion correction led to the same scaling factors, which probably points to the equal performance of these corrections. A similar trend was observed for the harmonic frequency scaling factors in ref. [34]. However, the most unexpected yet predictable result is that the optimal scaling factors for the B3LYP-Dn/def2-TZVP levels of theory were equal to one within the margins of error. This means that the scaling did not significantly improve the predicted rotational constant at this approximation. At the same time, the rRMSD values for the B3LYP-Dn/def2-TZVP levels are amongst the best in the dataset. Such behavior matches the popularity of these levels of theory for quantum–chemical computations among the rotational spectroscopy community.

Table 1. Optimal scaling factors for rotational constants (s_{opt}) and their uncertainties (σ_{opt}) were determined for the listed quantum–chemical approximations with Equations (6) and (7), respectively. rRMSD denotes the values of the relative root mean square deviation (Equation (5)) computed for the training dataset without scaling ($s = 1$) and with an optimized scaling factor ($s = s_{\text{opt}}$).

Method			$s_{\text{opt}} \pm \sigma_{\text{opt}}$	rRMSD(s) × 100%	
DF	**Dn**	**Basis**		$s = 1$	$s = s_{\text{opt}}$
B3LYP	D3(BJ)	def2-SVP	1.008 ± 0.001	2.748	2.645
		def2-TZVP	1.000 ± 0.001	2.225	2.225
	D4	def2-SVP	1.008 ± 0.001	2.765	2.656
		def2-TZVP	1.000 ± 0.001	2.244	2.244
PBE0	D3(BJ)	def2-SVP	0.995 ± 0.001	2.633	2.579
		def2-TZVP	0.988 ± 0.001	2.530	2.202
	D4	def2-SVP	0.995 ± 0.001	2.637	2.583
		def2-TZVP	0.988 ± 0.001	2.535	2.206
	PBEh-3c		0.987 ± 0.001	3.092	2.778
	r^{2}SCAN-3c		1.008 ± 0.001	2.689	2.577

4. Illustrative Cases

The simplest way to illustrate the robustness and generality of the scaling procedure is to apply this procedure to cases outside the training dataset. For this, we demonstrate the applicability of the obtained scaling factors for a popular quantum–chemical approximation, namely PBE0-D3(BJ)/def2-TZVP [40–42]. The geometries of the molecules discussed here were optimized at this level of theory using the ORCA 5 [35,36] software, and then their rotational constants were taken from the calculations. In the case of isotopic substitutions, the rotational constants of the isotopologues were re-computed from the optimized geometries using the UNEX 1.6 software [47]. To demonstrate the numerical performance of the scaling factor, we compared the rRMSD (Equation (5)) and mean absolute deviations (MADs) of the rotational constants from the experimentally determined values for the given molecules in the case of scaled and unscaled theoretical equilibrium rotational constants. The MAD values were calculated according to the following expression:

$$\text{MAD}(s) = \sum_{k=1}^{N_{\text{mol}}} \left(\left| s \cdot A_{e,k}^{\text{QC}} - A_{0,k}^{\text{exp}} \right| + \left| s \cdot B_{e,k}^{\text{QC}} - B_{0,k}^{\text{exp}} \right| + \left| s \cdot C_{e,k}^{\text{QC}} - C_{0,k}^{\text{exp}} \right| \right). \quad (8)$$

The first illustrative set of molecules included 15 linear top molecules from di- to pentatomic molecules. Since linear molecules have only one rotational constant, they were excluded from the training set, and the scaling effect on these systems will be the most clearly visible. The calculation of the rRMSD and MAD values (Equations (5) and (8)) for the linear molecules, thus, included only the B rotational constant. The second illustrative set was the case of isotopologues, which, for the simplicity of the analysis, were not included in the training dataset. We chose single-substituted isotopologues of imidazole ($C_3H_4N_2$), which had rotational constants of three singly substituted ^{13}C and two singly substituted ^{15}N isotopologues available from the literature [48]. The last example was a set of non-covalently bound molecular systems, namely, water–hydrochloric acid clusters $HCl(H_2O)_n$ ($n = 2$–$5, 7$), which are examples of hydrogen bond network structures. The rotational constants for these species were taken from refs. [33,49].

We can first take a look at a few exemplary cases of molecular systems from our test dataset, including two linear molecules (HCN and HCCCN), one imidazole ^{15}N-substituted isotopologue, namely imidazole-^{15}N(1) (nomenclature adapted from ref. [48]), and also the largest of our hydrochloric acid clusters, $HCl(H_2O)_7$. The structures of these molecules and their rotational constants are given in Figure 1 and Table 2. As one can see, B3LYP-D3(BJ)/def2-TZVP provided a reasonable estimation of the rotational constants, closer to the experimental values than the unscaled constants at the PBE0-D3(BJ)/def2-TZVP level of theory. However, PBE0-D3(BJ)/def2-TZVP after scaling became as accurate or even more accurate than the B3LYP-D3(BJ)/def2-TZVP-based results. This can be seen by comparing the deviations within the datasets. By looking at the rRMSD and MAD values for the scaled and unscaled rotational constants of these systems at the PBE0-D3(BJ)/def2-TZVP level of theory (Table 3), we observe that the scaling indeed improved the agreement of the theoretical and experimental values. Note that the rRMSD and MAD deviations for the non-covalently bound cluster are larger than for the covalently bound linear molecules and imidazole isotopologues. This might be either due to the less stiff nature of the intermolecular bonds, which allows for larger systematic deviations of the obtained numerical structures from the actual potential energy surface minimum, or due to a less accurate description of the intermolecular interactions in general, which leads to larger systematic errors, or both.

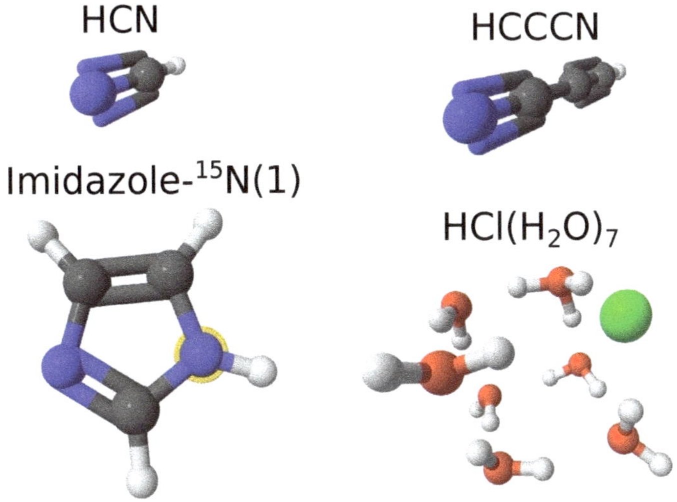

Figure 1. Examplary molecular systems to demonstrate the effect of scaling factors (see main text for details). The atomic color scheme is as follows: hydrogen—white; carbon—gray; nitrogen—blue; oxygen—red; and chlorine—green. The position of isotopic substitution in imidazole is shown with a yellow halo around the atom.

Table 2. A comparison of experimental and theoretical rotational constants for a few examples of molecular systems: HCN and HCCCN, an imidazole-^{15}N(1) isotopologue, and a $HCl(H_2O)_7$ cluster (see Figure 1). "B3LYP" denotes results at the B3LYP-D3(BJ)/def2-TZVP level of theory, and PBE0 denotes results at the PBE0-D3(BJ)/def2-TZVP level of theory. A comparison for all other molecular systems can be found in an Excel spreadsheet in the Supplementary Materials.

Molecular System	*A*/*B*/*C*	Rotational Constant Value [MHz]			
		Experimental	Theoretical		
			B3LYP	PBE0	
			$s = 1$	$s = 1$	$s = 0.988$
HCN [50]	*B*	44,316	44,941	44,969	44,415
HCCCN [51]	*B*	4549	4591	4593	4537
Imidazole-^{15}N(1) [48]	*A*	9695	9756	9850	9729
	B	9188	9218	9271	9157
	C	4716	4740	4776	4717
$HCl(H_2O)_7$ [33]	*A*	914	935	947	935
	B	737	739	750	740
	C	689	709	720	711

Table 3. Relative root mean square deviation (rRMSD, Equation (5)) and mean absolute deviation (MAD, Equation (8)) values for rotational constants from three illustrative test sets of molecular systems using the PBE0-D3(BJ)/def2-TZVP level of theory. The optimal scaling factor for this method ($s_{opt} = 0.988$) is taken from Table 1.

Dataset	rRMSD(s) × 100%		MAD(s) [MHz]	
	$s = 1$	$s = 0.988$	$s = 1$	$s = 0.988$
Linear molecules	1.0	0.3	33	16
Isotopologues	1.4	0.4	110	26
$HCl(H_2O)_n$	7.9	6.8	132	115

5. Conclusions

In this work, we introduced the concept of scaling factors for rotational constants. Applying a single tabulated scaling factor for all rotational constants was effectively equivalent to scaling the molecular size to account for systematic errors in the equilibrium structure due to the quantum–chemical approximation and for absent anharmonic effects. Sets of scaling factors for ten different DFT approximations, namely DF-Dn/def2-mVP (DF = B3LYP, PBE0; $n = 3$(BJ), 4; m = S, TZ) and PBEh-3c and r^{2}SCAN-3c, were produced from the database of 174 non-linear molecules. The applicability of these scaling factors was illustrated for the PBE0-D3(BJ)/def2-TZVP level of theory in the case of linear molecules, isotopologues, and non-covalently bonded systems. Thus, the application of such scaling factors can be recommended for the more accurate identification of species in rotational spectra and to support the assignment of specific molecular species in complicated broadband rotational spectra.

Supplementary Materials: The following supporting information can be downloaded at https://www.mdpi.com/article/10.3390/molecules29245874/s1.

Author Contributions: Conceptualization, D.S.T. and M.S.; methodology, D.S.T.; validation, D.S.T.; formal analysis, D.S.T., C.J.S., W.S., F.X., M.K., E.G., J.L., F.B., H.S. and C.M.T.; investigation, D.S.T., C.J.S., W.S., F.X., M.K., E.G., J.L., F.B., H.S. and C.M.T.; resources, M.S.; data curation, M.S.; writing—original draft preparation, D.S.T.; writing—review and editing, M.S.; supervision, M.S.; project administration, M.S.; funding acquisition, M.S. All authors have read and agreed to the published version of the manuscript.

Funding: C.J.S. acknowledges the Charles Hamilton Houston Internship Program at Amherst College that sponsored his internship at Deutsches Elektronen-Synchrotron DESY.

Institutional Review Board Statement: Not applicable

Informed Consent Statement: Not applicable

Data Availability Statement: The Excel sheet containing the data and computations for obtaining the scaling factors is provided in the Supplementary Materials.

Acknowledgments: All authors acknowledge DESY (Hamburg, Germany), a member of the Helmholtz Association HGF. In particular, D.S.T.'s calculations were enabled through the Maxwell computational resources operated at DESY.

Conflicts of Interest: The authors declare no conflicts of interest.

Abbreviations

The following abbreviations are used in this manuscript:

MW	microwave
MMW	millimeter wave
THZ	terahertz
QC	quantum–chemical
DFT	density functional theory
rRMSD	relative root mean square deviation
MAD	mean absolute deviation

References

1. Morino, Y.; Hirota, E. Microwave Spectroscopy. *Annu. Rev. Phys. Chem.* **1969**, *20*, 139–166. [CrossRef]
2. Park, G.B.; Field, R.W. Perspective: The first ten years of broadband chirped pulse Fourier transform microwave spectroscopy. *J. Chem. Phys.* **2016**, *144*, 200901. [CrossRef] [PubMed]
3. Prozument, K.; Barratt Park, G.; Shaver, R.G.; Vasiliou, A.K.; Oldham, J.M.; David, D.E.; Muenter, J.S.; Stanton, J.F.; Suits, A.G.; Barney Ellison, G.; et al. Chirped-pulse millimeter-wave spectroscopy for dynamics and kinetics studies of pyrolysis reactions. *Phys. Chem. Chem. Phys.* **2014**, *16*, 15739–15751. [CrossRef] [PubMed]
4. Cuisset, A.; Hindle, F.; Mouret, G.; Bocquet, R.; Bruckhuisen, J.; Decker, J.; Pienkina, A.; Bray, C.; Fertein, E.; Boudon, V. Terahertz Rotational Spectroscopy of Greenhouse Gases Using Long Interaction Path-Lengths. *Appl. Sci.* **2021**, *11*, 1229. [CrossRef]
5. Swearer, D.F.; Gottheim, S.; Simmons, J.G.; Phillips, D.J.; Kale, M.J.; McClain, M.J.; Christopher, P.; Halas, N.J.; Everitt, H.O. Monitoring Chemical Reactions with Terahertz Rotational Spectroscopy. *ACS Photonics* **2018**, *5*, 3097–3106. [CrossRef]
6. Giesen, T.; Brünken, S.; Caris, M.; Neubauer-Guenther, P.; Fuchs, U.; Lewen, F. Terahertz Rotational Spectroscopy. *Proc. Int. Astron. Union* **2005**, *231*, 87–96. [CrossRef]
7. Pinacho, P.; Quesada-Moreno, M.M.; Schnell, M. Conformations of borneol and isoborneol in the gas phase: Their monomers and microsolvation clusters. *J. Chem. Phys.* **2023**, *159*, 194305. [CrossRef]
8. Loru, D.; Cabezas, C.; Cernicharo, J.; Schnell, M.; Steber, A.L. Detection of ethynylbenzene in TMC-1 and the interstellar search for 1,2-diethynylbenzene. *A & A* **2023**, *677*, A166. [CrossRef]
9. Xie, F.; Sun, W.; Hartwig, B.; Obenchain, D.A.; Schnell, M. Hydrogen-Atom Tunneling in a Homochiral Environment. *Angew. Chem. Int. Ed.* **2023**, *62*, e202308273. [CrossRef] [PubMed]
10. Loru, D.; Steber, A.L.; Pérez, C.; Obenchain, D.A.; Temelso, B.; López, J.C.; Schnell, M. Quantum Tunneling Facilitates Water Motion across the Surface of Phenanthrene. *J. Am. Chem. Soc.* **2023**, *145*, 17201–17210. [CrossRef] [PubMed]
11. Li, W.; Tikhonov, D.S.; Schnell, M. Double Proton Transfer Across a Table: The Formic Acid Dimer–Fluorobenzene Complex. *Angew. Chem. Int. Ed.* **2021**, *60*, 25674–25679. [CrossRef] [PubMed]
12. Nguyen, H.V.L.; Caminati, W.; Grabow, J.U. The LAM of the Rings: Large Amplitude Motions in Aromatic Molecules Studied by Microwave Spectroscopy. *Molecules* **2022**, *27*, 3948. [CrossRef]
13. Carlotti, M.; Trombetti, A.; Velino, B.; Vrbancich, J. The rotation-inversion spectrum of 15NH3. *J. Mol. Spectrosc.* **1980**, *83*, 401–407. [CrossRef]
14. Krin, A.; Quesada Moreno, M.M.; Pérez, C.; Schnell, M. A Scent of Peppermint—A Microwave Spectroscopy Analysis on the Composition of Peppermint Oil. *Symmetry* **2022**, *14*, 1262. [CrossRef]
15. Sun, W.; Pinacho, P.; Obenchain, D.A.; Schnell, M. Gas-Phase Characterization of Adipic Acid, 6-Hydroxycaproic Acid, and Their Thermal Decomposition Products by Rotational Spectroscopy. *J. Phys. Chem. Lett.* **2024**, *15*, 817–825. [CrossRef] [PubMed]
16. Tretyakov, M. Spectroscopy underlying microwave remote sensing of atmospheric water vapor. *J. Mol. Spectrosc.* **2016**, *328*, 7–26. [CrossRef]
17. Muhleman, D.O.; Clancy, R.T. Microwave spectroscopy of the Mars atmosphere. *Appl. Opt.* **1995**, *34*, 6067–6080. [CrossRef]

18. Guélin, M.; Cernicharo, J. Organic Molecules in Interstellar Space: Latest Advances. *Front. Astron. Space Sci.* **2022**, *9*, 787567. [CrossRef]
19. Gordy, W.; Cook, R. *Microwave Molecular Spectra*, 3rd ed.; Wiley: New York, NY, USA, 1984.
20. Atkins, P.; Friedman, R. *Molecular Quantum Mechanics*; OUP: Oxford, UK, 2011.
21. Cristina Puzzarini, J.F.S.; Gauss, J. Quantum-chemical calculation of spectroscopic parameters for rotational spectroscopy. *Int Rev. Phys. Chem.* **2010**, *29*, 273–367. [CrossRef]
22. Gottschalk, H.C.; Poblotzki, A.; Fatima, M.; Obenchain, D.A.; Pérez, C.; Antony, J.; Auer, A.A.; Baptista, L.; Benoit, D.M.; Bistoni, G.; et al. The first microsolvation step for furans: New experiments and benchmarking strategies. *J. Chem. Phys.* **2020**, *152*, 164303. [CrossRef] [PubMed]
23. Tikhonov, D.S.; Vishnevskiy, Y.V. Describing nuclear quantum effects in vibrational properties using molecular dynamics with Wigner sampling. *Phys. Chem. Chem. Phys.* **2023**, *25*, 18406–18423. [CrossRef] [PubMed]
24. Mata, R.A.; Suhm, M.A. Benchmarking Quantum Chemical Methods: Are We Heading in the Right Direction? *Angew. Chem. Int Ed.* **2017**, *56*, 11011–11018. [CrossRef]
25. Pulay, P.; Fogarasi, G.; Pongor, G.; Boggs, J.E.; Vargha, A. Combination of theoretical ab initio and experimental information to obtain reliable harmonic force constants. Scaled quantum mechanical (QM) force fields for glyoxal, acrolein, butadiene, formaldehyde, and ethylene. *J. Am. Chem. Soc.* **1983**, *105*, 7037–7047. [CrossRef]
26. Irikura, K.K.; Johnson, R.D.; Kacker, R.N. Uncertainties in Scaling Factors for ab Initio Vibrational Frequencies. *J. Phys. Chem. A* **2005**, *109*, 8430–8437. [CrossRef] [PubMed]
27. Kesharwani, M.K.; Brauer, B.; Martin, J.M.L. Frequency and Zero-Point Vibrational Energy Scale Factors for Double-Hybrid Density Functionals (and Other Selected Methods): Can Anharmonic Force Fields Be Avoided? *J. Phys. Chem. A* **2015**, *119*, 1701–1714. [CrossRef] [PubMed]
28. Alecu, I.M.; Zheng, J.; Zhao, Y.; Truhlar, D.G. Computational Thermochemistry: Scale Factor Databases and Scale Factors for Vibrational Frequencies Obtained from Electronic Model Chemistries. *J. Chem. Theory Comput.* **2010**, *6*, 2872–2887. [CrossRef] [PubMed]
29. Laury, M.L.; Carlson, M.J.; Wilson, A.K. Vibrational frequency scale factors for density functional theory and the polarization consistent basis sets. *J. Comput. Chem.* **2012**, *33*, 2380–2387. [CrossRef] [PubMed]
30. Merrick, J.P.; Moran, D.; Radom, L. An Evaluation of Harmonic Vibrational Frequency Scale Factors. *J. Phys. Chem. A* **2007**, *111*, 11683–11700. [CrossRef] [PubMed]
31. Pople, J.A.; Scott, A.P.; Wong, M.W.; Radom, L. Scaling Factors for Obtaining Fundamental Vibrational Frequencies and Zero-Point Energies from HF/6–31G* and MP2/6–31G* Harmonic Frequencies. *Isr. J. Chem.* **1993**, *33*, 345–350. [CrossRef]
32. Khaikin, L.S.; Grikina, O.E.; Vogt, N.; Stepanov, N.F. Interpreting the vibrational spectra of uracil molecules and their deuterated isotopomers using a scaled quantum-chemical quadratic force field. *Russ. J. Phys. Chem. A* **2012**, *86*, 1855–1861. [CrossRef]
33. Xie, F.; Tikhonov, D.S.; Schnell, M. Electric nuclear quadrupole coupling reveals dissociation of HCl with a few water molecules. *Science* **2024**, *384*, 1435–1440. [CrossRef]
34. Tikhonov, D.S.; Gordiy, I.; Iakovlev, D.A.; Gorislav, A.A.; Kalinin, M.A.; Nikolenko, S.A.; Malaskeevich, K.M.; Yureva, K.; Matsokin, N.A.; Schnell, M. Harmonic scale factors of fundamental transitions for dispersion-corrected quantum chemical methods. *ChemPhysChem* **2024**, *25*, e202400547. [CrossRef]
35. Neese, F.; Wennmohs, F.; Becker, U.; Riplinger, C. The ORCA quantum chemistry program package. *J. Chem. Phys.* **2020**, *152*, 224108. [CrossRef]
36. Neese, F. Software update: The ORCA program system—Version 5.0. *WIREs Comput. Mol. Sci.* **2022**, *12*, e1606. [CrossRef]
37. Becke, A.D. Density-functional thermochemistry. III. The role of exact exchange. *J. Chem. Phys.* **1993**, *98*, 5648–5652. [CrossRef]
38. Becke, A.D. Density-functional exchange-energy approximation with correct asymptotic behavior. *Phys. Rev. A* **1988**, *38*, 3098–3100. [CrossRef]
39. Lee, C.; Yang, W.; Parr, R.G. Development of the Colle-Salvetti correlation-energy formula into a functional of the electron density. *Phys. Rev. B* **1988**, *37*, 785–789. [CrossRef]
40. Adamo, C.; Barone, V. Toward reliable density functional methods without adjustable parameters: The PBE0 model. *J. Chem. Phys.* **1999**, *110*, 6158–6170. [CrossRef]
41. Weigend, F.; Ahlrichs, R. Balanced basis sets of split valence, triple zeta valence and quadruple zeta valence quality for H to Rn: Design and assessment of accuracy. *Phys. Chem. Chem. Phys.* **2005**, *7*, 3297–3305. [CrossRef] [PubMed]
42. Grimme, S.; Ehrlich, S.; Goerigk, L. Effect of the damping function in dispersion corrected density functional theory. *J. Comput. Chem.* **2011**, *32*, 1456–1465. [CrossRef] [PubMed]
43. Caldeweyher, E.; Ehlert, S.; Hansen, A.; Neugebauer, H.; Spicher, S.; Bannwarth, C.; Grimme, S. A generally applicable atomic-charge dependent London dispersion correction. *J. Chem. Phys.* **2019**, *150*, 154122. [CrossRef] [PubMed]
44. Grimme, S.; Brandenburg, J.G.; Bannwarth, C.; Hansen, A. Consistent structures and interactions by density functional theory with small atomic orbital basis sets. *J. Chem. Phys.* **2015**, *143*, 054107. [CrossRef]
45. Grimme, S.; Hansen, A.; Ehlert, S.; Mewes, J.M. r2SCAN-3c: A "Swiss army knife" composite electronic-structure method. *J. Chem. Phys.* **2021**, *154*, 064103. [CrossRef] [PubMed]

46. Fokin, A.A.; Zhuk, T.S.; Blomeyer, S.; Pérez, C.; Chernish, L.V.; Pashenko, A.E.; Antony, J.; Vishnevskiy, Y.V.; Berger, R.J.F.; Grimme, S.; et al. Intramolecular London Dispersion Interaction Effects on Gas-Phase and Solid-State Structures of Diamondoid Dimers. *J. Am. Chem. Soc.* **2017**, *139*, 16696–16707. [CrossRef] [PubMed]
47. Vishnevskiy, Y.V. UNEX Version 1.6. Available online: https://unex.vishnevskiy.group (accessed on 13 October 2023).
48. Arenas, B.E.; Batra, G.; Steber, A.L.; Bizzocchi, L.; Pietropolli Charmet, A.; Giuliano, B.M.; Caselli, P.; Harris, B.J.; Pate, B.H.; Guillemin, J.C.; et al. Rotational spectroscopy of imidazole: Accurate spectroscopic information for three vibrationally excited states and the heavy-atom isotopologues up to 295 GHz. *J. Mol. Spectrosc.* **2021**, *378*, 111452. [CrossRef]
49. Kisiel, Z.; Białkowska-Jaworska, E.; Pszczółkowski, L.; Milet, A.; Struniewicz, C.; Moszynski, R.; Sadlej, J. Structure and properties of the weakly bound trimer $(H_2O)_2HCl$ observed by rotational spectroscopy. *J. Chem. Phys.* **2000**, *112*, 5767–5776. [CrossRef]
50. Winnewisser, G.; Maki, A.G.; Johnson, D.R. Rotational constants for HCN and DCN. *J. Mol. Spectrosc.* **1971**, *39*, 149–158. [CrossRef]
51. Creswell, R.; Winnewisser, G.; Gerry, M.C.L. Rotational spectra of the ^{13}C and ^{15}N isotopic species of cyanoacetylene. *J. Mol. Spectrosc.* **1977**, *65*, 420–429. [CrossRef]

molecules

MDPI

Article

Equilibrium Values for the Si-H Bond Length and Equilibrium Structures of Silyl Iodide and Halosilylenes

Jean Demaison [1,*] and Jacques Liévin [2]

[1] Physique des Lasers, Atomes et Molécules, Université de Lille, Bât. P5, 59655 Villeneuve d'Ascq cedex, France
[2] Spectroscopy, Quantum Chemistry and Atmospheric Remote Sensing (SQUARES), CP160/09, Faculté des Sciences, Université Libre de Bruxelles (U.L.B.), Ave. Roosevelt, 50, B-1050 Brussels, Belgium; jacky.lievin@ulb.be
* Correspondence: jean.demaison@gmail.com

Abstract: The equilibrium structures of silyl iodide, SiH_3I, and silylene halides, SiHX (X = F, Cl, Br, I), were determined by using the mixed regression method, where approximate values of the rotational constants are supplemented by the structural parameters of a different origin. For this goal, it is shown that the r(Si-H) bond length can be determined by using the isolated SiH stretching frequency and that an accurate estimation of the bond angles is obtained by an MP2 calculation with a basis set of triple zeta quality. To check the accuracy of the experimental structures, they were also optimized by means of all electron CCSD(T) calculations using basis sets of quadruple zeta quality.

Keywords: equilibrium structure; Si-H bond; silyl halides; silylene halides; ab initio

1. Introduction

Equilibrium structures are known for silyl fluoride, SiH_3F [1], silyl chloride, SiH_3Cl [2], silyl bromide, SiH_3Br [3], SiH_3CN [4], and silane, SiH_4 [5]. It is not easy to determine an accurate equilibrium structure (corresponding to the minimum of the potential hypersurface) because, in addition to the ground state rotational constants, it requires the knowledge of at least the vibration–rotation interaction constants of all fundamental vibrational states or, in other words, the rotational constants of all fundamental vibrational states. This is often a difficult problem because some of these vibrational states are perturbed by nearby states (through anharmonic resonances or Coriolis interactions) and reliable analysis of the corresponding rovibrational spectra is not straightforward. This is particularly true for a molecule such as silyl iodide, SiH_3I, whose rovibrational spectrum is complicated by the large quadrupole hyperfine structure due to the iodine nucleus. This is furthermore complicated by the small value of the B rotational constant, making the rovibrational spectrum very dense. For these reasons, approximate structures are often obtained by empirical methods. This is the case for SiH_3I for which only empirical effective structures (r_0) are available [6,7]. For this kind of empirical structure, all the rovibrational effects are neglected assuming that the equilibrium rotational constants are identical to the ground state rotational constants. These two r_0-structures are significantly different and none of them is expected to be close to the equilibrium structure. This is unfortunate because no equilibrium value is known for the Si-I bond length. Actually, a r_0 structure is available for monoiodosilylene, HSiI [8,9], but it is not expected to be a good approximation of the equilibrium structure.

The goal of this work is to present a set of consistent and accurate values for the Si-H bond length and to determine the equilibrium structure of SiH_3I and of the silylene halides, HSiX (X = F, Cl, Br, I).

The Section 2 describes the used quantum chemical methods. The Section 3 tries to determine an accurate value for the Si-H bond length. The Section 4 estimates the accuracy of the ∠(HSiH) and ∠(HSiX) bond angles. The Section 5 is dedicated to the structure of silyl iodide and the Section 6 to the structure of halosilylenes. A general conclusion is provided in Section 7.

2. Quantum Chemical Calculations

The structure of the molecules was optimized using three different methods: (i) the second-order Møller–Plesset perturbation theory (MP2) [10], (ii) the hybrid density functional B3LYP [11,12], and (iii) the coupled-cluster method including single and double excitations (CCSD) [13] augmented with a perturbational estimate of the effects of connected triple excitations, CCSD(T) [14]. Several basis sets were used for frozen core calculations: Dunning's correlation consistent polarized valence triple-zeta basis set: cc-pVTZ for first-row atoms [15], cc-pV(T+d)Z for second-row atoms [16], and for larger atoms, cc-pVTZ-PP with the corresponding small core relativistic pseudopotential [17]. The def2-SVP basis set [18] was also used as well as the 6-311G** basis set [19] for iodine and the 6-311+G(3df, 2pd) [20] for all other atoms.

CCSD(T) calculations correlating all electrons were also performed with the cc-pwCVQZ basis set for first- and second-row atoms [21] and with the cc-pwCVQZ-PP basis set and the corresponding small core pseudopotential [22] for larger atoms.

The CCSD(T) calculations were performed with Molpro [23,24] and the MP2 and B3LYP calculations with Gaussian16 [25].

Different Types of Structures

The equilibrium structure corresponding to the minimum of the potential hypersurface is obtained by high-level ab initio optimizations. It can also be determined experimentally from a fit of the equilibrium rotational constants (or the corresponding moments of inertia). The equilibrium rotational constants, B_e, are obtained by correcting the ground state rotational constants, B_0, from the rovibrational contribution: $B_e = B_0 + \sum_i \alpha_i d_i/2$ where α_i are the vibration–rotation interaction constants and d_i the degeneracies of these states. The summation is over all the normal modes. The α_i are derived from the rotational constants of the fundamental vibrational states. Alternatively, they can be calculated from an ab initio cubic force field giving the semi-experimental equilibrium structure.

As it is not always easy to determine an equilibrium structure, many empirical methods have been devised.

The simplest one is the effective structure, r_0, which assumes that the equilibrium rotational constants are identical to the ground state rotational constants. Although the rovibrational correction is only a few percent of the rotational constants, it can be a poor approximation of the equilibrium structure.

A slightly better method is the substitution structure, r_s, which assumes that the rovibrational correction is isotopically independent. In this case, the difference between the moment inertia of an isotopologue and the parent species is used.

The r_z distance is the distance between the average nuclear positions in the vibrational ground state at 0 K. It has a clear physical meaning permitting comparisons between molecules. In this case, it is assumed that the anharmonic part of the rovibrational correction is zero. The r_z bond length may be used to estimate the equilibrium using the "diatomic approximation" where it is assumed that only the stretching cubic force constant of the bond is different from zero (it is often assumed to be equal to the value of the corresponding diatomic molecule). The accuracy of the r_e-bond lengths derived from the r_z structure is limited due to the diatomic approximation, and it cannot be better than the accuracy of the r_z structure, which can be rather poor, in particular, for large molecules. Its accuracy is perhaps not better than a few thousandths of an Å.

3. Si-H Bond Length

It is well established that the MP2 method with a basis set of triple zeta quality gives a satisfactory approximation of the C-H bond length [26]. The same conclusion is valid for the Si-H bond, as shown in Table 1 where our ab initio values are compared to the equilibrium values [1–5,27–36].

Table 1. Equilibrium values for the Si-H bond length (in Å).

	r_e	MP2 [a]	r_e—MP2	ν_{is} [b]	Calc [c]	r_e—calc	Refs. [d]
$SiHD_3$	1.4742	1.4746	−0.0004	2182	1.4740	0.0002	[5], [37]
SiH_3F	1.4695	1.4703	−0.0008	2207.4	1.4692	0.0003	[1], [6]
SiH_3Cl	1.4688	1.4698	−0.0010	2206.7	1.4693	−0.0006	[2], [6]
SiH_3Br	1.4696	1.4699	−0.0003	2205.61	1.4695	0.0001	[3], [6]
SiH_3CN	1.468	1.4671	0.0009	2217.4	1.4673	0.0007	[4], [38]
SiH_2F_2	1.4615	1.4625	−0.0010	2248	1.4616	−0.0001	[33], [39]
$SiHF_3$	1.4487	1.4503	−0.0016	2316.9	1.4487	0.0000	[34]
SiH_2	1.513	1.5115	0.0015	1973.3	1.5130	0.0000	[28], [40]
SiHCl	1.5140	1.5137	0.0003	1968.7	1.5139	0.0001	[29], [28]
SiH_3SiH_3	1.4769	1.4779	−0.0010	2162.6	1.4776	−0.0007	[35], [6]
$CH_2{=}SiH_2$	1.4671	1.4685	−0.0014				[36]
Mediane			−0.0008			0.0000	
Mean			−0.0004			0.0000	
MAD			0.0005			0.0002	
s(MAD)			0.0008			0.0003	
Predictions [e]							
SiH_2Cl_2	1.4671	1.4643	0.0028	2231.1	1.4648	0.0023	[27], [27]
CH_3SiH_3		1.4780		2166.6	1.4768		-, [6]
SiHF	1.529	1.5230	0.006	1930	1.5211	0.008	[30], [41]
SiHBr	1.503	1.5126	−0.010	1976	1.5125	−0.009	[31], [42]
SiHI	1.5151	1.5114	0.0037	1983	1.5112	0.0039	[32], [8]
SiH_3I		1.4704		2201.1	1.4704		-, [6]

[a] This work, cc-pVTZ on H, C, N, cc-p(V+d)Z on Si, Cl, cc-pVTZ-PP on Br, I. [b] Isolated stretching frequency for the Si-H bond; partially deuterated compounds are used when Si is bonded to more than one H atom. [c] Calculated from the isolated stretching frequency. See text. [d] The first number gives the reference for r_e and the second one, the reference for the isolated stretching frequency for the Si-H bond. [e] Whereas there are published values for most molecules listed below, they are not accurate as can be seen by comparing with columns MP2 and calc.

The median of residuals (r_e–r_{MP2}) is −0.0008 Å and the mean is −0.0004 Å. The corresponding MAD (mean absolute deviation) is 0.0005 Å corresponding to a standard deviation of 0.0008 Å. In other words, the MP2/cc-pV(T+d)Z level of theory gives a value that is slightly too large (by about 0.0008 Å) but with an excellent accuracy (standard deviation of 0.0008 Å). However, it has to be noted that there are large residuals for most silylene derivatives (up to 0.0095 Å for HSiBr). This discrepancy will be discussed later.

There is another way to estimate the Si-H bond length. It was noticed by Bernstein [43] and developed by McKean [44] that there is a correlation between the isolated C-H stretching frequency (i.e., when only one hydrogen atom is present in the molecule) and the C-H bond length. This correlation permits us to determine the C-H bond length with an accuracy of 0.001 Å [45]. McKean extended this method to the Si-H bond [46,47]; however, to establish the correlation, he used empirical (r_0 or r_s) bond lengths. Using instead equilibrium values, this correlation is confirmed, as shown in Table 1, where most ν_{is}(Si–H) values are taken from [6], see also Table 1 for specific references [27,28,32,34,37–42]. A fit of 11 data gives the following linear expression:

$$r_e(\text{Si-H})\ [\text{Å}] = 1.8823(27) - 1.871(12) \times 10^{-4} \nu_{is}(\text{Si-H})\ [\text{cm}^{-1}] \quad (1)$$

with a correlation coefficient $\rho = 0.9997$ and a standard deviation $\sigma = 0.0004$ Å.

Using Equation (1) and the ab initio results allows us to confirm that the published Si-H bond lengths of most silylene halides are probably inaccurate, see Table 1. This conclusion also applies to dichlorosilane, SiH_2Cl_2 [27]. The origin of this discrepancy is easy to explain: the r_e structures of these molecules were estimated from the r_z structure and it is known that, in some cases, the method can be inaccurate when the difference r_z–r_e is large. This is particularly true for the X-H bond lengths. Furthermore, the r_z structure used for the extrapolation is already not accurate enough.

4. (HSiH) Bond Angles

The MP2 method with a basis set of at least triple zeta quality is able to predict the bond angles with an accuracy generally better than 0.4° [48]. This is well verified for the ∠(HSiX) (X = F, Cl, Br, I, CN, SiH_3, $=CH_2$) bond angles as shown in Table 2.

Table 2. Equilibrium ∠(HSiH) and ∠(HSiX) bond angles in degrees (X = F. Cl. Br. I. . . .).

	Experimental			MP2/cc-pV(T+d)Z [a]		
Molecule	**HSiH**	**HSiX**	**Ref.**	**HSiH**	**HSiX**	**Exp–Calc**
SiH_3F	110.63	108.28	[1]	110.44	108.48	−0.20
SiH_3Cl	110.47	108.46	[2]	110.39	108.53	−0.07
SiH_3Br	110.54	108.38	[3]	110.43	108.50	−0.12
SiH_2F_2	114.39	108.61	[33]	114.07	108.60	0.01
SiH_2Cl_2	112.45	108.67	[27]	112.52	108.56	0.11
$SiHF_3$		110.69	[34]		110.49	0.19
SiH_2	92.04		[28]	92.17	92.17	−0.13
SiH_3CN	111.43	107.43	[4]	111.23	107.65	−0.22
SiH_3SiH_3	115.22	110.23	[35]	108.70	110.24	−0.01
SiHCl		94.66	[29]		95.23	−0.57
$CH_2{=}SiH_2$	115.22	122.39	[36]	114.51	122.75	−0.36
Mediane						−0.12
Mean						−0.12
MAD						0.11
s(MAD)						0.17
Predictions						
CH_3SiH_3	110.563	108.357	[6]	110.56	108.36	−2.21
SiHF		92.8	[30]		96.90	−4.10
SiHBr		96.9	[31]		94.70	2.20
SiHI		92.5	[8]		93.77	−1.27
SiH_3I				110.42	108.51	

[a] cc-pVTZ-PP on Br and I.

The median as well as the mean of residuals is −0.12°. The corresponding MAD is 0.11° corresponding to a standard deviation of 0.17°, which may be considered as small. Again, the silylene compounds show large deviations.

5. Structure of SiH_3I

The method to determine the Si-I bond length is to use a well-conditioned least-squares fit. In such a case, the result is not sensitive to the accuracy of the rovibrational correction as shown for instance in the case of difluoroethane, $CH_2{=}CF_2$ [49]. One way to obtain a well-conditioned system is to use the method of mixed regression [50]. In this method, structures are fitted concurrently to the moments of inertia and bond parameters obtained from another method. The structures obtained in this work are determined by only three parameters: the Si-H and Si-X bond lengths, and the ∠(HSiX) (or ∠(HSiH)) bond angle. As shown in Sections 3 and 4, two of these parameters can be determined accurately and can, therefore, be used as predicate observations in the least-squares fit. Everything happens as if there is only one parameter to determine. However, an approximate value of the rovibrational correction is required.

For SiH_3I, it is relatively easy because it is known that the rovibrational correction $\varepsilon = I_0 - I_e$ is approximately proportional to the square root of the moment of inertia [51]. However, empirically, the exponent of I_0 is often found to be different from ½. To determine it, the ε of two sets of molecules was used: CH_3Br and CH_3I, and SiH_3Cl and SiH_3Br. For both sets, an exponent of 0.777 was found. This allowed us to deduce an ε value of 0.561 $\mu Å^2$ for SiH_3I from that of 0.452 $\mu Å^2$ for SiH_3Br. A one-dimensional fit using Si-H = 1.471 Å and ∠(HSiH) = 110.54° gives Si-I = 2.4327 Å. When only one parameter is fitted, the least-squares system is well conditioned (condition number = 1). Therefore, there is no amplification of errors. In this case, $\partial r(\text{Si-I})/\partial \varepsilon = 0.0082\ \mu^{-1} Å^{-1}$ to be compared with the value of ε which is about 0.66 $\mu Å^2$. Assuming an error of 0.001 Å for Si-H, 0.3° for ∠(HSiH), and a rather large error of 0.1 $\mu Å^2$ for ε, gives an uncertainty of 0.0012 Å, see Table 3. Using the isotopologues

^{28}Si and ^{30}Si for the moments of inertia gives identical results. This value is significantly smaller than the r_0 values: 2.437 Å or 2.438 Å. Finally, a mixed regression was performed using three predicate values: r(Si-I) = 2.435(5) Å, r(Si-H) = 1.470(2) Å; ∠(HSiI) = 108.51(30)° and the three estimated equilibrium B_e rotational constants of $^{28}SiH_3I$, $^{29}SiH_3I$, and $^{30}SiH_3I$. The uncertainty used to weigh these rotational constants was 0.1 MHz. The fit is rather well-conditioned and the residuals are small. The used semi-experimental rotational constants and the residuals of the fit are given in Table S1 of the Supplementary Material. However, the number of data is small: three rotational constants and three predicate values. Therefore, the resulting standard deviations are probably too small.

Table 3. Structure of SiH_3I (bond lengths in Å, angles in degrees).

Method	r(Si-H)	r(Si-I)	∠(HSiI)	Ref.
r_0	1.487(8)	2.437(3)	108.4	[7]
r_0	1.474(1)	2.4384(6)	108.16(17)	[6]
MP2 [a]	1.4704	2.4352	108.507	This work
r_e	1.471 [b]	2.4327(12)	108.51 [b]	This work
r_e [c]	1.4700(7)	2.4325(2)	108.49(8)	This work

[a] cc-pVTZ on H, cc-p(V+d)Z on Si, cc-pVTZ-PP on I. [b] Fixed value, see text. [c] Mixed regression with r(Si-I) = 2.435(5) Å, r(Si-H) = 1.470(2) Å; ∠(HSiI) = 108.51(30)° and an uncertainty of 0.1 MHz on the B rotational constants.

6. Structure of Halosilylenes

For the Si-H bond lengths, the equilibrium values for SiH_2 [28] and SiHCl [29] are in good agreement with the ab initio optimizations and with the prediction using the isolated stretching frequency, see Tables 1 and 2. However, for SiHF [30], SiHBr [31], and SiHI [8], there are significant discrepancies. It is important to determine whether these discrepancies are due either to the different bonding or to the poor accuracy of the equilibrium structures. The first information is obtained by comparing the ab initio optimizations and the prediction from the isolated stretching frequencies which are in perfect agreement with the Si-H bond length. However, a more stringent test is welcomed. For this reason, the structures of SiHF, SiHBr, and SiHI have been redetermined.

6.1. Structure of SiHF

The equilibrium structure of SiHF has been determined twice [30,52]. In each case, it has been derived from the average (r_z) structure. First, it has to be noted that the Si-H bond length is extremely imprecise and quite different from the predictions of Table 1.

There is an obvious explanation for this poor result. As the molecule is planar, the number of independent rotational constants is only four for SiHF and SiDF whereas there are three parameters to determine. In other words, the system is ill-conditioned and quite sensitive to small errors in the data.

This a typical example where the mixed regression should improve the situation. To calculate a semi-experimental equilibrium structure, the rovibrational corrections were obtained from an ab initio anharmonic force field calculated at the MP2/6-311+G(3df,2pd) level of theory. The ground state rotational constants were taken from Refs. [30,52] and the semi-experimental rotational constants were corrected for a small electronic effect. The values of the diagonal elements of the rotational g-tensor were calculated at the B3LYP/6-311+G(3df,2pd) level of theory. The results are $g_{aa} = -4.29$; $g_{bb} = -0.099$; $g_{cc} = -1.70$. The electronic correction is small for B and C, of the order of magnitude of one MHz, i.e., similar to the experimental accuracy. On the other hand, it is extremely large for the A rotational constants: 534 MHz for SiHF. However, this is not a problem because the A rotational constants were not used in the fit. Three predicate values were used: r(Si-H) = 1.523(2) Å, r(Si-F) = 1.602(5) Å, and ∠(HSiF) = 96.9(5)° (CCSD(T)/cc-pwCVQZ values, all electrons correlated). The estimates of the standard deviations are larger than their probable values to avoid a preponderant role of the predicate values. The predicate value of the Si-F bond length was estimated assuming that the error of the MP2 method is constant for a given bond. In other words:

$$r_e(\text{Si-F[SiHF]}) = r_{MP2}(\text{Si-F[SiHF]}) + r_e(\text{Si-F[SiH}_3\text{F]}) - r_{MP2}(\text{Si-F[SiH}_3\text{F]}) \quad (2)$$

The resulting structure is given in Table 4 and the semi-experimental rotational constants and their residuals in Table S2 of the Supplementary Material. From the residuals and the standard deviations of the fitted parameters, it appears that the predicate values are compatible with the semi-experimental rotational constants. In other words, the structure is expected to be reliable.

Table 4. Structure of HSiF (bond lengths in Å, angle in degrees).

Method	r(Si-H)	r(Si-F)	∠(HSiF)	Refs.
r_e^z [a]	1.528(5)	1.603(3)	96.9(5)	[52]
r_e [a]	1.529(6)	1.603(1)	96.9(3)	[30]
MP2/6-311+G(3df,2pd)	1.5201	1.6250	96.59	This work
MP2/cc-pVTZ [b]	1.5230	1.6132	96.90	This work
r_e^{SE}	1.5227(8)	1.6028(2)	96.75(8)	This work

[a] Derived from the r_z structure. [b] cc-pV(T+d)Z on Si.

As the number of data is quite small, the standard deviations of the parameters are not expected to be a reliable estimate of their accuracy. To palliate this inconvenient, the standard deviations were also calculated using the law of propagation of errors assuming an error of 2 MHz for the rotational constants and for the predicates: 0.002 Å for r(Si-H), 0.005 Å for r(Si-F) and 0.5° for ∠HSiF).

It is interesting to note that the Si-F bond length at 1.603 Å is close to the value found for the radical SiF, 1.601 Å [53].

6.2. Structure of SiHCl

There are already equilibrium structures available for SiHCl. The equilibrium structure was first derived from the average (r_z) structure [54]. There are also two high-level ab initio structures [29,55] optimized at the CCSD(T) level of theory. Finally, there are two semi-experimental structures [29] using two different force fields [29], see Table 5. However, there are small differences between these two semi-experimental structures. It might be explained by the fact that rotational constants are available for only three isotopologues, i.e., six independent data (as the molecule is planar). To improve the accuracy, we have repeated the semi-experimental determination by adding the three CCSD(T)_ae/cc-pwCVQZ parameters with conservative uncertainties as predicates, see Table 5. This new structure is in agreement with the previous ones, but it is believed to be slightly more accurate.

Table 5. Structure of HSiCl (bond lengths in Å, angle in degrees).

Method	r(Si-H)	r(Si-Cl)	∠(HSiCl)	Refs.
CCSD(T)_ae/cc-pwCVQZ	1.5138	2.0697	95.302	[29]
CCSD(T)_ae/cc-pCVQZ	1.51469	2.07122	95.303	[55]
CCSD(T)_ae/cc-pwCVQZ	1.5138	2.0697	95.31	This work
r_e^z [a]	1.525(5)	2.067(3)	96.9(5)	[54]
r_e^{SE} [b]	1.5146	2.0697	94.78	[29]
r_e^{SE} [c]	1.5140	2.0724	94.66	[29]
Predicates [d]	1.5147(20)	2.0712(50)	95.303(500)	This work
r_e^{SE} [e]	1.5145(7)	2.0719(2)	95.22(11)	This work

[a] Derived from the r_z structure. [b] Rovibrational corrections from a CCSD(T)/cc-pVTZ cubic force field. [c] Rovibrational corrections from a CCSD(T)/cc-pCVTZ cubic force field. [d] Predicates from the CCSD(T)/cc-pwCVQZ structure [55]. [e] Semi-experimental rotational constants from [29] and predicates from the previous line.

6.3. Structure of SiHBr

The structure of SiHBr was first determined in 1964 by Herzberg and Varma [56]. They recorded the spectrum of SiHBr in the region 6000 to 4100 Å obtained in the flash photolysis of SiH_3Br. More recently, Harjanto et al. [42] examined the ground and first excited singlet electronic states of SiHBr through analysis of the 500–400 nm band system, using pulsed discharge jet and laser-induced fluorescence techniques. Rotational analysis of the electronic ground state bands of SiHBr and SiDBr yielded average ground state rotational constants because the bromine isotope splittings could not be resolved. A structural fit using the average of the two bromine isotopologues yielded an effective (r_0) structure. Hostutler et al. [31] determined the ground state harmonic frequencies of gas phase H/DSi^{79}Br by exciting single vibronic bands of the $\tilde{A}^1A''–\tilde{X}^1A'$ electronic transition and recorded the dispersed fluorescence. The derived harmonic force constants and the ground state rotational constants of Harjanto et al. [42] were used to calculate an average (r_z) structure which permitted us to estimate an equilibrium (r_e) structure. Finally, Tackett et al. [57] measured the microwave spectrum by Fourier transform microwave spectroscopy and were able to determine the ground state values of $\overline{B} = B + C$ for the ^{79}Br and ^{81}Br isotopologues. We used these values to calculate a semi-experimental equilibrium structure. The rovibrational corrections were obtained from an ab initio anharmonic force field calculated at the MP2/6-311+G(3df,2pd) level of theory. As a check, the calculation was repeated with the def2-TZVP basis set. Almost identical values were obtained. The two experimental data are not enough to determine a complete structure. A sensitivity analysis was first performed and it was found that $\overline{B}$ is mainly sensitive to the value of the r(Si-Br) bond length: $\partial\overline{B}/\partial r$(Si-Br) = 4168 MHz Å^{-1}, $\partial\overline{B}/\partial r$(Si-H) = 80 MHz Å^{-1} and $\partial\overline{B}/\partial(\angle$SiHBr) = −4 MHz degree^{-1}. For this reason, the Si-H bond length derived from the isolated Si-H stretching frequencies (see Table 1) was fixed as well as the $\angle$(SiHBr) bond angle calculated at the MP2 level of theory (see Table 2). Only the SiBr bond length was fitted. To estimate the accuracy of the result, it was assumed that the standard deviation of r(Si-H) is 0.002 Å, that of ($\angle$SiHBr) was 0.4° (in agreement with the results of Sections 3 and 4), and that the accuracy of the rovibrational correction is 10% [58]. It gives r_e(Si-Br) = 2.2321(4) Å. The structures are given in Table 6.

Table 6. Structure of HSiBr (bond lengths in Å, angle in degrees).

Method	r(Si-H)	r(Si-Br)	∠(HSiBr)	Ref.
r_0	1.518(1)	2.237(1)	93.4(3)	[42]
r_e^z [a]	1.503(9)	2.235(1)	92.8(4)	[31]
MP2/6-311+G(3df,2pd)	1.5111	2.2439	94.55	This work
MP2/cc-pVTZ [b]	1.5126	2.2333	94.70	This work
MP2/def2-TZVP	1.5134	2.2390	94.64	This work
r_e^{SE}	1.5125(20)	2.2321(4)	94.70(40)	This work

[a] Derived from the r_z structure. [b] cc-pV(T+d)Z on Si, cc-pVTZ-PP on Br.

As a check the Si-Br bond length was estimated from the MP2 optimizations, see Equation (2). It gives r(Si-Br) 2.230 Å, in excellent agreement with the previous result.

6.4. Structure of SiHI

The analysis of single vibronic level emission spectra of jet-cooled SiHI and SiDI permitted us to determine a harmonic force field [8]. Using this force field and previously determined ground state rotational constants [32], Tackett and Clouthier were able to determine an average (r_z) structure which was used to derive an approximate equilibrium structure. Later, the accuracy of the rotational constants was improved by measuring the Fourier transform microwave spectra of SiHI and SiDI and a new effective structure was determined [9].

To determine a semi-experimental structure, an anharmonic force field was calculated at the MP2/6-311G** level of theory. This permitted us to estimate the rovibrational corrections that were used to correct the ground state rotational constants of Kang et al. [9] and obtain semi-experimental equilibrium moments of inertia. These moments of inertia were fitted by the mixed regression method. Three predicate values were used: (i) the Si-H bond length derived from the isolated Si-H stretching frequencies (see Table 1), (ii) the ∠(HSiI) bond angle calculated at the MP2 level of theory (see Table 2), (iii) an estimate of the Si-I bond length which was estimated assuming that the error of the MP2 method is constant for a given bond, see Equation (2) where F is replaced by I.

As a check, the structure was calculated with anharmonic force fields computed at the MP2/def2-TZVP and B3LYP/def2-TZVP levels of theory. Nearly identical results were obtained. The results are given in Table 7.

Table 7. Structure of HSiI (bond lengths in Å, angle in degrees).

	r(Si-H)	r(Si-I)	∠(HSiI)	Refs.
r_0	1.534(1)	2.463(1)	92.4(1)	[8]
r_e^z [a]	1.5151(2)	2.4610(1)	92.5(1)	[8]
r_0	1.5405(16)	2.46143(9)	92.68(6)	[9]
MP2/6-311G**	1.5084	2.4967	93.70	This work
MP2/cc-pVTZ [b]	1.5114	2.4567	93.77	This work
B3LYP/def2-TZVP	1.5230	2.4906	93.41	This work
MP2/def2-TZVP	1.5128	2.4495	93.90	This work
Predicates	1.5112(40) [c]	2.455(5)	93.8(5) [d]	This work
r_e^{SE} [e]	1.5138(8)	2.45746(6)	93.26(5)	This work
r_e^{SE} [f]	1.5128(11)	2.45684(8)	93.254(7)	This work
r_e^{SE} [g]	1.5125(7)	2.45760(5)	93.185(4)	This work

[a] Derived from the r_z structure. [b] cc-pV(T+d)Z on Si and cc-pVTZ-PP on I. [c] From isolated stretching frequency, see text. [d] MP2 value, see Table 2. [e] Cubic force field calculated at the MP2/6-311G** level of theory. [f] Cubic force field calculated at the B3LYP/def2-TZVP level of theory. [g] Cubic force field calculated at the MP2/def2-TZVP level of theory.

7. Discussion

Accurate equilibrium structures were determined for SiH_3I, SiHF, SiHBr, and SiHI. The method was to use well-conditioned least-squares fits, i.e., not sensitive to the uncertainties of the input data. For all these molecules, the structure is defined by three parameters: the r(Si-H) and r(Si-X) bond lengths, as well as the ∠(HSiX) bond angle (with X = F, Cl, Br, I). It was shown that it is possible to obtain an accurate value of the r(Si-H) bond length using either isolated stretching frequencies or MP2 calculations with a triple-zeta basis set. Furthermore, the ∠(HSiX) bond angle can be accurately estimated with MP2 calculations. Finally, using the method of mixed estimation, the third parameter, r(Si-X), is determined by the semi-experimental equilibrium moments of inertia. In such a case, it was confirmed that the results are not much sensitive to the accuracy of the rovibrational corrections.

There is another way to check the accuracy of the results by plotting the r(Si-X) bond of SiHX molecules as a function of the r(Si-X) bond of SiH_3X molecules [59]. As shown in Figure 1, a nice linear correlation is obtained.

Our predictions are also found to be in good agreement with the results of high-level CCSD(T)/cc-pwCVQZ calculations correlating with all electrons (see Table 8). The mean absolute deviation is of 0.0006 and 0.0013 Å for the r(Si-H) and r(Si-X) bond lengths, respectively, and 0.13° for the HSiX bond angle.

As already observed by Duncan et al. [6], the structures of the halides form a remarkably consistent set, Si-H bond length and the ∠(HSiX) bond angle remaining almost constant for the silyl halides. For the silylene halides, this conclusion remains true if we exclude SiHF. For the silylene halides, both the bond angle ∠(HSiX) and the bond length r(Si-H) increase with the electronegativity of the halogen X.

Table 8. Comparison of CCSD(T)[a] and equilibrium structures (bond lengths in Å, angles in degrees).

Molecule	Parameter	CCSD(T) [a]	r_e [b]	r_e—CCSD(T)	Ref. [c]
HSiF	r(Si-H)	1.5235	1.5227	0.0008	This work
	r(Si-F)	1.6020	1.6028	−0.0008	
	∠(HSiF)	96.96	96.75	0.21	
HSiCl	r(Si-H)	1.5138	1.5145	−0.0007	This work
	r(Si-Cl)	2.0697	2.0719	−0.0022	
	∠(HSiCl)	95.31	95.22	0.09	
HSiBr	r(Si-H)	1.5127	1.5125	0.0002	This work
	r(Si-Br)	2.2323	2.2321	0.0002	
	∠(HSiBr)	94.66	94.70	−0.04	
HSiI	r(Si-H)	1.5117	1.5128	−0.0011	This work
	r(Si-I)	2.4547	2.4568	−0.0021	
	∠(HSiI)	93.65	93.25	0.40	
SiH_3F	r(Si-H)	1.4694	1.4695	−0.0001	[1]
	r(Si-F)	1.5903	1.5915	−0.0012	
	∠(HSiF)	108.41	108.28	0.13	
SiH_3Cl	r(Si-H)	1.4686	1.4688	−0.0002	[2]
	r(Si-Cl)	2.0469	2.0458	0.0011	
	∠(HSiCl)	108.52	108.46	0.06	
SiH_3Br	r(Si-H)	1.4687	1.4696	−0.0009	[3]
	r(Si-Br)	2.2080	2.207	0.001	
	∠(HSiBr)	108.43	108.38	0.05	
SiH_3I	r(Si-H)	1.4693	1.4700	−0.0007	This work
	r(Si-I)	2.4312	2.4325	−0.0013	
	∠(HSiI)	108.46	108.49	−0.03	

[a] cc-pVQZ on H, cc-pwCVQZ on Si, Cl, cc-pwCVQZ-PP on Br, I, all electrons correlated. [b] Experimental or semi-experimental equilibrium structure. [c] Reference for r_e.

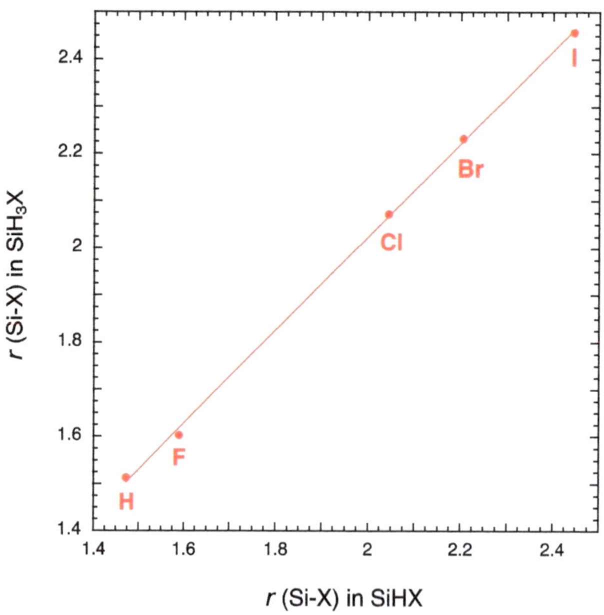

Figure 1. Plot of r(Si-X) in SiH_3X as a function of r(Si-X) in SiHX.

Supplementary Materials: The following supporting information can be downloaded at: https://www.mdpi.com/article/10.3390/molecules29133101/s1. Table S1. Semi-experimental values for the *B* rotational constants of SiH_3I and residuals of the fit (MHz). Table S2. Semi-experimental values for the rotational constants of HSiF and residuals of the fit (MHz).

Author Contributions: Conceptualization, J.D. and J.L. Methodology, J.D. and J.L. Software, J.D. and J.L. Validation, J.D. and J.L. Formal Analysis, J.D. and J.L. Investigation, J.D. and J.L. Resources, J.L. Data Curation, J.D. and J.L. Writing—Original Draft Preparation, J.D. Writing—Review and Editing, J.L. Visualization, J.L. Supervision, J.D. Project Administration, J.D. Funding Acquisition, J.L. All authors have read and agreed to the published version of the manuscript.

Funding: This research received no external funding.

Institutional Review Board Statement: Not applicable.

Informed Consent Statement: Not applicable.

Data Availability Statement: The data presented in this study are available in article and Supplementary Materials.

Acknowledgments: J.L. thanks the Fonds de la Recherche Scientifique-FNRS through IISN (Grant No. 4.4504.10) for financial support and the Consortium des Equipements de Calcul Intensif (CECI), funded by the Fonds de la Recherche Scientifique de Belgique (F.R.S.-FNRS) under grant number 2.5020.11 and by the Walloon Region, for computational resources.

Conflicts of Interest: The authors declare no conflict of interest.

References

1. Boulaftali, N.; Ben Sari-Zizi, N.; Graner, G.; Demaison, J. The equilibrium structure of silyl fluoride. *J. Mol. Spectrosc.* **2002**, *216*, 284–291. [CrossRef]
2. Demaison, J.; Šormova, H.; Bürger, H.; Margulès, L.; Constantin, F.L.; Ceausu-Velescu, A. Pure rotational spectrum, ab initio anharmonic force field, and equilibrium structure of silyl chloride. *J. Mol. Spectrosc.* **2005**, *232*, 323–330. [CrossRef]
3. Demaison, J.; Cosléou, J.; Bürger, H.; Mkadmi, E.B. High-Resolution Infrared Spectrum of $D_3Si^{79}Br$ in the ν_1/ν_4 Region: The Structure of Silyl Bromide. *J. Mol. Spectrosc.* **1997**, *185*, 384–391. [CrossRef] [PubMed]
4. Priem, D.; Cosléou, J.; Demaison, J.; Merke, I.; Stahl, W.; Jerzembeck, W.; Bürger, H. Analysis of the Rotational Spectra of SiH_3CN and Its Isotopomers: Experimental and Ab Initio Determinations of the dipole moment and structure. *J. Mol. Spectrosc.* **1998**, *191*, 183–198. [CrossRef] [PubMed]
5. Ohno, K.; Matsuura, H.; Endo, Y.; Hirota, E. The microwave spectra of deuterated silanes, germanes, and stannanes. *J. Mol. Spectrosc.* **1986**, *118*, 1–17. [CrossRef]
6. Duncan, J.L.; Harvie, J.L.; McKean, D.C.; Cradock, S. The ground state structures of disilane, methyl silane and the silyl halides, and an SiH bond length correlation with stretching frequency. *J. Mol. Struct.* **1986**, *145*, 225–242. [CrossRef]
7. Kewley, R.; McKinney, P.M.; Robiette, A.G. The microwave spectra and molecular structures of the silyl halides. *J. Mol. Spectrosc.* **1970**, *34*, 390–398. [CrossRef]
8. Tackett, B.S.; Clouthier, D.J. Structural and spectroscopic trends in the ground states of the monohalosilylenes: Emission spectroscopy of jet-cooled HSiI and DSiI. *J. Chem. Phys.* **2003**, *118*, 2612–2619. [CrossRef]
9. Kang, L.; Gharaibeh, M.A.; Clouthier, D.J.; Novick, S.E. Fourier transform microwave spectroscopy of the reactive intermediate monoiodosilylene, HSiI and DSiI. *J. Mol. Spectrosc.* **2012**, *271*, 33–37. [CrossRef]
10. Møller, C.; Plesset, M.S. Note on an approximation treatment for many-electron systems. *Phys. Rev.* **1934**, *46*, 618–622. [CrossRef]
11. Becke, A.D. Density functional thermochemistry. *III. The role of exact change. J. Chem. Phys.* **1993**, *98*, 5648–5652.
12. Lee, C.; Yang, W.; Parr, R.G. Development of the Colle-Salvetti correlation-energy formula into a functional of the electron density. *Phys. Rev. B* **1988**, *37*, 785–789. [CrossRef] [PubMed]
13. Purvis III, G.D.; Bartlett, R.J. Full Coupled-Cluster Singles and Doubles Model: The Inclusion of Disconnected Triples. *J. Chem. Phys.* **1982**, *76*, 1910–1918. [CrossRef]
14. Raghavachari, K.; Trucks, G.W.; Pople, J.A.; Head-Gordon, M.A. Fifth-Order Perturbation Comparison of Electron Correlation Theories. *Chem. Phys. Lett.* **1989**, *157*, 479–483. [CrossRef]
15. Dunning, T.H., Jr. Gaussian basis sets for use in correlated molecular calculations. I. The atoms boron through neon and hydrogen. *J. Chem. Phys.* **1989**, *90*, 1007–1023. [CrossRef]
16. Dunning, T.H., Jr.; Peterson, K.A.; Wilson, A.K. Gaussian basis sets for use in correlated molecular calculations. X. The atoms aluminum through argon revisited. *J. Chem. Phys.* **2001**, *114*, 9244–9253. [CrossRef]
17. Peterson, K.A.; Shepler, B.C.; Figgen, D.; Stoll, H. On the spectroscopic and thermochemical properties of ClO, BrO, IO, and their anions. *J. Phys. Chem. A* **2006**, *110*, 13877–13883. [CrossRef]

18. Weigend, F.; Ahlrichs, R. Balanced basis sets of split-valence, triple zeta valence and quadruple zeta valence quality for H to Rn: Design and assessment of accuracy. *Phys. Chem. Chem. Phys.* **2005**, *7*, 3297–3305. [CrossRef]
19. Glukhovtsev, M.N.; Pross, A.; McGrath, M.P.; Radom, L. Extension of Gaussian-2 (G2) theory to bromine- and iodine-containing molecules: Use of effective core potentials. *J. Chem. Phys.* **1995**, *103*, 1878–1885. [CrossRef]
20. Frisch, M.J.; Pople, J.A.; Binkley, J.S. Self-consistent molecular orbital methods. 25. Supplementary functions for Gaussian basis sets. *J. Chem. Phys.* **1984**, *80*, 3265–3269. [CrossRef]
21. Peterson, K.A.; Dunning, T.H., Jr. Accurate Correlation Consistent Basis Sets for Molecular Core–Valence Correlation Effects: The Second Row Atoms Al–Ar, and the First Row Atoms B–Ne Revisited. *J. Chem. Phys.* **2002**, *117*, 10548–10560. [CrossRef]
22. Peterson, K.A.; Yousaf, K.E. Molecular core-valence correlation effects involving the post-d elements Ga-Rn: Benchmarks and new pseudopotential-based correlation consistent basis sets. *J. Chem. Phys.* **2010**, *133*, 174116. [CrossRef]
23. Werner, H.-J.; Knowles, P.J.; Knizia, G.; Manby, V.; Schütz, M.; Celani, P.; Györffy, W.; Kats, D.; Korona, T.; Lindh, R.; et al. *MOLPRO* version 2022.1; A Package of Ab Initio Programs. Available online: https://www.molpro.net (accessed on 1 January 2022).
24. Werner, H.-J.; Knowles, P.J.; Manby, F.R.; Black, J.A.; Doll, K.; Heßelmann, A.; Kats, D.; Köhn, A.; Korona, T.; Kreplin, D.A.; et al. The Molpro quantum chemistry package. *J. Chem. Phys.* **2020**, *152*, 144107. [PubMed]
25. Frisch, M.J.; Trucks, G.W.; Schlegel, H.B.; Scuseria, G.E.; Robb, M.A.; Cheeseman, J.R.; Scalmani, G.; Barone, V.; Petersson, G.A. Nakatsuji, H.; et al. *Gaussian 16, Revision C.01*; Gaussian, Inc.: Wallingford, CT, USA, 2016.
26. Demaison, J.; Craig, N.C. Semiexperimental equilibrium structure for cis, trans-1,4- difluorobutadiene by the mixed estimation method. *J. Phys. Chem. A* **2011**, *115*, 8049–8054. [CrossRef] [PubMed]
27. Davis, R.W.; Robiette, A.G.; Gerry, M.C.L. The harmonic force fields of methylene chloride and dichlorosilane from combined microwave and infrared data. *J. Mol. Spectrosc.* **1981**, *85*, 399–415. [CrossRef]
28. Kokkin, D.L.; Ma, T.; Steimle, T.; Sears, T.J. Detection and characterization of singly deuterated silylene, SiHD, via optical spectroscopy. *J. Chem. Phys.* **2016**, *144*, 244304/1–244304/13. [CrossRef]
29. Vázquez, R.; Stanton, J.F. Theoretical Investigation of the Structure and Vibrational Spectrum of the Electronic Ground State $\tilde{X}(^1A')$ of HSiCl. *J. Phys. Chem. A* **2002**, *106*, 4429–4434. [CrossRef]
30. Hostutler, D.A.; Clouthier, D.J.; Judge, R.H. Single vibronic level emission spectroscopy of jet-cooled HSiF and DSiF. *J. Chem. Phys.* **2001**, *114*, 10728–10732. [CrossRef]
31. Hostutler, D.A.; Ndiege, N.; Clouthier, D.J.; Pauls, S.W. Emission spectroscopy, harmonic vibrational frequencies, and improved ground state structures of jet-cooled monochloro- and monobromosilylene (HSiCl and HSiBr). *J. Chem. Phys.* **2001**, *115*, 5485–5491
32. Clouthier, D.J.; Harper, W.W.; Klusek, C.M.; Smith, T.C. The electronic spectrum of monoiodosilylene (HSiI) revisited. *J. Chem. Phys.* **1998**, *109*, 7827–7834. [CrossRef]
33. D'Eu, J.-F.; Demaison, J.; Buerger, H. Millimeter-wave and high-resolution FTIR spectroscopy of SiH_2F_2: The ground and v4 = 1 states. *J. Mol. Spectrosc.* **2003**, *218*, 12–21.
34. Demaison, J.; Margules, L.; Breidung, J.; Thiel, W.; Bürger, H. Ab initio anharmonic force field, spectroscopic parameters and equilibrium structure of trifluorosilane. *Mol. Phys.* **1999**, *97*, 1053–1067. [CrossRef]
35. Puzzarini, C.; Taylor, P.R. An ab initio study of the structure, torsional potential energy function, and electric properties of disilane, ethane, and their deuterated isotopomers. *J. Chem. Phys.* **2005**, *122*, 054315. [CrossRef] [PubMed]
36. Bailleux, S.; Bogey, M.; Demaison, J.; Bürger, H.; Senzlober, M.; Breidung, J.; Thiel, W.; Fajgar, R.; Pola, J. The equilibrium structure of silene $H_2C{=}SiH_2$ from millimeter wave spectra and from ab initio calculations. *J. Chem. Phys.* **1997**, *106*, 10016–10026. [CrossRef]
37. Shimanouchi, T. *Tables of Molecular Vibrational Frequencies Consolidated Volume I*; National Bureau of Standards: Gaithersburg, MD, USA, 1972; pp. 1–160.
38. McKean, D.C. On the assignment of SiH and SiD stretching frequencies: A reanalysis of the ν_7 bands of Si_2H_6 and Si_2D_6 and a harmonic local mode force field for disilane. *Spectrochim. Acta A* **1992**, *48*, 1335–1345.
39. Bunnell, J.; Ford, T.A. Intramolecular coupling of vibrational modes and the assignments of the partially deuterated dihalogenomethanes, -silanes, and -germanes. *J. Mol. Spectrosc.* **1983**, *100*, 215–233.
40. Fredin, L.; Hauge, R.H.; Kafafi, Z.H.; Margrave, J.L. Matrix isolation studies of the reactions of silicon atoms with molecular hydrogen. The infrared spectrum of silylene. *J. Chem. Phys.* **1985**, *82*, 3542–3545. [CrossRef]
41. Suzuki, T.; Hakuta, K.; Saito, S.; Hirota, E. Doppler limited dye laser excitation spectroscopy of the $\tilde{A}^1A''(000)$–$\tilde{X}^1A'(000)$ band of HSiF. *J. Chem. Phys.* **1985**, *82*, 3580–3583. [CrossRef]
42. Harjanto, H.; Harper, W.W.; Clouthier, D.J. Resolution of anomalies in the geometry and vibrational frequencies of monobromosilylene (SiHBr) by pulsed discharge jet spectroscopy. *J. Chem. Phys.* **1996**, *105*, 10189–10200. [CrossRef]
43. Bernstein, H.J. The average XH stretching frequency as a measure of XH bond properties. *Spectrochim. Acta* **1962**, *18*, 161–170. [CrossRef]
44. McKean, D.C. Individual CH bond strengths in simple organic compounds: Effects of conformation and substitution. *Chem. Soc. Rev.* **1978**, *7*, 399–422. [CrossRef]
45. Demaison, J.; Rudolph, H.D. Ab initio anharmonic force field and equilibrium structure of propene. *J. Mol. Spectrosc.* **2008**, *248*, 66–76. [CrossRef]
46. McKean, D.C. New light on the stretching vibrations, lengths and strengths of CH, SiH and GeH bonds. *J. Mol. Struct.* **1981**, *113*, 251–266. [CrossRef]

47. McKean, D.C.; Torto, I.; Boggs, J.E.; Fan, K. Infrared and ab initio studies of molecules containing SiH bonds. *J. Mol. Struct. (Theochem.)* **1992**, *260*, 27–46. [CrossRef]
48. Margulès, L.; Demaison, J.; Boggs, J.E. Ab initio and equilibrium bond angles. Structures of HNO and H_2O_2. *J. Mol. Struct. (Theochem.)* **2000**, *500*, 245–258. [CrossRef]
49. Vogt, N.; Demaison, J.; Vogt, J.; Rudolph, H.D. Why it is sometimes difficult to determine the accurate position of a hydrogen atom by the semiexperimental method: Structure of molecules containing the OH or the CH_3 group. *J. Comput. Chem.* **2014**, *35*, 2333–2342. [CrossRef]
50. Belsley, D.A. *Conditioning Diagnostics. Collinearity and Weak Data in Regression*; Wiley: New York, NY, USA, 1991.
51. Demaison, J.; Nemes, L.L. A correlation between the vibrational correction and the effective moment of inertia. *J. Mol. Struct.* **1979**, *55*, 295–299. [CrossRef]
52. Harper, W.W.; Hostutler, D.A.; Clouthier, D.J. Pulsed discharge jet spectroscopy of DSiF and the equilibrium molecular structure of monofluorosilylene. *J. Chem. Phys.* **1997**, *106*, 4367–4375. [CrossRef]
53. Tanaka, T.; Tamura, M.; Tanaka, K. Infrared diode laser spectroscopy of the SiF radical. Analysis of hot bands up to ν = 9–8. *J. Mol. Struct.* **1997**, *413–414*, 153–166. [CrossRef]
54. Harper, W.W.; Clouthier, D.J. Reinvestigation of the HSiCl electronic spectrum: Experimental reevaluation of the geometry, rotational constants, and vibrational frequencies. *J. Chem. Phys.* **1997**, *106*, 9461–9473. [CrossRef]
55. Coriani, S.; Marchesan, D.; Gauss, J.; Hättig, C.; Helgaker, T.; Jørgensen, P. The accuracy of ab initio molecular geometries for systems containing second-row atoms. *J. Chem. Phys.* **2005**, *123*, 184107. [CrossRef]
56. Herzberg, G.; Verma, R.D. Spectra and structures of the free SiHCl and SiHBr radicals Can. *J. Phys.* **1964**, *42*, 395–432.
57. Tackett, B.S.; Clouthier, D.J.; Landry, J.N.; Jaeger, W. Fourier transform microwave spectroscopy of SiHBr: Exploring the Si-Br bond through quadrupole hyperfine coupling. *J. Chem. Phys.* **2005**, *122*, 214314. [CrossRef]
58. Vogt, N.; Vogt, J.; Demaison, J. Accuracy of the rotational constants. *J. Mol. Struct.* **2011**, *988*, 119–127. [CrossRef]
59. Demaison, J.; Vogt, N. *Accurate Structure Determination of Free Molecules*; Springer: Cham, Switzerland, 2020; pp. 205–232.

molecules

MDPI

Article

Perfluoropropionic Acid ($CF_3CF_2C(O)OH$): Three Conformations and Dimer Formation

Carlos O. Della Védova [1], Rosana M. Romano [1,*], Hans-Georg Stammler [2] and Norbert W. Mitzel [2,*]

1 Centro de Química Inorgánica "Dr. Pedro J. Aymonino", CEQUINOR (Universidad Nacional de La Plata, UNLP, Centro Científico y Tecnológico, Consejo Nacional de Investigaciones Científicas y Técnicas, CCT-CONICET La Plata, Associated with Comisión de Investigaciones Científicas de la Provincia de Buenos Aires, CIC-PBA), Departamento de Química, Facultad de Ciencias Exactas, Universidad Nacional de La Plata, Boulevard 120 N° 1465, La Plata CP 1900, Argentina; carlosdv@quimica.unlp.edu.ar

2 Lehrstuhl für Anorganische Chemie und Strukturchemie, Center for Molecular Materials CM2, Bielefeld University, Universitätsstraße 25, 33615 Bielefeld, Germany; georg.stammler@uni-bielefeld.de

* Correspondence: romano@quimica.unlp.edu.ar (R.M.R.); nmitzel@uni-bielefeld.de (N.W.M.)

Abstract: Perfluoropropionic acid ($CF_3CF_2C(O)OH$) has been investigated with a focus on its complex structural properties. As a formal derivative of propanoic acid, the incorporation of fluorine atoms imparts unique structural features, including three distinct monomeric conformations and a dimeric structure. This study presents experimental findings, supported by computational modeling, to explore these characteristics. The analysis includes an FTIR study of the isolated species in an Ar-cryogenic matrix and the low-temperature determination of its crystalline structure using single-crystal X-ray diffraction.

Keywords: perfluoropropionic acid; low-temperature crystal structure; cryogenic Ar matrix study; conformers; monomers; dimer; computational calculations

1. Introduction

The chemistry of fluorine has long been, and continues to be, with renewed momentum, one of the most captivating fields for chemists worldwide. Fluorine, in its compounds, cannot be simply regarded as either a "larger hydrogen" or a "smaller chlorine". The properties of its compounds often prove to be unexpected, unpredictable, fascinating, and truly unique. For example, the energy transition has once again positioned fluorine at the forefront of the discipline. It is a key component of the salt used in lithium battery electrolytes, $LiPF_6$, and in its elemental form plays a vital role in chemically eliminating trace water by oxidizing it into OF_2 and HF [1].

In particular, fluoro- and perfluoro-organic compounds have found a wide range of applications. Specifically, perfluoroorganic compounds, one of which is the focus of this work, exhibit remarkable stability due to the presence of C-F bonds within their molecular structure. This stability—thermal, chemical, biological, and, to some extent, photochemical—confers upon them a significant degree of environmental persistence [2–13].

Perfluoropropionic acid, $CF_3CF_2C(O)OH$ (PFPA), the title compound, may not accumulate in the environment to the same extent as the longer chain perfluorinated carboxylic acids [7,8], and its natural sources have not been identified so far. That it has been detected in rainwater [14–16], however, stresses its role as an environmentally active molecule. To fully understand how this and other long-lived products are formed under the complex environmental conditions present in a given reactive matrix, it is essential to acquire a detailed knowledge of the photochemical evolution of the species, their association equilibria at various temperatures, and to experimentally determine the existence of all conformers present at room temperature.

In the context of the anhydride acid molecule central to this study, it is worth noting that, from a structural perspective, the hydrate and dihydrate of anhydrous acid were investigated using Chirped-Pulse Fourier Transform Microwave (CP-FTMW) Spectroscopy. The study revealed that the complexation of the -OH group of the acid with one or two water molecules occurs on the plane of the carboxylic acid group, resulting in the formation of a six- or eight-membered ring structure [17].

Important for this work is a vibrational study of $CF_3CF_2C(O)OH$, published by Crowder in 1972, in which he detailed the partial and total association of the species in the vapor and liquid states, respectively. The use of fundamental vibrational concepts, such as evaluating group electronegativity, allowed him to understand, for example, the shift to higher wavenumbers of the carbonyl stretch when a CF_3CF_2- group is formally replaced by a CF_3- group, and its connection with the hydrogen bonding comparison between the two species [18,19].

Another study explored the far-infrared spectra of a set of 27 carboxylic acids in aqueous solution, including the compound examined in this work. The analysis of the spectra, particularly in the OH stretching region, enables the determination of conformational isomerism [20]. Computational studies using DFT-B3LYP/6-311+G** and ab initio MP2/6-311+G** calculations on perfluoropropionic acid revealed the existence of an equilibrium between two conformations: the *cis* form (where the C=O group eclipses the C-C bond) and the *gauche* conformer. The calculations predict the *gauche* conformer to be the lower-energy form at ambient temperature, with an abundance of 76% *gauche* and 24% *cis* at 298.15 K [21]. The structural properties of perfluoropropionic acid have been resolved through the study of its rotational spectrum using a pulsed nozzle, chirped-pulse Fourier transform microwave spectrometer within the frequency range of 8–14 GHz. Combined quantum chemical calculations and spectroscopic analysis supports the assignment of the *gauche* form, with a C−C−C=O dihedral angle of 106–107°, and variations depending on the level of approximation used [22].

In another work, the chemistries of perfluoropropionic acid and its close derivatives were studied, described, and compared [23]. Perfluoropropionic acid was also included in an early study utilizing neutron spectroscopy to examine its vibrational spectrum, alongside a broader group of related organic acids [24]. The title compound was also investigated using gas-phase mid-IR, near-IR, and visible vibrational spectroscopy, alongside perfluorooctanoic and perfluorononanoic acid, employing Fourier transform and cavity ring-down spectroscopy. The authors of this work concluded that these compounds exhibit more harmonic O–H bonds, lower transition wavenumbers, and reduced intensities compared to shorter-chain hydrocarbon acids, alcohols, and peroxides [25]. The vibrational spectra of the title compound were also found within the range of 11,000–1000 cm^{-1} and were compared with those of its hydrocarbon homolog, propionic acid [26]. Perfluoropropionic acid was one of the compounds used to study a simple drop-coating deposition using Raman spectroscopy methods to concentrate perfluoroalkyl substances and subsequently design an accessible and reliable spectral library [27].

A family of polyfluorinated compounds, including $CF_3CF_2C(O)OH$, was analyzed from the perspective of the fragmentation process of the deprotonated species. It was demonstrated that the relative energy of the transition state of the formed CCFC ring, which leads to the FCO_2^- anion, is directly linked to the subsequently observed dissociation [28]. In another work, we also employed perfluoropropionic acid to study details related to its photoexcitation, photoionization, and photofragmentation using synchrotron light energies in the range between 11.7 and 715.0 eV. At low energies, the detected fragments were

COH^+, $C_2F_4^+$, and the parent M^+ ion. In this work, and in line with the experimental variables used—for instance, very low pressures on the order of 10^{-6} mm Hg—there was no evidence of the existence of a dimer of perfluoropropionic acid [29].

$CF_3CF_2C(O)OH$ was also part of a very recent study that evaluated the decomposition products of this family of compounds with the aim of providing more information about the thermal evolution process [30]. In this context, and in connection with this evidence, the degradation of perfluoropropionic acid and related compounds was investigated using an argon plasma under various conditions [31].

A previous analysis of this type reported that after decomposition in an N_2 atmosphere at temperatures between 200 and 780 °C, the identified products were $CF_2{=}CF_2$, CF_3CF_2H, and $CF_3C(O)F$. In an O_2 atmosphere at below 400 °C, the main product is OCF_2, accompanied by the inevitable formation of SiF_4 due to the use of a quartz reactor [32]. From a computational perspective, and to understand transport behavior of relatively stable substances over considerable distances, such as perfluoroalkyl and polyfluoroalkyl compounds, this work aims to comparatively determine the gas-phase thermochemical properties of the compounds, which includes perfluoropropionic acid [33]. The use of perfluoropropionic acid for studying its role in the nucleation of atmospheric molecules under ambient conditions is computationally analyzed in order to understand, at a molecular level, the composition and formation mechanism of secondary organic aerosols [34]. In connection with the above-mentioned decomposition processes, we highlight that pentafluoropropionate salts (salts of Li, Na, K, Cs, Mg, Ca and Ba) were also examined. In that study, the principal pyrolysis product of the pentafluoropropionate salts under dynamic vacuum was tetrafluoroethylene ($CF_2{=}CF_2$) [35].

To fully comprehend how this and other long-lived products are formed under the intricate environmental conditions present in a given reactive matrix, it is crucial to gain a detailed understanding of the photochemical evolution of the species and their association equilibria at various temperatures and to experimentally confirm the existence of all conformers present at room temperature. This comprehensive approach ensures a deeper insight into the mechanisms and interactions that drive the formation and stability of these products.

2. Results and Discussion

2.1. Quatum Chemical Calculations

2.1.1. Monomer

With the aim of determining which conformations of $CF_3CF_2C(O)OH$ coexist in the gas phase at room temperature, a potential energy surface was calculated as a function of the dihedral angles φ(C−C−C=O) and φ(O−C−O−H), using the B3LYP/6-311+G(D) approximation (Figure S1).

The *gauche–syn*, *gauche–anti*, and *syn–syn* conformations (Figure 1) correspond to minima on the aforementioned potential energy surface, while the *syn–anti* structure corresponds to a saddle point. This represents a notable difference between $CF_3CF_2C(O)OH$ and its hydrogenated analog, which admits four stable conformations. Despite the stability of three $CF_3CF_2C(O)OH$ conformations, only one (*gauche–syn*) has been properly detected experimentally and studied.

The structures corresponding to the three mentioned conformers were optimized and their harmonic vibrational frequencies were calculated using different approximations. Table 1 presents the values of the dihedral angles φ(C−C−C=O) and φ(O−C−O−H) for the different conformations, their relative energies, and their populations determined using the Boltzmann equation at room temperature, taking into account the double degeneracies for the *gauche* species due to symmetric considerations, using the MP2/6-311+G(D) level of approximation.

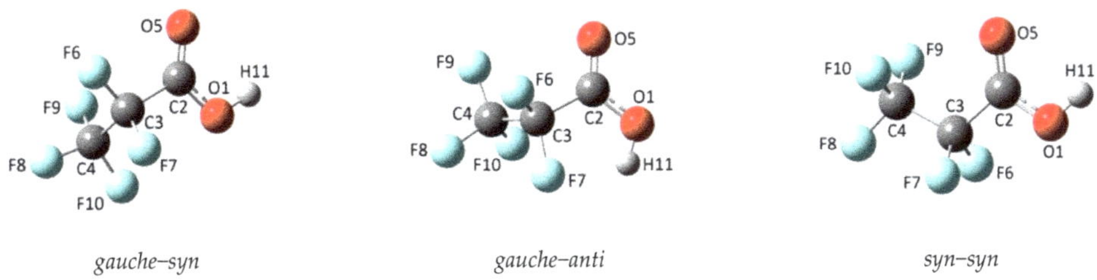

Figure 1. Optimized structures (MP2/6-311+G(D)) of the *gauche–syn*, *syn–syn*, and *gauche–anti* conformers of $CF_3CF_2C(O)OH$.

Table 1. Energy and Gibbs free energy differences among *gauche–syn*, *gauche–anti*, and *syn–syn* conformers of $CF_3CF_2C(O)OH$; dihedral angles φ(C−C−C=O)) and φ(O−C−O−H); and conformational population χ calculated at 298 K using the MP2/6-311+G(D) approximation.

Conformer	φ(C−C−C=O)	φ(O−C−O−H)	ΔE (kcal/mol)	ΔG (kcal/mol)	χ (%)
gauche–syn	101.2	−0.3	0.00	0.00	85.1
syn–syn	−0.1	0.0	0.43	0.62	14.7
gauche–anti	82.3	176.6	3.37	3.64	0.2

The obtained and tabulated results indicate that the *gauche–syn* conformer has the highest conformational population percentage at 298 K (85.1%), followed by the *syn–syn* rotamer (14.7%) and finally the *gauche–anti* form (0.2%).

Table S1 (Supplementary Information) lists the theoretically calculated vibrational wavenumbers for each of the three conformers, obtained at the MP2/6-311+G(d) level of theory, which correspond to the experimental wavenumber range. A tentative spectral assignment is also provided. These data will later facilitate the interpretation of the FTIR spectra of matrix-isolated $CF_3CF_2C(O)OH$.

2.1.2. Dimer

The structure of the $CF_3CF_2C(O)OH$ dimer (Figure 2) was calculated using the MP2/6-311+G(D) approximation, taking into account that the structure determined by X-ray diffraction reproduces these data (see X-ray diffraction section). The dimer consists of two enantiomeric monomeric units (*gauche–syn*) that are properly oriented and linked to each other through two hydrogen bonds. The calculated geometry belongs to the C_i point group.

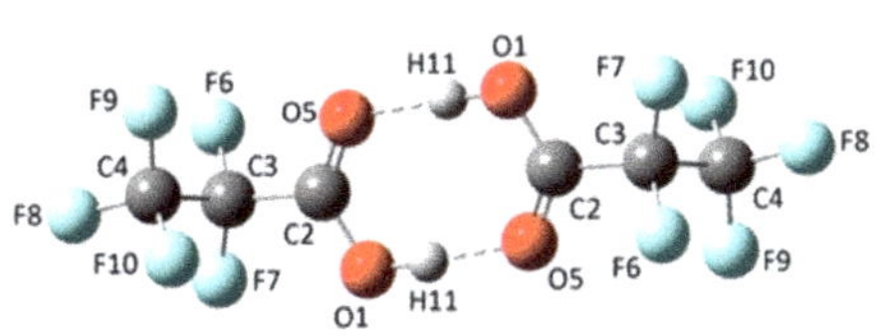

Figure 2. Optimized structures (MP2/6-311+G(D)) of the dimer of $CF_3CF_2C(O)OH$.

The IR spectrum of the dimeric form was computed at the MP2/6-311+G(d) level of theory. The wavenumbers of the IR-active vibrational modes, along with their tentative assignments, are compiled in Table S1. Similarly to the simulated spectra of the monomeric conformers, the theoretical dimer spectrum serves as a key reference for interpreting and assigning the experimental gas-phase and matrix-isolation IR spectra.

2.2. *Experimental Results*

2.2.1. Gas-Phase FTIR Spectra

In 1972, Crowder reported for the first time the infrared spectrum of the gas and liquid phases of $CF_3CF_2C(O)OH$ [18]. The experimental gas-phase FTIR spectrum of $CF_3CF_2C(O)OH$ demonstrates clear evidence for the simultaneous presence of monomeric and dimeric forms (Figures 3 and 4). Despite the good resolution of the acquired spectra (0.5 cm^{-1}), the different conformational contributions of the monomer remain unclear. The gas-phase infrared spectrum assignment is detailed in Table S1.

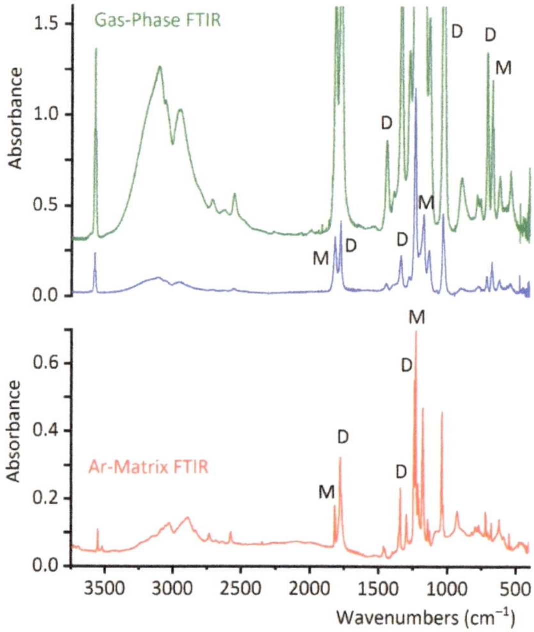

Figure 3. Gas-phase FTIR spectra of $CF_3CF_2C(O)OH$ (optical path, 10 cm; resolution, 0.5 cm^{-1}; and pressures, 40 torr (top, green line) and 8 torr (middle, blue line)) and the Ar matrix FTIR spectrum (bottom, red line) with a $CF_3CF_2C(O)OH$:Ar ratio at 1:500 and with a resolution of 0.5 cm^{-1}, in the 3740–400 cm^{-1} wavenumber region. Some of the bands of the monomer and dimer species are indicated by M and D, respectively.

The gas-phase IR spectra (Figures 3 and 4) exhibit six different absorptions attributable to the dimeric species. Notably, the ν(O−H) stretching vibration undergoes a significant redshift from 3576 cm^{-1} (monomer) to ~3100 cm^{-1} (dimer), consistent with strong hydrogen bonding between subunits. Due to the dimer's C_i symmetry, only the antisymmetric (out-of-phase) O−H stretching fundamental mode is IR-active.

Similarly, the carbonyl stretching vibration shifts from 1821 cm^{-1} (monomer) to 1779 cm^{-1} (dimer), indicating substantial intermolecular interaction via the C=O groups. This observation aligns with the formation of the characteristic cyclic structure of carboxylic acid dimers, as predicted computationally (Figure 2). The experimental wavenumber shifts for these modes show excellent agreement with theoretical calculations (Table S1), validating the proposed dimeric structure.

The wavenumbers and tentative assignments of the four additional dimer absorptions observed in the gas-phase IR spectra are detailed in Table S1. Notably, a distinct band at 900 cm^{-1}—assigned to the out-of-phase HCO deformation mode of the dimer—appears in a spectral region devoid of monomer absorptions (Table S1). This feature provides strong spectroscopic evidence for dimer formation.

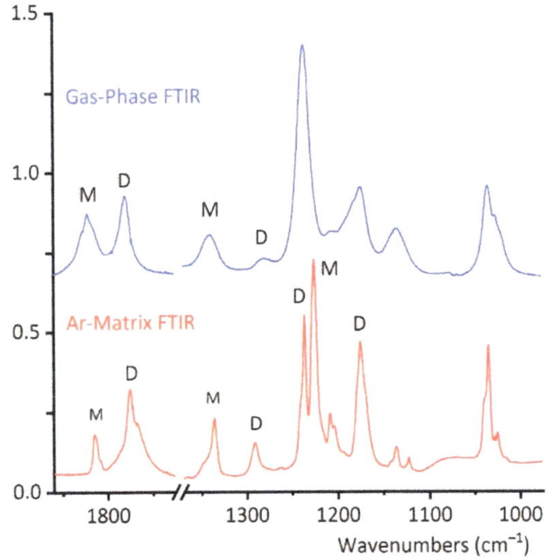

Figure 4. Selected regions of the gas-phase FTIR spectrum of $CF_3CF_2C(O)OH$ (optical path, 10 cm; resolution, 0.5 cm^{-1}; pressure, 8 torr (top, blue line)) and the Ar matrix FTIR spectrum (bottom, red line) with a $CF_3CF_2C(O)OH$:Ar ratio at 1:500 and a resolution of 0.5 cm^{-1}. Some of the bands of the monomer and dimer species are indicated by M and D, respectively.

Next, the coexistence of monomers and a dimer of $CF_3CF_2C(O)OH$ will be confirmed through experiments conducted in the vapor phase. It is important to have this information to ensure that, during the preparation of the $CF_3CF_2C(O)OH$ matrix in Ar for deposition and measurement at cryogenic temperature, the dimer concentration is minimized. Figure 5 shows a selected section of the FTIR spectra in the vapor phase for $CF_3CF_2C(O)OH$ samples measured at different temperatures. In this region, two distinct bands clearly appear: one at 714 cm^{-1}, corresponding to the out-of-phase CF_2 deformation mode of the dimer, and another at 676 cm^{-1}, corresponding to the $\delta(CF_2)$ mode of the monomer. Figures 5 and 6 describe the results of experiments conducted with the vapor phase of the species, aiming to determine the optimal experimental conditions for ensuring that monomeric species predominate over the dimer in the matrix isolation experiments of perfluoropropionic acid. The remaining dimer bands discussed in this section exhibit consistent spectral shifts relative to the monomer absorptions, further supporting their proposed assignments.

According to the experimental design used to record these spectra, the rise in temperature is directly associated with an increase in the vapor pressure of perfluoropropionic acid, promoting monomer interactions and resulting in a higher proportion of dimeric species.

This experiment should not be confused with the one originally conducted by Crowder, who recorded infrared spectra of $CF_3CF_2C(O)OH$ at different temperatures while keeping the pressure constant. As expected, the increase in temperature favors the growth of the entropic term associated with the system's evolution toward the formation of a greater number of monomeric species [18]. These monomeric species, predicted by Crowder, have now been determined with the help of computational calculations.

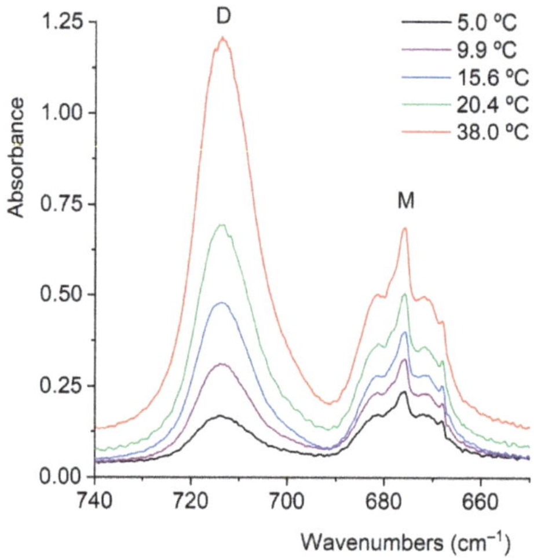

Figure 5. Gas-phase FTIR spectra of $CF_3CF_2C(O)OH$ (optical path: 10 cm; resolution: 0.5 cm^{-1}) recorded from the liquid phase at different temperatures.

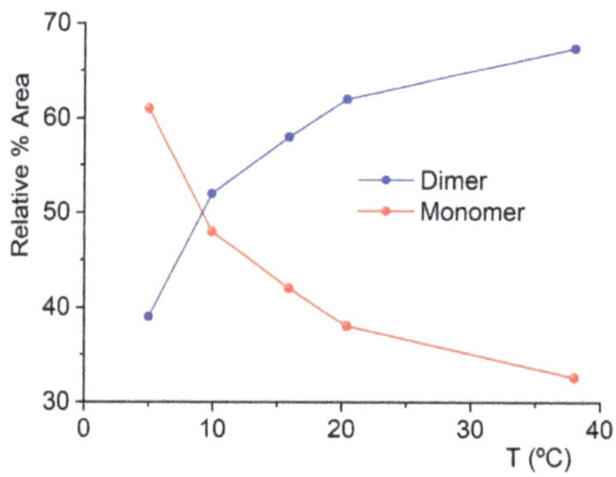

Figure 6. Relative percentage areas of the IR bands at 714 cm^{-1}, arising from the bending vibration of the O−C=O group in the dimeric $CF_3CF_2C(O)OH$ species, and at 676 cm^{-1}, corresponding to the same mode in the monomeric acid in its *gauche–syn* conformation, as a function of the liquid temperature in equilibrium with the vapor phase.

2.2.2. FTIR Spectrum of $CF_3CF_2C(O)OH$ Isolated in Solid Argon

The study of species isolated using cryogenic matrices enhances the resolution of infrared spectra by eliminating the contributions of rotational broadening at low temperatures, typically around 15 K. Thus, the close inspection of the infrared spectrum of the $CF_3CF_2C(O)OH$:Ar (1:500) mixture, in combination with the aid of the computational predictions, reveals the contributions of the *gauche–syn* and *syn–syn* conformers of $CF_3CF_2C(O)OH$, which coexist in the gas phase. No clear contribution of the less abundant *gauche–anti* conformer to the IR spectrum is observed.

Four distinct absorptions assigned to the *syn–syn* conformer exhibit measurable wavenumbers shifts relative to the dominant *gauche–syn* conformer, with sufficient intensity for detection in the matrix–isolation IR spectrum (Table S1). These bands correspond to ν(C=O) stretching, ν(C_2-C_3) stretching, ν(C-O) stretching, and $\nu_S(CF_3)$ symmetric stretching fundamental modes. The observed shifts align quantitatively with theoretical predictions reported in this study, confirming the spectral assignment of the *syn–syn* conformer.

The presence of a second conformation represents a difference between propionic acid, $CH_3CH_2C(O)OH$, and its perfluorinated counterpart: the matrix FTIR spectrum of propionic acid at low temperatures only shows the existence of a single conformer [36]. Figure 4 compares the FTIR spectra of $CF_3CF_2C(O)OH$ in the gas phase and in the matrix, highlighting the respective conformational contributions in the latter.

2.2.3. Matrix FTIR Spectra of $CF_3CF_2C(O)OH$ After Broadband UV–Vis Irradiation

The matrix of $CF_3CF_2C(O)OH$ diluted in argon in a 1:500 ratio at cryogenic temperatures was exposed to UV–vis broadband irradiation in the range of $200 \leq \lambda \leq 800$ nm. Spectra were acquired before irradiation and at different irradiation times (0.5, 1.5, 3, 6, 12, 30, and 60 min). The irradiation resulted in a decrease in the population of the lowest energy and most abundant conformer in the gas phase, the *gauche–syn* conformer, and an increase in the *syn–syn* conformer and the dimeric species. A significant finding was that after 30 min of irradiation, signals of the *gauche–anti* form appeared in the spectrum. The ν(O-H) and ν(C=O) vibrational modes of the *syn–syn* conformer were observed at higher wavenumbers (3574 and 1834 cm^{-1}, respectively) compared to those of the more abundant *gauche–syn* conformer. Two additional bands exhibiting similar spectral shifts were detected at 682 and 616 cm^{-1}, which were assigned to $\delta(CF_2)$ and $\delta(CF_3)$ deformation modes, respectively. The *gauche–syn* conformer had previously been elusive due to its relatively low concentration in the gas phase at room temperature. Thus, we obtained the first experimental evidence for the existence of this conformer. Figure 7 depicts the FTIR spectra of $CF_3CF_2C(O)OH$ isolated in Ar, recorded immediately after deposition and after 60 min of broad band irradiation in the carbonyl stretching vibrational region, which is highly sensitive to conformation. For clarity, the spectra were normalized to the carbonyl absorption of the lowest-energy conformer, the *gauche–syn* rotamer.

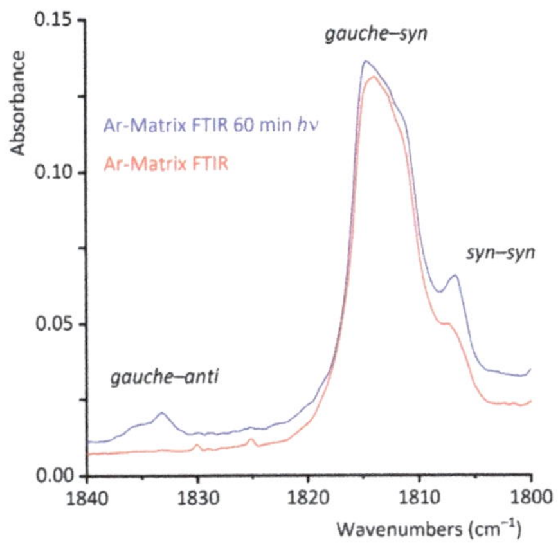

Figure 7. Ar matrix FTIR spectra of $CF_3CF_2C(O)OH$ (resolution, 0.5 cm^{-1}; $CF_3CF_2C(O)OH$:Ar ratio, 1:500) between 1840 and 1800 cm^{-1}, taken immediately after deposition (bottom, red line) and after 60 min of broad band UV–vis irradiation (top, blue line).

Figure 8 shows the variation in absorbance, measured as the integrated area of the IR bands, as a function of irradiation time. The features assigned to the *syn–syn* and *gauche–syn* forms increase their intensities at the expense of the bands corresponding to the *gauche–syn* rotamer. Additionally, the absorptions of the dimer also increase upon photolysis, presumably due to some monomer diffusion during irradiation.

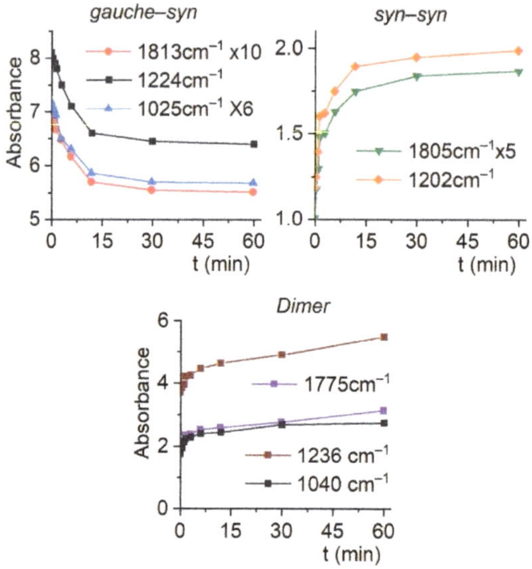

Figure 8. Absorbances of the IR bands of the $CF_3CF_2C(O)OH$:Ar (1:500) matrix for the *gauche–syn* (**top left**) and *syn–syn* (**top right**) conformers, as well as for the dimer (**bottom**), as a function of irradiation time.

2.2.4. Solid State Structure

The solid-state structure of $CF_3CF_2C(O)OH$ has been studied from an in situ grown crystal. $CF_3CF_2C(O)OH$ crystallizes in the space group $P2_1/c$, forming dimers in which both monomers adopt a *gauche–syn* conformation, related to each other by a crystallographic center of inversion (Figure 9). Table S2 presents crystallographic information obtained from the structural analysis and refinement of $CF_3CF_2C(O)OH$.

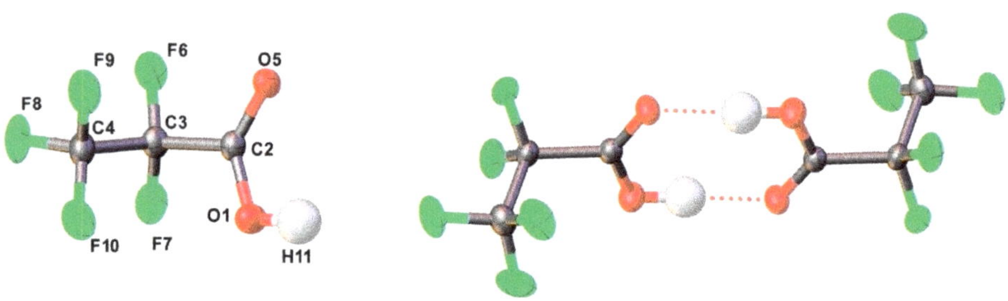

Figure 9. Left: structure of $CF_3CF_2C(O)OH$ in the solid state. Thermal ellipsoids are shown at a probability level of 50%. Right: dimer of $CF_3CF_2C(O)OH$; symmetry code used: $-x, 1-y, -z$.

The geometric parameters of $CF_3CF_2C(O)OH$ obtained by X-ray diffraction are listed below in Table 2, where they are compared with the values obtained by quantum chemistry calculations showing that the calculated values reproduce the experimentally obtained fairly well, even though the computational values would be closer to those determined in the gas phase due to their intrinsic nature.

According to NBO quantum chemical calculations at the B3LYP/6-311+G(D) level of the theory, the aforementioned hydrogen bonds, which are responsible for dimer formation, arise from electronic transfer from a lone pair of the carbonyl oxygen to the anti-bonding molecular orbital σ*(O−H). This interaction leads to a second-order perturbation stabilization energy $E^{(2)}$ of 9.20 kcal mol^{-1}.

Table 2. Experimental structure parameters obtained by X-ray diffraction and computed parameters (distances in Å; angles in degrees) corresponding to the *gauche–syn* conformer of $CF_3CF_2C(O)OH$.

Parameter	X-Ray Diffraction	MP2/6-311+G(D)
r(F6−C3)	1.338(2)	1.344
r(F7−C3)	1.336(2)	1.351
r(F9−C4)	1.319(2)	1.336
r(F8−C4)	1.317(2)	1.330
r(F10−C4)	1.304(2)	1.332
r(O5=C2)	1.215(2)	1.203
r(O1−H11)	0.97(3)	0.971
r(O1−C2)	1.286(2)	1.337
r(C2−C3)	1.545(2)	1.542
r(C4−C3)	1.542(2)	1.542
α(H11−O1−C2)	112.6(17)	108.3
α(O5−C2−O1)	127.9(2)	126.9
α(O5−C2−C3)	120.0(2)	123.0
α(O1−C2−C3)	112.0(2)	110.0
α(F9−C4−F8)	108.9(2)	109.0
α(F9−C4−F10)	108.8(2)	108.8
α(F9−C4−C3)	109.4(2)	109.4
α(F8−C4−F10)	109.1(2)	108.9
α(F8−C4−C3)	110.1(2)	110.1
α(F10−C4−C3)	110.6(2)	110.5
α(F6−C3−C7)	108.7(2)	108.9
α(F6−C3−C2)	109.0(2)	108.8
α(F6−C3−C4)	108.1(2)	107.7
α(F7−C3−C2)	110.5(2)	110.6
α(F7−C3−C4)	107.9(2)	108.0
α(C2−C3−C4)	112.6(2)	112.8
τ(H11−O1−C2−O5)	−0.5(18)	−0.3
τ(H11−O1−C2−C3)	−178.2(17)	178.8
τ(O5−C2−C3−F6)	21.2(2)	−18.2
τ(O5−C2−C3−F7)	140.6(2)	−137.8
τ(O5−C2−C3−C4)	−98.7(2)	101.2
τ(O1−C2−C3−F6)	−160.9(2)	162.6
τ(O1−C2−C3−F7)	−41.5(2)	43.1
τ(O1−C2−C3−C4)	79.2(2)	−78.0
τ(F9−C4−C3−F6)	−65.5(2)	65.1
τ(F9−C4−C3−F7)	177.2(2)	−177.4
τ(F9−C4−C3−C2)	54.9(2)	−54.9
τ(F8−C4−C3−F6)	54.1(2)	−54.7
τ(F8−C4−C3−F7)	−63.3(2)	62.8
τ(F8−C4−C3−C2)	174.5(2)	−174.7
τ(F10−C4−C3−F6)	174.7(2)	−175.0
τ(F10−C4−C3−F7)	57.3(2)	−57.6
τ(F10−C4−C3−C2)	−64.9(2)	65.0

The hydrogen bond H11···O5′ (symmetry code $-x, 1-y, -z$) has a length of 1.70(3) Å, and the distance O1···O5′ is 2.665(2) Å. Additionally, the only appreciable distance below the van der Waals distance (r_{vdW}) [37] is the contact F6···C2′ (symmetry code: $+x, 3/2-y, 1/2+z$) with 3.113(2) Å. However, the stabilization energy calculated for these interactions is not appreciable. All intermolecular F···O and F···F distances are longer than their van der Waals distances. A section of the crystal lattice, viewing roughly along the *b* axis, is shown in Figure 10.

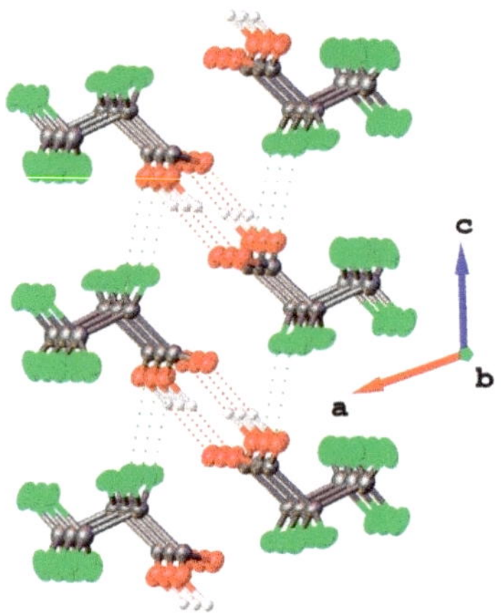

Figure 10. Section of the crystal lattice. Intermolecular contacts below the van der Waals distances are shown as dotted lines.

3. Materials and Methods

3.1. $CF_3CF_2C(O)OH$

The $CF_3CF_2C(O)OH$ 97% was purchased from Sigma Aldrich (Saint Louis, MO, USA) and subsequently purified by distillation through a series of U-shaped cold traps immersed in cold baths at −50, −80, and −110 °C, respectively. The acid was collected in the trap cooled to −50 °C.

3.2. *Quantum Chemical Calculations*

The Gaussian 03 program [38] was used to perform quantum chemical calculations, including the calculation of the potential energy function for a specific dihedral angle, followed by the geometry optimization of the corresponding minima and the calculation of their harmonic wavenumbers. For these purposes, the DFT [39] and MP2 [40] methods were chosen in conjunction with the 6-311+G(D) basis set. NBO [41] calculations were performed with the NBO 5.G package [42] incorporated in Gaussian 03. Additionally, the dimeric structure was computed using Gaussian 03. In this case, a potential energy curve was generated, followed by the optimization of the obtained minimum, employing the B3LYP/6-311+G(D) level of approximation.

3.3. *Infrared Spectroscopy*

Infrared spectra were recorded using a Nicolet™ 6700 spectrometer (Thermo Electron Corporation, Madison, WI, USA) with a double-wall cell featuring a 10 cm optical path length and 0.5 mm thick Si windows. The spectral resolution was 0.5 cm^{-1}, and each spectrum was obtained by averaging 64 scans. To optimize the equilibrium between monomers and dimers of perfluoropropionic acid in the matrix study at cryogenic temperatures, spectra were recorded at various temperatures (5.0, 9.9, 15.6, 20.4, and 38.0 °C) to determine the best experimental conditions.

3.4. *Matrix Isolation Experiments*

The gas mixture was deposited on a 15 K CsI window using the pulse deposition technique [43–45]. Low temperatures were achieved using a Displex closed-cycle refrigerator SHI-APD Cryogenics, model DE-202(AS Scientific Products, Abingdon, UK). The corresponding FTIR spectra were acquired with the previously described instrument. A Spectra-Physics Hg-Xe arc lamp operating at 1000 W was used to irradiate the matrix within the 200–800 nm broad band range. To prevent matrix heating, a water filter was placed between the lamp and the matrix. Several spectra were recorded at different irradiation times.

3.5. X-Ray Diffraction Analysis

A single crystal of $CF_3CF_2C(O)OH$ was grown in situ within a capillary. The sample was filled into a capillary, cooled with liquid nitrogen forming a polycrystalline material. At 180 K, a solid/liquid equilibrium near the melting point was established by melting the solid, leaving only a tiny crystal seed intact, using a thin copper wire as an external heat source. The temperature was then gradually lowered at 1 K/h to 176 K, at which point the entire capillary was filled with the crystalline specimen, followed by cooling to 146 K at a rate of 44 K per hour.

The crystal was maintained at 146.0(1) K during data collection, which was performed using an Agilent SuperNova diffractometer (Agilent, Santa Clara, CA, USA) at Bielefeld University. Using Olex2 [46], the structure was solved by direct methods with SHELX-97 [47], and the refinement was carried out using Olex2.refine [48] and spherical scattering factors were calculated with NoSpherA2 [49].

4. Conclusions

The evaluation of the results obtained through various spectroscopic techniques—vibrational IR, matrix IR spectra with UV–vis broad band irradiation, and X-ray diffraction analysis—provide complementary data that allow for the study of conformations and equilibria in different families. It is worth noting that conformational properties are crucial for understanding the chemical behavior of macromolecules and are responsible for the fundamental chemical behavior of biological molecules [50,51].

In the present case, and in relation to the above-mentioned data, the structure of $CF_3CF_2C(O)OH$, its dimer, and the existence of three conformations in equilibrium in the vapor phase have been conclusively determined. The combined analysis of matrix–isolation IR spectra and computationally predicted vibrational wavenumbers provides a powerful approach for identifying rotamers and investigating conformational equilibria. This is particularly evident when monitoring IR spectral changes induced by irradiation, which serve as distinctive markers for different conformers.

It is worth noting that the accurate assessment of the number of real conformers is a key tool for precisely approximating the electronic spectra of molecules. A purely computational approach requires evaluating thousands of geometries. In a recently published study, this computational effort is drastically reduced by using effective conformers for the calculation of UV–vis spectra [52].

Dedication

With all our gratitude, deepest affection, and boundless admiration, we dedicate this work to Prof. Jaan Laane. His unwavering devotion to knowledge—pursued with quiet resilience and constant passion—has been a shining inspiration in our lives. Beyond his brilliance as a scientist and mentor, it is his kindness, his generosity of spirit, and the profound care he has shown us that we will carry with us always. From the unforgettable days in College Station to the laughter-filled, sunny moments in La Plata, every memory made with him has been a gift.

Supplementary Materials: The following supporting information can be downloaded at: https://www.mdpi.com/article/10.3390/molecules30091887/s1. Figure S1: Potential energy surface of $CF_3CF_2C(O)OH$ as a function of the dihedral angles φ(O−C−O−H) and φ(C−C−C=O), displayed from different perspectives; Table S1: Comparison between experimental and computed IR vibrational modes for $CF_3CF_2C(O)OH$ and their proposed assignment; Table S2: Crystallographic information for $CF_3CF_2C(O)OH$.

Author Contributions: All authors contributed to the results presented in this manuscript. In particular, H.-G.S. was responsible for the crystal structure resolution of the studied species. All authors have read and agreed to the published version of the manuscript.

Funding: This article was supported by Consejo Nacional de Investigaciones Científicas y Técnicas (CONICET), PIP 0352 and PUE-2017-22920170100053; the Agencia Nacional de Promoción Científica y Tecnológica (ANPCyT), PICT-2018-04355 and PICT-2020-03746; and the Universidad Nacional de La Plata, UNLP-11/X971. The German authors acknowledge the Deutsche Forshungsgemeinschaft (DFG) for their financial support.

Institutional Review Board Statement: Not applicable.

Informed Consent Statement: Not applicable.

Data Availability Statement: The original contributions presented in this study are included in the article/Supplementary Materials. Further inquiries can be directed to the corresponding author.

Acknowledgments: The authors thank Yanina Berrueta Martínez for her contribution to the manuscript. This time, as she is working in a different area, she prefers not to be included as one of the authors.

Conflicts of Interest: The authors declare no conflicts of interest.

References

1. Susarla, N.; Ahmed, S. Estimating cost and energy demand in producing lithium hexafluorophosphate for Li-ion battery electrolyte. *Ind. Eng. Chem. Res.* **2019**, *58*, 3754–3766. [CrossRef]
2. Moody, C.A.; Martin, J.W.; Kwan, W.C.; Muir, D.C.G.; Mabury, S.A. Monitoring perfluorinated surfactants in biota and surface water samples following an accidental release of fire-fighting foam into Etobicoke Creek. *Environ. Sci. Technol.* **2002**, *36*, 545–551. [CrossRef] [PubMed]
3. Zhang, C.; Yan, H.; Li, F.; Zhou, Q. Occurrence and fate of perfluorinated acids in two wastewater treatment plants in Shanghai, China. *Environ. Sci. Pollut. Res.* **2015**, *22*, 1804–1811. [CrossRef] [PubMed]
4. Yamashita, N.; Kannan, K.; Taniyasu, S.; Horii, Y.; Petrick, G.; Gamo, T. A global survey of perfluorinated acids in oceans. *Mar. Pollut. Bull.* **2005**, *51*, 658–668. [CrossRef]
5. So, M.K.; Miyake, Y.; Yeung, W.Y.; Ho, Y.M.; Taniyasu, S.; Rostkowski, P.; Yamashita, N.; Zhou, B.S.; Shi, X.J.; Wang, J.X.; et al. Perfluorinated compounds in the Pearl River and Yangtze River of China. *Chemosphere* **2007**, *68*, 2085–2095. [CrossRef]
6. Ellis, D.A.; Mabury, S.A.; Martin, J.W.; Muir, D.C.G. Thermolysis of fluoropolymers as a potential source of halogenated organic acids in the environment. *Nature* **2001**, *412*, 321–324. [CrossRef]
7. Tian, Y.; Yao, Y.; Chang, S.; Zhao, Z.; Zhao, Y.Y.; Yuan, X.J.; Wu, F.C.; Sun, H.W. Occurrence and phase distribution of neutral and ionizable per- and polyfluoroalkyl substances (PFASs) in the atmosphere and plant leaves around landfills: A case study in Tianjin, China. *Environ. Sci. Technol.* **2018**, *52*, 1301–1310. [CrossRef]
8. Martin, J.W.; Mabury, S.A.; Solomon, K.R.; Muir, D.C.G. Dietary accumulation of perfluorinated acids in juvenile rainbow trout (*Oncorhynchus mykiss*). *Environ. Tox. Chem.* **2003**, *22*, 189–195. [CrossRef]
9. Martin, J.W.; Mabury, S.A.; Solomon, K.R.; Muir, D.C.G. Bioconcentration and tissue distribution of perfluorinated acids in rainbow trout (*Oncorhynchus mykiss*). *Environ. Tox. Chem.* **2003**, *22*, 196–204. [CrossRef]
10. U.S. Environmental Protection Agency. *Preliminary Risk Assessment of the Developmental Toxicity Associated with Exposure to Perfluorooctanoic Acid and Its Salts*; Office of Pollution Prevention and Toxics, RiskAssessment Division: Washington, DC, USA, 2003.
11. Berthiaume, J.; Wallace, K.B. Perfluorooctanoate, perflourooctanesulfonate, and N-ethyl perfluorooctanesulfonamido ethanol; peroxisome proliferation and mitochondrial biogenesis. *Toxicol. Lett.* **2002**, *129*, 23–32. [CrossRef]
12. Upham, B.L.; Deocampo, N.D.; Wurl, B.; Trosko, J.E. Inhibition of gap junctional intercellular communication by perfluorinated fatty acids is dependent on the chain length of the fluorinated tail. *Int. J. Cancer* **1998**, *78*, 491–495. [CrossRef]
13. Biegel, L.B.; Hurtt, M.E.; Frame, S.R.; Connor, J.O.; Cook, J.C. Mechanisms of extrahepatic tumor induction by peroxisome proliferators in male CD rats. *Toxicol. Sci.* **2001**, *60*, 44–55. [CrossRef] [PubMed]
14. Jackson, D.A.; Young, C.J.; Hurley, M.D.; Wallington, T.J.; Mabury, S.A. Atmospheric degradation of perfluoro-2-methyl-3-pentanone: Photolysis, hydrolysis and hydration. *Environ. Sci. Technol.* **2011**, *45*, 8030–8036. [CrossRef]

15. Scott, B.F.; Spencer, C.; Mabury, S.A.; Muir, D.C.G. Environ. Poly and perfluorinated carboxylates in North American precipitation. *Sci. Technol.* **2006**, *40*, 7167–7174. [CrossRef]
16. Young, C.J.; Hurley, M.I.D.; Wallington, T.J.; Mabury, S.A. Atmospheric chemistry of CF_3CF_2H and $CF_3CF_2CF_2CF_2H$: Kinetics and products of gas-phase reactions with Cl atoms and OH radicals, infrared spectra, and formation of perfluorocarboxylic acids. *Chem. Phys. Lett.* **2009**, *473*, 251–256. [CrossRef]
17. Grubbs, G.S., II; Obenchain, D.A.; Frank, D.S.; Novick, S.E.; Cooke, S.A.; Serrato, A., III; Lin, W. A study of the monohydrate and dihydrate complexes of perfluoropropionic acid using Chirped-Pulse Fourier Transform Microwave (CP-FTMW) spectroscopy. *J. Phys. Chem. A* **2015**, *119*, 10475–10480. [CrossRef]
18. Crowder, G.A. Infrared and Raman spectra of pentafluoropropionic acid. *J. Fluorine Chem.* **1972**, *1*, 385–389. [CrossRef]
19. Kagarise, R.E. Relation between the electronegativities of adjacent substitutents and the stretching frequency of the carbonyl group. *J. Am. Chem. Soc.* **1955**, *77*, 1377–1379. [CrossRef]
20. Statz, G.; Lippert, E. Far infrared spectroscopic studies on carboxylic acid solutions. *Ber. Bunsen-Ges.* **1967**, *71*, 673–679.
21. Badawi, H.M.; Al-Khaldi, M.A.A.; Al-Sunaidi, Z.H.A.; Al-Abbad, S.S.A. Conformational properties and vibrational analyses of monomeric pentafluoropropionic acid CF_3CF_2COOH and pentafluoropropionamide $CF_3CF_2CONH_2$. *Can. J. Anal. Sci. Spectrosc.* **2007**, *52*, 252–269.
22. Grubbs, G.S., II; Serrato, A., III; Obenchain, D.A.; Cooke, S.A.; Novick, S.E.; Lin, W. The rotational spectrum of perfluoropropionic acid. *J. Mol. Spectrosc.* **2012**, *275*, 1–4. [CrossRef]
23. Husted, D.R.; Ahlbrecht, A.H. The chemistry of the perfluoro acids and their derivatives. V. Perfluoropropionic acid. *J. Am. Chem. Soc.* **1953**, *75*, 1605–1608. [CrossRef]
24. Collins, M.F.; Haywood, B.C. Vibration spectra of carboxylic acids by neutron spectroscopy. *J. Chem. Phys.* **1970**, *52*, 5740–5745. [CrossRef]
25. Rontu, N.; Vaida, V. Vibrational spectroscopy of perfluorocarboxylic acids from the infrared to the visible regions. *J. Phys. Chem. B* **2008**, *112*, 276–282. [CrossRef] [PubMed]
26. Rontu, N.; Vaida, V. Vibrational spectroscopy of perfluoropropionic acid in the region between 1000 and 11000 cm^{-1}. *J. Mol. Spectros.* **2006**, *237*, 19–26. [CrossRef]
27. Cho, S.; Remucal, C.K.; Wei, H. Common and distinctive Raman spectral features for the identification and differentiation of per- and polyfluoroalkyl substances. *ACS EST Water* **2025**, *5*, 300–309. [CrossRef]
28. Lee, E.; Featherstone, J.; Martens, J.; McMahon, T.B.; Hopkins, W.S. Fluorinated propionic acids unmasked: Puzzling fragmentation phenomena of the deprotonated species. *J. Phys. Chem. Lett.* **2024**, *15*, 3029–3036. [CrossRef]
29. Berrueta Martínez, Y.; Bava, Y.B.; Cavasso Filho, R.L.; Erben, M.F.; Romano, R.M.; Della Védova, C.O. Valence and inner electronic excitation, ionization, and fragmentation of perfluoropropionic acid. *J. Phys. Chem. A* **2018**, *122*, 9842–9850. [CrossRef]
30. Litvanová, K.; Klemetsrud, B.; Xiao, F.; Kubátová, A. Investigation of real-time gaseous thermal decomposition products of representative per- and polyfluoroalkyl substances (PFAS). *J. Am. Soc. Mass Spectrom.* **2025**, *36*, 108–118. [CrossRef]
31. Wiens, J.P.; Miller, T.H.; Ard, S.G.; Viggiano, A.A.; Shuman, N.S. Elementary reactions leading to perfluoroalkyl substance degradation in an Ar^+/e^- Plasma. *J. Phys. Chem. A* **2022**, *126*, 9076–9086. [CrossRef]
32. Wang, W.; Song, M.; Abusallout, I.; Hanigan, D. Thermal decomposition of two gaseous perfluorocarboxylic acids: Products and mechanisms. *Environ. Sci. Technol.* **2023**, *57*, 6179–6187. [CrossRef]
33. Melin, T.R.L.; Harell, P.; Ali, B.; Loganathan, N.; Wilson, A.K. Thermochemistry of per- and polyfluoroalkyl substances. *J. Comput. Chem.* **2023**, *44*, 570–580. [CrossRef] [PubMed]
34. Medeiros, F.S., Jr.; Mota, C.; Chaudhuri, P. Perfluoropropionic acid-driven nucleation of atmospheric molecules under ambient conditions. *J. Phys. Chem. A* **2022**, *126*, 8449–8458. [CrossRef] [PubMed]
35. Hercules, D.A.; Parrish, C.A.; Sayler, T.S.; Tice, K.T.; Williams, S.M.; Lowery, L.E.; Brady, M.E.; Coward, R.B.; Murphy, J.A.; Hey, T.A.; et al. Preparation of tetrafluoroethylene from the pyrolysis of pentafluoropropionate salts. *J. Fluor. Chem.* **2017**, *196*, 107–116. [CrossRef]
36. Sablinskas, V.; Pucetaite, M.; Ceponkus, J.; Kimtys, L. Structure of propanoic acid dimers as studied by means of MIR and FIR spectroscopy. *J. Mol. Struct.* **2010**, *976*, 263–269.
37. Alvarez, S. A cartography of the van der Waals territories. *Dalton Trans.* **2013**, *42*, 8617–8636. [CrossRef]
38. Frisch, M.J.; Trucks, G.W.; Schlegel, H.B.; Scuseria, G.E.; Robb, M.A.; Cheeseman, J.R.; Montgomery, J.A., Jr.; Vreven, T.; Kudin, K.N.; Burant, J.C.; et al. *Gaussian 03, Rev C.02*; Gaussian, Inc.: Wallingford, CT, USA, 2004.
39. Parr, R.G.; Yang, W. *Density-Functional Theory of Atoms and Molecules (International Series of Monographs on Chemistry)*; Oxford University Press: New York, NJ, USA, 1989.
40. Møller, C.; Plesset, M.S. Note on an approximation treatment for many-electron systems. *Phys. Rev.* **1934**, *46*, 618–622. [CrossRef]
41. Foster, J.P.; Weinhold, F. Natural hybrid orbitals. *J. Am. Chem. Soc.* **1980**, *102*, 7211–7218. [CrossRef]
42. Glendening, E.D.; Badenhoop, J.K.; Reed, A.E.; Carpenter, J.E.; Bohmann, J.A.; Morales, C.M.; Weinhold, F. *NBO 5.0*; Theoretical Chemistry Institute, University of Wisconsin-Madison: Madison, WI, USA, 2004.

43. Almond, M.J.; Downs, A.J. Spectroscopy of matrix isolated species. *Adv. Spectrosc.* **1989**, *17*, 1–505. [CrossRef]
44. Dunkin, I.R. *Matrix-Isolation Techniques: A Practical Approach;* Oxford University Press: New York, NJ, USA, 1998.
45. Perutz, R.N.; Turner, J.J. Pulsed matrix isolation. A comparative study. *J. Chem. Soc. Faraday Trans. 2 Mol. Chem. Phys.* **1973**, *69* 452–461. [CrossRef]
46. Dolomanov, O.V.; Bourhis, L.J.; Gildea, R.J.; Howard, J.A.K.; Puschmann, H. OLEX2: A complete structure solution, refinement and analysis program. *J. Appl. Cryst.* **2009**, *42*, 339–341. [CrossRef]
47. Sheldrick, G.M. A short history of SHELX. *Acta Cryst.* **2008**, *A64*, 112–122. [CrossRef]
48. Bourhis, L.J.; Dolomanov, O.V.; Gildea, R.J.; Howard, J.A.K.; Puschmann, H. The anatomy of a comprehensive constrained restrained refinement program for the modern computing environment—Olex2 dissected. *Acta Crystallogr.* **2015**, *A71*, 59–75.
49. Kleemiss, F.; Dolomanov, O.V.; Bodensteiner, M.; Peyerimhoff, N.; Midgley, L.; Bourhis, L.J.; Genoni, A.; Malaspina, L.A. Jayatilaka, D.; Spencer, J.L.; et al. Accurate crystal structures and chemical properties from NoSpherA2. *Chem. Sci.* **2021**, *12* 1675–1692. [CrossRef] [PubMed]
50. Hogg, P.J. Disulfide bonds as switches for protein function. *Trends Biochem. Sci.* **2003**, *28*, 210–214. [CrossRef] [PubMed]
51. Hilal, R.; El-Aaser, A.M. A comparative quantum chemical study of methyl acetate and S-methyl thioacetate Toward an understanding of the biochemical reactivity of esters of coenzyme A. *Biophys. Chem.* **1985**, *2*, 145–150. [CrossRef]
52. Beck, I.T.; Mitchell, E.C.; Webb Hill, A.; Turney, J.M.; Rotavera, B.; Schaefer, H.F., III. Evaluating the importance of conformers for understanding the vacuum-ultraviolet spectra of oxiranes: Experiment and theory. *J. Phys. Chem. A* **2024**, *128*, 10906–10920 [CrossRef]

molecules

MDPI

Article

Infrared Matrix-Isolation and Theoretical Studies of the Reactions of Bis(benzene)chromium with Ozone

Roger W. Kugel and Bruce S. Ault *

Department of Chemistry, University of Cincinnati, Cincinnati, OH 45221, USA; kugelrr@ucmail.uc.edu
* Correspondence: bruce.ault@uc.edu

Abstract: Reactions of bis(benzene)chromium (Bz_2Cr) and ozone (O_3) were studied using low-temperature argon matrix-isolation infrared spectroscopy with supporting DFT calculations. When Bz_2Cr and O_3 were co-deposited, they reacted upon matrix deposition to produce two new prominent peaks in the infrared spectrum at 431 cm^{-1} and 792 cm^{-1}. These peaks increased upon annealing the matrix to 35 K and decreased upon UV irradiation at λ = 254 nm. The oxygen-18 and mixed oxygen-16,18 isotopic shift pattern of the peak at 792 cm^{-1} is consistent with the antisymmetric stretch of a symmetric ozonide species. DFT calculations of many possible ozonide products of this reaction were made. The formation of a hydrogen ozonide (H_2O_3) best fits the original peaks and the oxygen-18 isotope shift pattern. Energy considerations lead to the conclusion that the chromium-containing product of this reaction is the coupled product benzene-chromium-biphenyl-chromium-benzene (BzCrBPCrBz). $2Bz_2Cr + O_3 \rightarrow H_2O_3 + BzCrBPCrBz$, $\Delta E_{calc} = -52.13 kcal/mol$.

Keywords: matrix isolation; organometallic complex; infrared spectroscopy; theoretical calculations; ozonolysis

1. Introduction

Organometallic compounds have drawn interest in recent years as precursors for the vapor phase deposition of metal-containing thin films, either by chemical vapor deposition (CVD) or by atomic layer deposition (ALD). In particular, volatile metallocene and metallo-arene compounds react in the gas phase with oxidants like oxygen or ozone to produce metal oxide or pure metal thin films that find applications in the electronics industry [1–6]. In spite of their utility in industry, little is understood about the detailed mechanisms of such reactions. We undertook a study of some of these reactions using the low-temperature matrix-isolation technique in an attempt to produce and isolate possible reaction intermediates that escape detection in higher temperature kinetic studies.

Our previous work in this area involved the reactions of ozone with ferrocene [7] and of ozone with ruthenocene [8] in argon matrices. These studies revealed that virtually no thermal reaction occurred between ozone and these metallocenes in the gas phase under twin-jet deposition conditions when the reactants encountered each other only briefly in the dark. When the matrices were irradiated with low-energy, red, or near infrared light from an LED (λ = 625 nm or λ = 880 nm), however, different photochemical reactions were initiated. In the case of ferrocene, the red light photolyzed ozone to produce ground state $O(^3P)$ atoms that inserted into one of the cyclopentadienyl rings to form a coordinated, six-membered ring: a pyranyl anion [7]. In the case of ruthenocene, the infrared light photolyzed ozone to produce ground state $O(^3P)$ atoms that reacted with one of the cyclopentadiene rings to form a coordinated cyclopentadienone displacing one H atom to the ruthenium center, forming a ruthenium hydride [8]. In neither case was a reaction of ozone with the π-system of the ring observed to produce a primary ozonide, secondary ozonide, or Criegee intermediate typical of the ozone reactions with cyclopentene or cyclopentadiene previously observed [9,10].

In this work, we extend our research to the metallo-bisarene compound bis(benzene)chromium (Bz_2Cr). We report here our observations of the argon matrix reactions of O_3 with Bz_2Cr and propose an interpretation of the results.

2. Results and Discussion

1. Bis(benzene)chromium (Bz_2Cr) blank

"Blank" spectra of bis(benzene)chromium (Bz2Cr) isolated in an argon matrix (Ar/Bz2Cr $\cong$ 200) at 15 K showed infrared absorbance peaks that agreed with those reported earlier [11] and with the DFT calculations carried out in this work. The observed and calculated absorbance peaks are listed in Table 1. The calculated structure of Bz2Cr is shown in Figure 1 below.

Table 1. Observed and calculated infrared absorbance peaks for bis(benzene)chromium (Bz_2Cr) isolated in an argon matrix at 15 K.

Experiment [a]		Literature [b]		Calculations [a]		
Bz_2Cr Vaporized at 85 C and Co-Deposited with Argon at 15 K		$Cr+2C_6H_6 \overset{Armatrix}{\rightarrow} Bz_2Cr$ Bz_2Cr Formed in the Ar Matrix on Deposition		DFT/B3LYP 6-311G++(d,2p)		
$\tilde{\nu}$ (cm^{-1})	*A* (a.u.) [c]	$\tilde{\nu}$ (cm^{-1})	*A* (a.u.)	$\tilde{\nu}$ (cm^{-1})	*I*(km/mol) [d]	Assignment
467	0.65	466	0.81	435	177 d	A_{2u}
491 494	0.16 0.20	492	0.23	478 478	42 42	E_{1u} E_{1u}
793	0.22	793	0.15	798	58	A_{2u}
861	0.04	862	0.02	876 876	7 7	E_{1u} E_{1u}
974	0.10	974	0.13	986	25	A_{2u}
1003	0.03	1001	0.03	1021 1021	9 9	E_{1u} E_{1u}
1433	0.07	1433	0.06	1466 1466	2 2	E_{1u} E_{1u}
3060	0.19	3062	0.11	3179 3179	83 83	E_{1u} E_{1u}
				3185	18	A_{2u}

[a] This work; [b] Reference [11]. [c] *A* (a.u.) is absorbance in absorbance units. [d] *I* (km/mol) is calculated intensity in kilometers per mole.

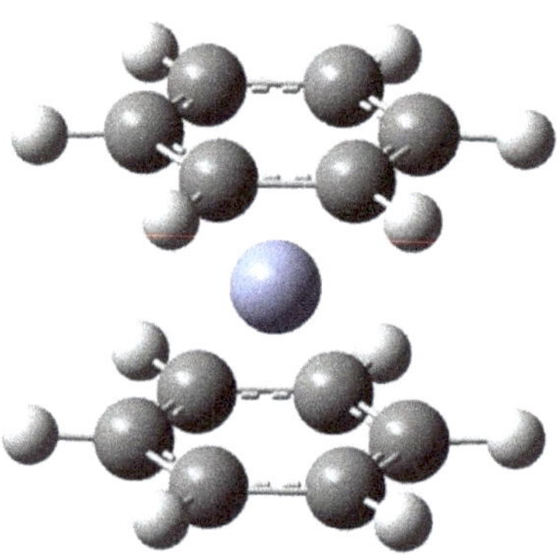

Figure 1. Calculated structure of Bz_2Cr.

A few observations should be noted about the spectra summarized in Table 1. First, Boyd, Lavoie, and Gruen [11] reported an additional weak, unassigned absorbance at 432.5 cm^{-1}. This peak was neither observed in the gas-phase spectrum of Bz_2Cr [12], nor in the argon matrix spectrum reported in this work, nor is it predicted from the DFT quantum calculations. It should be noted that Boyd et al. [11] produced Bz_2Cr by co-deposition of benzene and atomic chromium with argon onto a liquid-helium-cooled cesium iodide window. We believe that the peak they observed at 432.5 cm^{-1} might be due to a small amount of the less stable mono(benzene)chromium (BzCr) formed during the co-deposition of benzene and atomic chromium in the argon matrix. Secondly, the symmetry assignments given by Boyd et al. [11] for the two lowest frequency vibrations of Bz_2Cr are reversed from those predicted by the current quantum calculations, and from those previously reported for the gas phase spectrum [13]. We believe that these assignments should be 466 cm^{-1}(A_{2u}) and 492 cm^{-1}(E_{1u}), as shown in Table 1. Thirdly, the highest frequency reported in our spectrum at 3060 cm^{-1} is actually flanked by two unresolved shoulders at about 3055 and 3066 cm^{-1}. It is possible that the highest frequency vibration predicted by the quantum calculations at 3185 cm^{-1} is unresolved in this peak pattern. Finally, when the argon-matrix-isolated Bz_2Cr was irradiated with ultraviolet light, new peaks appeared in the infrared spectrum indicating that a photochemical reaction had occurred. The new product, whose most prominent peak occurs at 426 cm^{-1}, is likely mono(benzene)chromium, but it was not further characterized.

2. The reaction of Bz_2Cr with O_3 in an Ar matrix

When gas mixtures of Bz_2Cr plus Ar and O_3 plus Ar were co-deposited through twin jets onto the CsI cold window, new peaks were observed in the infrared spectra of the matrices that formed. These peaks were "new" in the sense that they were not observed in similar matrices made from separate "blank" mixtures of Bz_2Cr in Ar or of O_3 in Ar. The most significant of these peaks, at 431 and 792 cm^{-1}, increased in intensity upon warming the matrices to 35 K, and decreased in intensity upon irradiating the matrices with ultraviolet light from an unfiltered mercury pen lamp. These results suggest that O_3 reacts with Bz_2Cr rapidly and with little-to-no activation energy during deposition, or in the forming or annealing matrices. The results further suggest that the compound formed is readily decomposed by ultraviolet light. The new infrared peaks observed in these matrices are summarized in Table 2. Partial infrared spectra showing the major peaks are given in Figure 2.

Table 2. New infrared peaks observed in the twin-jet co-deposition of gas mixtures: $Ar/Bz_2Cr \cong 200$ and $Ar/O_3 \cong 200$ onto a CsI window at 15 K.

New Peaks Observed upon Deposition [a] of Bz_2Cr and O_3 in Solid Ar at 15 K		
$\tilde{\nu}$ (cm^{-1})	A^b (a.u.)	$\Delta\tilde{\nu}_{O^{18}}$ (cm^{-1})^c
431	0.16	0
472 [d]	0.03	0
622	0.02	−35
752	0.02	−42
758	0.02	−41
782	0.03	−42
792	0.12	−44
865 [d]	0.02	−23
890 [d]	0.03	−21
1007 [d]	0.03	-
1431	0.03	−2

[a] Note that all the peaks listed in this table appeared upon deposition, increased upon annealing to 35 K, and decreased upon UV irradiation at 254 nm. [b] The intensities listed were those measured after 18 h of deposition. [c] $\Delta\nu_O$(O-18) is the oxygen-18 isotope shift in wavenumbers (cm^{-1}). [d] The peaks found at 472, 865, 890, and 1007 cm^{-1} are due to trace amounts of unassigned byproducts that remain unknown.

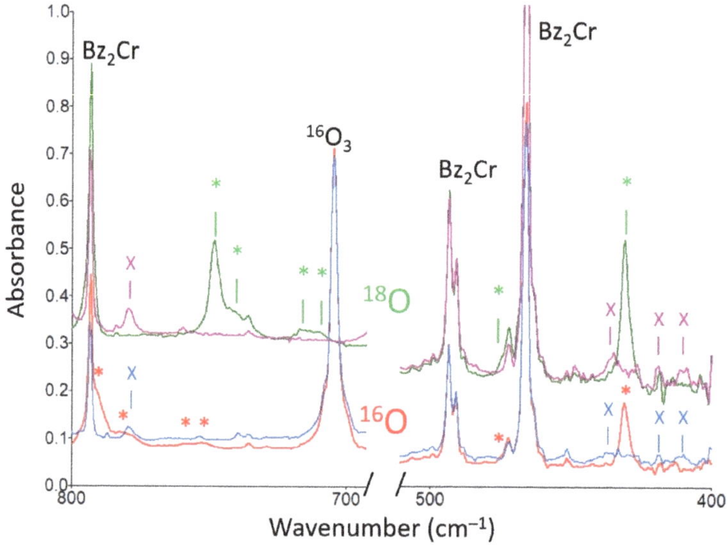

Figure 2. Partial infrared spectra from two experiments. Lower traces: Bz_2Cr plus $^{16}O_3$ in an argon matrix at 15 K; red trace: deposit with new peaks marked with a red asterisk * (431, 472, 752, 758, 782, and 792 cm^{-1}); blue trace: after 1 h of UV irradiation with the unfiltered light from a mercury pen lamp. New peaks marked with a blue **x** (411, 419, 436, and 779 cm^{-1}) are unidentified decomposition products. Upper traces: Bz_2Cr plus $^{18}O_3$ in an argon matrix at 15 K; green trace: deposit with new peaks marked with a green asterisk * (431, 472, 710, 717, 740, and 748 cm^{-1}); magenta trace after 1.25 h of UV irradiation with the unfiltered light from a mercury pen lamp. New peaks marked with a magenta **x** (411, 419, 436, and 779 cm^{-1}) are unidentified decomposition products. In both experiments, the new peaks appeared upon deposition, disappeared upon UV irradiation, and reappeared upon annealing the matrix to 35 K.

The most intense peak at 431 cm^{-1} in the infrared spectrum of the new species shows no significant oxygen-18 isotope shift. However, the second-most intense peak at 792 cm^{-1} shows an exceptionally large oxygen-18 isotope shift of 44 cm^{-1}. This large shift suggests that this vibrational mode involves significant motion of one or more oxygen atoms. To explore this isotope shift, a mixed isotope experiment was carried out with scrambled $^{16,18}O_3$ ozone. The results of this experiment revealed six isotopomer peaks between 748 and 792 cm^{-1}. These peaks and their relative intensities are listed in Table 3 and shown in Figure 3. The peak pattern shown by these peaks is suggestive of the antisymmetric stretch of a symmetric ozonide moiety. The pattern shows the second and fifth peaks in the isotopic progression are more intense than the third and fourth peaks, indicating that they correspond to the following isotope sequences: 16, 16, 18 and 16, 18, 18, which would be twice as intense for the antisymmetric stretch of a symmetric ozonide.

Another interesting feature of the intense infrared absorption of the new species at 792 cm^{-1} is that it has three less-intense "satellite" peaks associated with it: a shoulder at 782 cm^{-1} and (barely resolved) peaks at 758 and 752 cm^{-1}. This entire four-peak pattern appears upon deposition, increases upon annealing the matrix, and disappears upon UV irradiation. Furthermore, all four peaks show similar large oxygen-18 isotope shifts of 41 to 44 cm^{-1}, so that the entire peak pattern repeats at 748, 740, 717, and 710 cm^{-1}. That the behavior of the peaks in this pattern is so similar suggests that they are due to similar molecular structures possibly perturbed by environmental or conformational effects. The fact that the spread of the peak pattern for a given oxygen isotope is so large (~40 cm^{-1}) seems to rule out simple matrix site effects and suggests that the satellite peaks are caused by different isomers or conformers of the new species that is formed on deposition of the matrix.

Table 3. New isotopomer infrared peaks observed in the twin-jet co-deposition of gas mixtures: $Ar/Bz_2Cr \cong 200$ and $Ar/^{16,18}O_3 \cong 200$ onto a CsI window at 15 K. Isotopomer peaks for the $^{16,18}O_3$ scrambled, mixed isotope experiment.

Isotopomer Peaks for the $^{16,18}O_3$ Scrambled Mixed Isotope Experiment	
$\tilde{\nu}$ (cm^{-1})	A (a.u.)
748	0.040
756	0.056
766	0.052
774	0.048
782	0.084
792	? [a]

[a] Note that the intensity of the peak at 792 cm^{-1} could not be accurately determined due to its proximity to the intense parent peak for Bz_2Cr at 793 cm^{-1}.

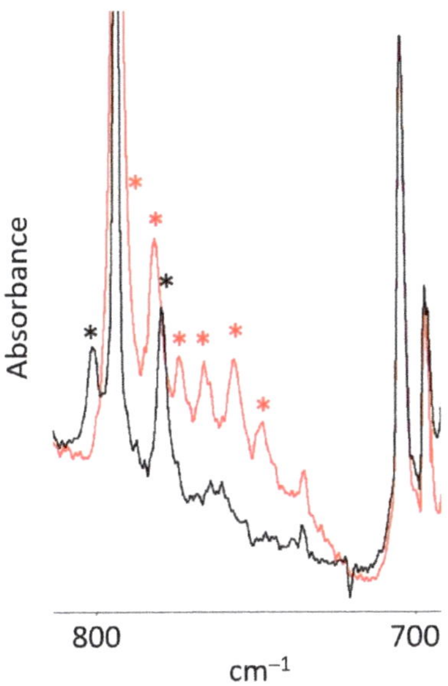

Figure 3. Partial infrared spectra from the mixed isotope experiment: Bz_2Cr plus $^{16,18}O_3$ in an argon matrix at 15 K: red trace: deposit with new peaks marked with a red asterisk * (748, 756, 766, 774, 782, and 792 cm^{-1}); black trace: deposit after 1.0 h of irradiation at 254 nm with the unfiltered light from a mercury pen lamp. The isotopomer peaks have disappeared and two new peaks that appeared with UV irradiation are marked with a black asterisk * (779 and 801 cm^{-1}).

Several symmetric ozonide molecules were calculated as potential candidates for the new product formed between ozone and bis(benzene)chromium in an argon matrix. Focus was placed on the calculated energy of reaction, ΔE, and the calculated frequency and intensity of the antisymmetric O-O-O stretching mode, $\tilde{\nu}_{asym\ O-O-O\ stretch}$ (I). A list of potential candidate molecules with appropriate calculated data is shown in Table 4 and the calculated Gaussian structures are shown in Figure 4.

Table 4. Calculated symmetric ozonide molecules as potential products of the reaction between Bz_2Cr and O_3 in an argon matrix. The calculated values of ΔE are relative to the separated molecules. The frequencies listed are calculated frequencies for the antisymmetric O-O-O stretch. The calculated frequency and intensity values for ozone are included at the end of the list for reference.

	Symmetric Ozonide	ΔE	$\tilde{\nu}_{asym\ O-O-O\ stretch}$ **(*I*)**
	Bz_2Cr+O_3 →	**(kcal/mol)**	**(cm^{-1}) (km/mol)**
(a)	1,2-ring ozonide (*exo*)	−8.47	721 (14)
(b)	1,2-ring ozonide (*endo*)	−8.62	687 (20)
(c)	1,4-ring ozonide	+10.15	599 (14)
(d)	Cr-ozonide (flat 1)	−40.88	621 (26)
(e)	Cr-ozonide (flat 2)	−56.46	457 (6)
(f)	Cr-ozonide (bent)	−52.93	695 (33)
(g)	1,2 van der Waals	−8.27	1089 (168)
(h)	1,1′ van der Waals	−9.16	1086 (155)
(i)	*syn*-H_2O_3 + BzByCr	+29.52	786 (113)
(j)	*syn*-H_2O_3 + Ph_2Cr	+41.31	786 (113)
	2 Bz_2Cr + O_3→		
(k)	*syn*-H_2O_3 + BzCrBPCrBz	−52.13	786 (113)
	2 Bz_2Cr + O_3→		
(l)	*anti*-H_2O_3 + BzCrBPCrBz	−54.76	780 (98)
(m)	Ozone, O_3 [a]	---	1183 (239)

[a] Note that the calculated frequency of ozone is included for reference.

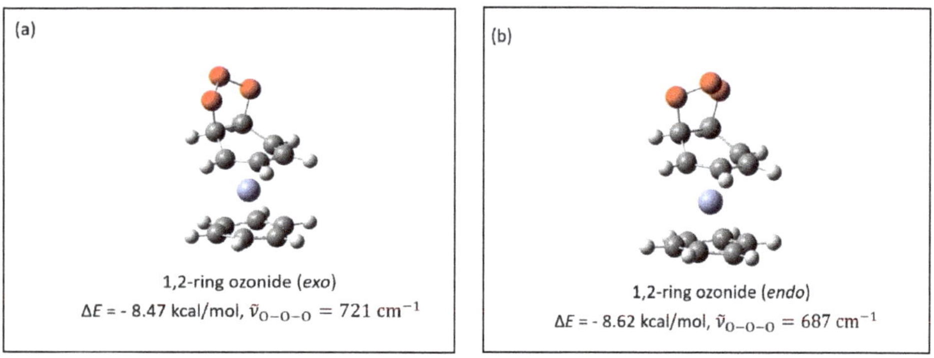

Figure 4. *Cont.*

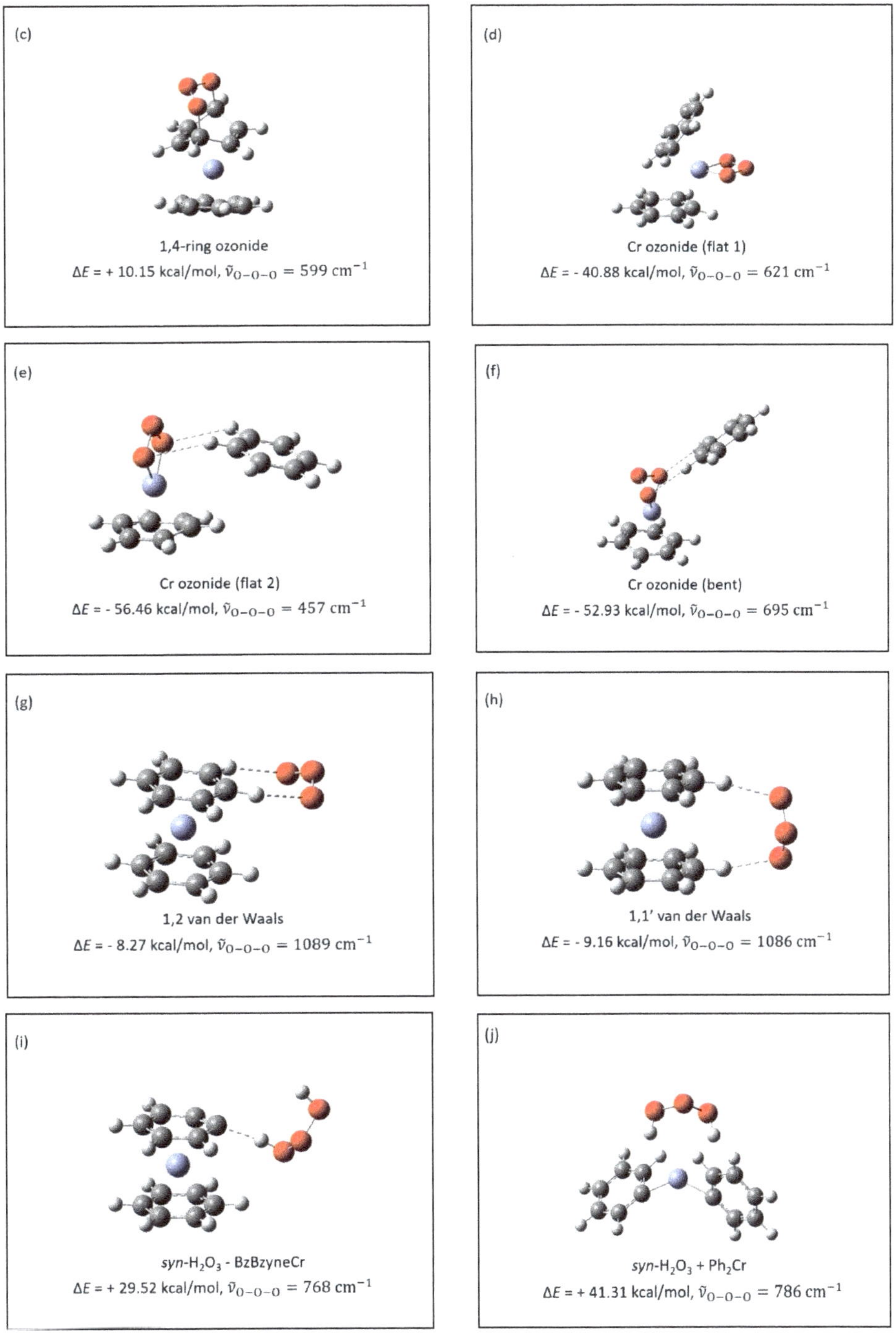

Figure 4. *Cont.*

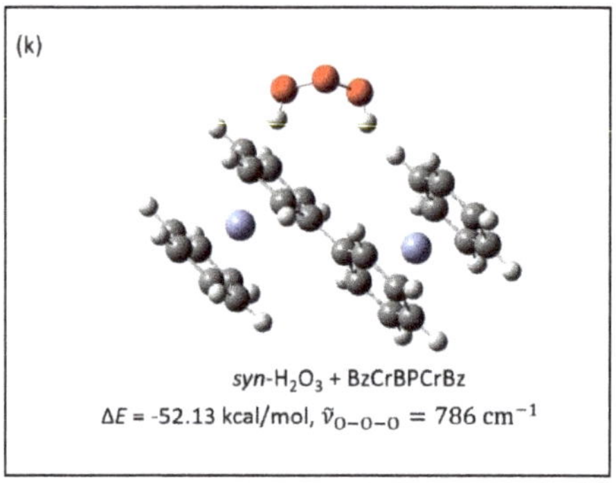

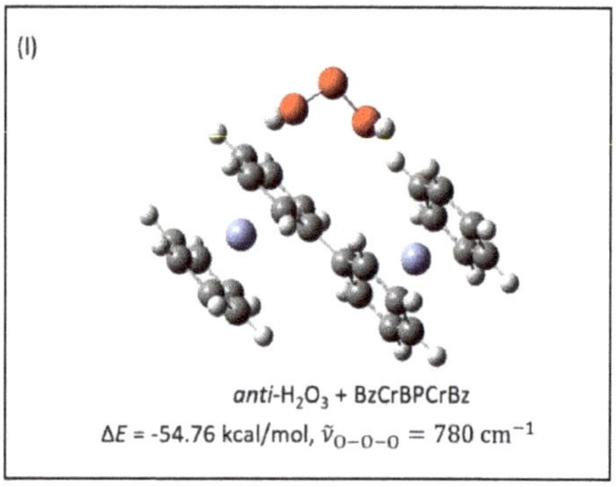

Figure 4. Gaussian structures with symmetric ozonides. (**a–j**): potential products from the reaction: $Bz_2Cr + O_3 \rightarrow$ product; (**k**,**l**): potential products from the reaction: $2\ Bz_2Cr + O_3 \rightarrow$ product. ΔE values represent the differences in energy between the structure represented and the Bz_2Cr + O_3 reactants; $\tilde{\nu}_{O-O-O}$ represents the antisymmetric vibrational frequency of the ozonide product. Reference for the calculated value for $\tilde{\nu}_{O-O-O}$ in ozone: 1183 cm^{-1} (239 km/mol).

By far, of all the symmetric ozonides calculated, the one that best fits the observed infrared data is hydrogen ozonide (H_2O_3). The two most prominent peaks in the calculated infrared spectrum of *syn*-H_2O_3 are at 444 cm^{-1} $\left(\Delta\tilde{\nu}_{^{18}_{8}O} = -3\ \text{cm}^{-1}\right)$ and 786 cm^{-1} $\left(\Delta\tilde{\nu}_{^{18}_{8}O} = -44\ \text{cm}^{-1}\right)$, and correspond closely to the two most prominent peaks in the experimental spectrum of the new species: 431 cm^{-1} $\left(\Delta\tilde{\nu}_{^{18}_{8}O} = 0\ \text{cm}^{-1}\right)$ and 792 cm^{-1} $\left(\Delta\tilde{\nu}_{^{18}_{8}O} = -44\ \text{cm}^{-1}\right)$. Not only is the calculated antisymmetric stretching frequency $\left(786\text{cm}^{-1}\right)$ closest to that observed $\left(792\text{cm}^{-1}\right)$, but the calculated oxygen-18 isotope shift $\left(-44\text{cm}^{-1}\right)$ is also exactly the same as that observed $\left(-44\text{cm}^{-1}\right)$. These results comparing experiment, the literature, and calculation are summarized in Table 5. Note that H_2O_3 has been observed in previous matrix experiments by Engdahl and Nelander, who isolated H_2O_3 from the argon matrix reaction of H_2O_2 with O(^{1}D) [14]. It was also observed recently in our laboratory by Pinelo et al., who isolated H_2O_3 from the argon matrix reaction of ozone with 1,4-cyclohexadiene to produce benzene and hydrogen ozonide (H_2O_3) [15]. Hydrogen ozonide (H_2O_3) can be formed in *syn*- and *anti*-conformations, with the *anti*-form more stable by 2.63 kcal/mol. The *syn*- and *anti*-calculated Gaussian structures for H_2O_3 are shown in Figure 5. The *syn/anti* calculated energy diagram for H_2O_3 is shown in Figure 6.

Table 5. The new peaks observed in the argon matrix reaction of O_3 + Bz_2Cr compared with those previously observed for H_2O_3 and with those calculated for H_2O_3.

Experiment [a]			Literature [b]	Calculations [a]		
$\tilde{\nu}$ (cm^{-1})	A (a.u.)	$\Delta\tilde{\nu}_{^{18}_{8}O}$ (cm^{-1})	$\tilde{\nu}$ (cm^{-1})	$\tilde{\nu}$ (cm^{-1})	I (km/mol)	$\Delta\tilde{\nu}_{^{18}_{8}O}$ (cm^{-1})
431	0.14	0	387	444	118	−3
622	0.02	−35	509	514	9	−29
792	0.10	−44	776	786	113	−44
1431	0.03	−2	1359	1377	45	−6
-	-	-	3530	3731	36	−13

[a] This work. [b] Reference [14].

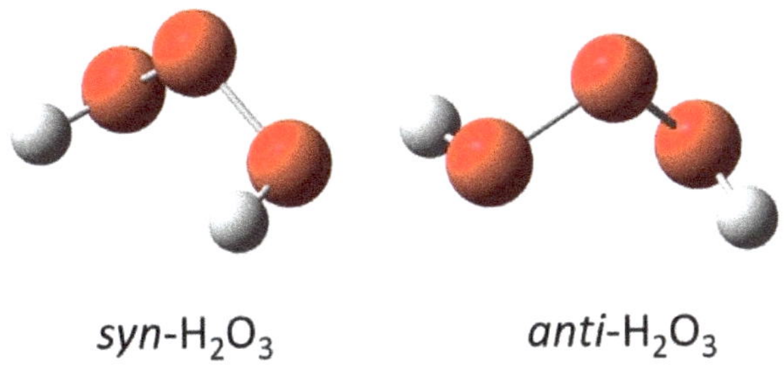

Figure 5. Hydrogen ozonide Gaussian calculated structures for *syn*-H_2O_3 and *anti*-H_2O_3.

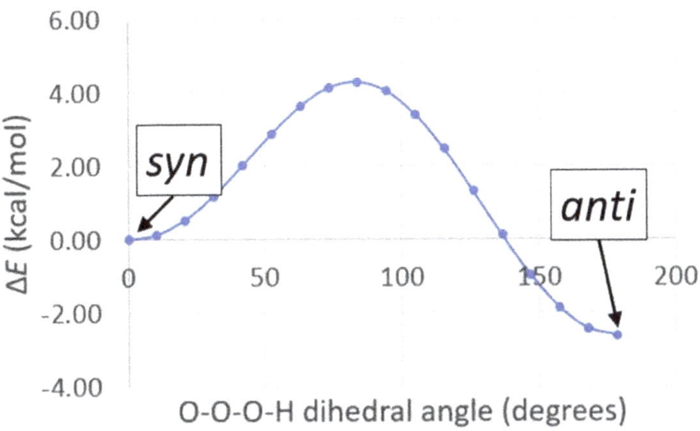

Figure 6. Energy diagram for *syn/anti* conversion of H_2O_3.

Furthermore, the peak pattern observed in the scrambled, mixed isotope experiment (See Figure 3) is almost exactly reproduced by the calculated peaks of all six of the H_2O_3 isotopomers. Figure 7 shows the observed and calculated peak patterns of the $^{16,18}O$ isotopomers of H_2O_3. Note that in Figure 7 the observed peaks in the experimental spectrum are recalculated for baseline correction. The isotopomer peak assignments are given in Table 6.

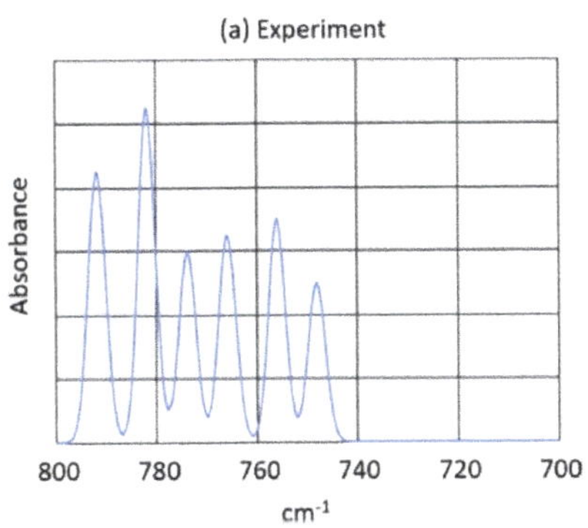

Figure 7. *Cont.*

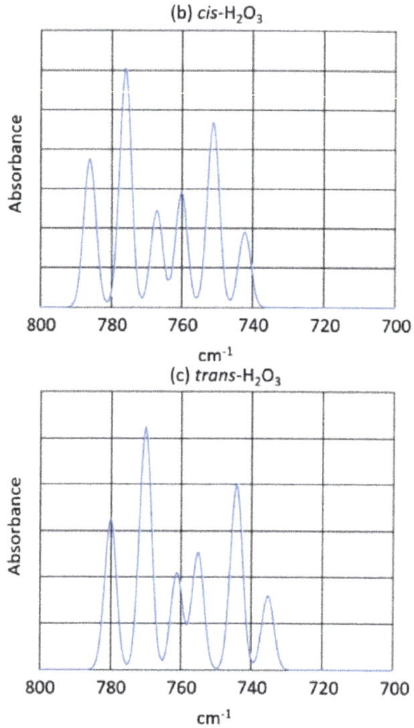

Figure 7. Isotopomer infrared peaks. (**a**) Experiment: Bz_2Cr + $^{16,18}O_3$. The experimental spectrum was baseline-corrected with the measured intensities and wavenumbers. (**b**) Calculated: *syn*-H_2O_3. (**c**) Calculated: *anti*-H_2O_3. The calculated spectra used the calculated wavenumbers and intensities and assumed 45% oxygen-18 in the $^{16,18}O_3$ mixture.

Table 6. Isotopomer peak assignments for the antisymmetric stretch of H_2O_3. In the assignment column, 666 represents H-^{16}O-^{16}O-^{16}O-H, 868 represents H-^{18}O-^{16}O-^{18}O-H, etc.

Experiment		***syn*-H_2O_3**		***anti*-H_2O_3**		**Assignment**
$\tilde{\nu}$	$\Delta\tilde{\nu}_{^{18}_{8}O}$	$\tilde{\nu}$	$\Delta\tilde{\nu}_{^{18}_{8}O}$	$\tilde{\nu}$	$\Delta\tilde{\nu}_{^{18}_{8}O}$	**6 Represents ^{16}O**
(cm^{-1})	(cm^{-1})	(cm^{-1})	(cm^{-1})	(cm^{-1})	(cm^{-1})	**8 Represents ^{18}O**
792	0	786	0	780	0	666
782	10	776	10	770	10	668 and 866
774	18	767	19	761	19	868
766	26	760	26	755	25	686
756	36	751	35	744	36	688 and 886
748	44	742	44	735	45	888

The peak patterns calculated for the *syn*- and *anti*-H_2O_3 are very similar to that observed in the Bz_2Cr + $^{16,18}O_3$ matrix experiment. The entire pattern is slightly shifted to lower frequencies by 6 cm^{-1} for the *syn*-isomer and by 12 cm^{-1} for the *anti*-isomer. Also, the individual oxygen-18 isotope shifts for the individual *syn*- and *anti*-H_2O_3 isotopomers are nearly identical to all the shifts observed in the experiment. Next, notice in the peak assignments that the 668/866 isotopomers and the 688/886 isotopomers are identical because H_2O_3 is a symmetrical ozonide. Finally, the isotope shift for the 686 isotopomer is greater than that for the 868 isotopomer because in the antisymmetric stretch of the

O-O-O group, the middle O atom moves a greater distance than do the other two. Thus, the isotope shift for one O atom in the middle is greater than that for two O atoms on the ends of the O-O-O group. Such would not be the case for the symmetric O-O-O stretch.

The formation of hydrogen ozonide, (H_2O_3) begs the following questions: Where do the H atoms come from? And what is the mechanism of this formation reaction? Calculations show that O_3 interacts strongly with hydrogen atoms of the benzene rings of Bz_2Cr. These intermolecular interactions are shown in Figure 4g,h. In Figure 4g, the terminal O atoms of ozone are simultaneously interacting with two H atoms on the same benzene ring of Bz_2Cr. This calculated interaction energy is $\Delta E = -8.27$ kJ/mol and the calculated antisymmetric vibrational frequency is 1089 cm^{-1}. In Figure 4h, the terminal O atoms of ozone are also simultaneously interacting with two H atoms, in this case one on each benzene ring of Bz_2Cr. This calculated interaction energy is $\Delta E = -9.16$ kJ/mol and the calculated antisymmetric vibrational frequency is 1086 cm^{-1}. These calculated intermolecular interaction energies are unusually large for this type of interaction, approaching the strength of a hydrogen bond for each C-H---O encounter. It should be noted that the calculated non-covalent interactions of ozone with benzene itself are no greater than $\Delta E = -1.65$ kJ/mol total, and the ozone O atoms are not particularly attracted to the benzene H atoms. The stronger interaction of ozone with the Bz_2Cr hydrogen atoms is understandable in terms of the fact that the Cr atom is pulling electron density from both benzene rings, leaving the H atoms with significant positive charge. As a result, we believe that co-deposition of O_3 with Bz_2Cr in an argon matrix will result in a significant number of Bz_2Cr---O_3 van der Waals complexes forming in the matrix. It should be noted that the strong, hydrogen-bond-like interactions between the oxygen atoms of ozone and the hydrogen atoms of Bz_2Cr are consistent with reports of similar interactions between O atoms of water and C-H bonds of Bz_2Cr^+ ions in crystals of hydrated salts from X-ray diffraction data [16]. In fact, the C-H---O interactions in the Bz_2Cr+ crystals are so strong that these researchers conclude the following:

"In summary this study provides evidence that hydrogen bonds (both of the conventional OH-O and controversial CH-O types) afford a pattern of interactions in common between organics, organometallics, and water that can be utilized to engineer crystalline materials on the basis of the complementarity between donors and acceptors".

If the O_3 moiety of the Bz_2Cr---O_3 complex were able to abstract two hydrogen atoms from Bz_2Cr, forming H_2O_3 + BzBzyneCr, it probably would. Unfortunately, that reaction is energetically unfavorable, as is the formation of Ph_2Cr with H_2O_3. Figure 4i,j show these structures and their calculated energies as $\Delta E = +29.52$ kJ/mol and $\Delta E = +41.31$ kJ/mol, respectively, relative to the Bz_2Cr + O_3 starting materials.

The Bz_2Cr---O_3 complexes and the uncomplexed O_3 and Bz_2Cr species will be fairly isolated at the beginning of the deposition. However, as deposition proceeds over a 20–24-h period, the pressures of Ar and O_3 gases decrease and so do their deposition rates. Bz_2Cr, on the other hand, is fed into the depositing Ar stream by its constant vapor pressure at 85 °C. So, the relative proportion of Bz_2Cr in the matrix increases toward the end of deposition. As a result, formation of the Bz_2Cr---O_3 complexes will be more numerous toward the end of the deposition process. In addition, the likelihood of these complexes encountering another molecule of Bz_2Cr will also increase. Consequently, during late-term deposition or during annealing to 35 K, three-way encounters like Bz_2Cr---O_3 + Bz_2Cr will more likely occur. These encounters of non-isolated molecules in the matrix could conceivably lead to a coupling reaction between the Bz_2Cr species, splitting out H_2O_3 and probably going through a BzBzyneCr-like intermediate. This coupling would result in a coupled benzene-chromium-biphenyl-chromium-benzene (BzCrBPCrBz) product, where BP = biphenyl. The energy of this overall reaction is very favorable at $\Delta E = -52.13$ kJ/mol for the *syn*-H_2O_3 and $\Delta E = -54.76$ kJ/mol for the *anti*-H_2O_3 product. The calculated energies and structures are shown in Figure 4k,l. There is evidence in the literature for the formation of biphenyl from the reaction of benzene with benzyne [17,18]. The calculated structure for the BzCrBPCrBz molecule is shown in Figure 8 and the proposed mechanism for this coupling reaction is shown in Scheme 1.

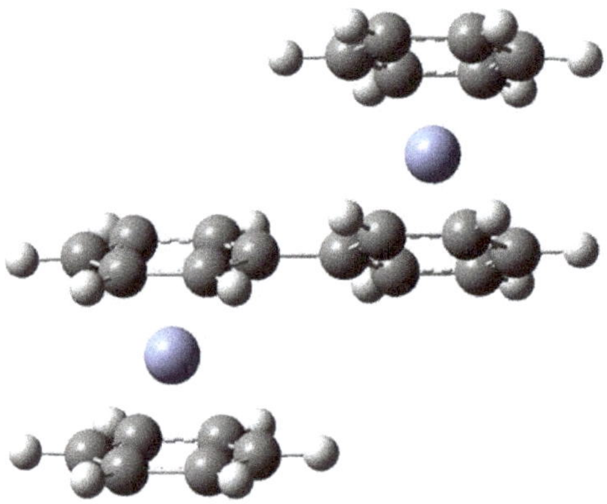

Figure 8. Calculated structure for BzCrBPCrBz. The calculated angle between the biphenyl rings is 27°, significantly less than that calculated for biphenyl itself (41°).

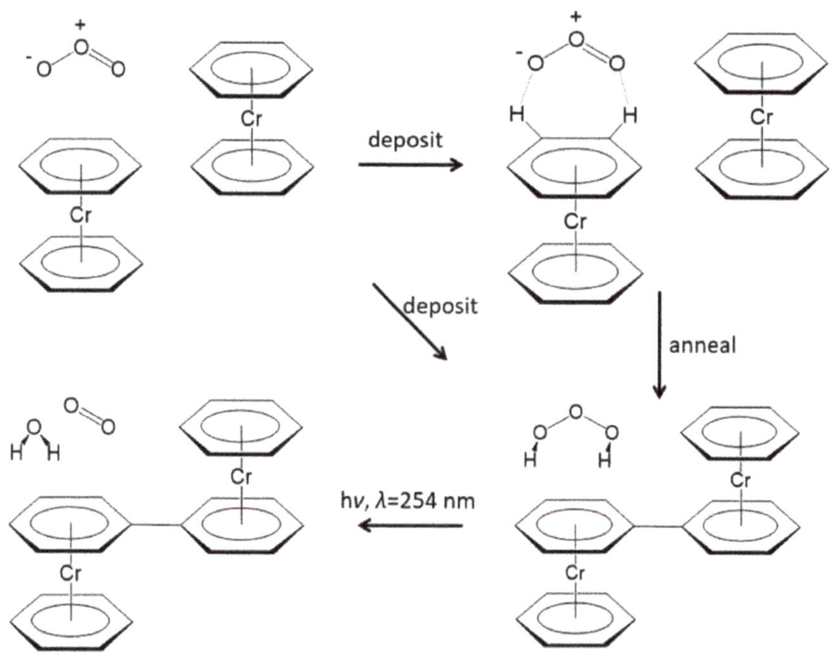

Scheme 1. O_3 interacts and reacts with Bz_2Cr upon deposition and annealing of the argon matrix. H_2O_3 and coordinated biphenyl form during deposition and annealing. H_2O_3 is decomposed by UV photolysis at λ = 254 nm.

Given this proposed mechanism, it seems likely that the *syn*-conformer of H_2O_3 is preferred in the reaction, since the interaction is a simultaneous frontal attack of ozone on two hydrogen atoms of the benzene ring of Bz_2Cr. The more stable *anti*-conformer of H_2O_3 would be less prevalent. This is also consistent with the experimental spectrum, since the observed "satellite" peaks are less intense and shifted to lower frequencies from the primary peak at 792 cm^{-1}. Also, the major peaks in the calculated spectrum of BzCrBPCrBz are indistinguishable from (within 2 or 3 wavenumbers of) those in Bz_2Cr.

It should be noted that biphenyl-coordinated chromium compounds were formed in very early syntheses of chromium phenyl compounds by the reaction of phenyl magnesium bromide (PhMgBr) with anhydrous chromium(III) chloride ($CrCl_3$) in benzene solution [19]. Although, at the time of this early work by Hein (1918–1936) [20], it was not known that biphenyl itself formed during the reaction to produce the metallocene compound, η^6-bis(biphenyl)chromium (I) bromide. In later work, the structure we propose in Figure 8 (BzCrBPCrBz) was actually produced and characterized (among many other chromium arenes) by the direct deposition of Cr(*g*) with benzene(*g*) and biphenyl(*g*) on cold surfaces [21].

3. Materials and Methods

Oxygen (O_2, 99.999%) and argon (Ar, 99.999%) gases were supplied by Wright Brothers, Inc., Cincinnati, OH, USA. Oxygen-18 isotopically enriched oxygen ($^{18}O_2$, 99-atom percent ^{18}O) was supplied by Sigma-Aldrich Chemical Co., St. Louis, MO, USA. Ozone (O_3 or $^{18}O_3$) was produced using a high-frequency Tesla coil discharge through oxygen gas (O_2 or $^{18}O_2$) while being condensed with liquid nitrogen. To produce a statistically scrambled mix of $^{16,18}O_3$ isotopomers, a 50–50 mixture of O_2 and $^{18}O_2$ was discharged. To minimize the ozone explosion hazard, the pressure of ozone gas at room temperature was never allowed to exceed 5.0 in. of Hg (127 torr). A stock supply of bis(benzene)chromium (Bz_2Cr) was obtained from Strem Chemicals, Inc., Newburyport, MA, USA, and purified by sublimation. Gas mixtures of ozone with argon (at mole ratios of approximately Ar/O_3 = 200) were prepared by standard manometric techniques. Ar and O_3 were premixed, and the Ar/Bz_2Cr mixtures were prepared by the sublimation of the bis(benzene)chromium at ~85 °C into a flowing stream of pure argon during deposition. Both mixtures were deposited using stainless steel needle valves onto a cryogenically cooled CsI window held at 10−15 K with a closed-cycle helium cryostat (CTI Cryogenics, now from Brooks Automation, Inc., Chelmsford, MA, USA). Matrix formation took place over 20−24 h at an average argon deposition rate of 2–3 mmol/h.

The infrared spectra of the argon matrices were scanned using FT-IR spectroscopy (PerkinElmer Spectrum One, Waltham, MA, USA) at 1 cm^{-1} resolution from 4000 to 400 cm^{-1}. After deposition, the matrices were subsequently either irradiated or annealed to observe by infrared spectroscopy any chemical changes that may have occurred. Irradiation was performed with ultraviolet light from a mercury pen lamp (Pen-Ray PS-1, UVP Cambridge, Cambridge, UK) whose primary output is the Hg line at a wavelength of λ = 254 nm. Annealing was accomplished with a small resistance heater attached to the brass mounting for the CsI cold window.

Quantum chemistry calculations (geometry optimizations, total energies, and infrared vibrational frequencies) were carried out with the Gaussian 09 suite of programs (Revision C.01) [22] using density functional theory (DFT) and the B3LYP functionals with the 6-311G++(d,2p) basis set. Gaussian calculations were carried out remotely at the Ohio Supercomputer Center (OSC) in Columbus, OH, USA.

4. Conclusions

1. Bis(benzene)chromium (Bz_2Cr) was isolated in an Ar matrix from the vapor pressure of $Bz_2Cr(s)$ at 85 °C fed into a stream of argon gas.
2. The infrared spectrum of the resulting matrix-isolated Bz_2Cr agrees with the infrared spectrum of matrix-isolated Bz_2Cr produced by co-deposition of C_6H_6(*g*) and Cr(*g*) with Ar(*g*) [11].
3. When bis(benzene)chromium (Bz_2Cr) was co-deposited in the dark with ozone (O_3), a new product was formed upon deposition that increased upon annealing to 35 K, and was destroyed by UV irradiation at 254 nm.

4. The new product showed two strong bands in the infrared spectrum, at 431 cm^{-1} and at 792 cm^{-1}, as well as some minor peaks. The 431 cm^{-1} peak showed virtually no oxygen-18 isotope shift, but the 792 cm^{-1} peak showed an exceptionally large oxygen-18 isotope shift consistent with the antisymmetric O-O-O stretch of a symmetric ozonide product.
5. The structures of several possible symmetric ozonide products of the reaction between O_3 and Bz_2Cr were calculated. The best fit of the antisymmetric stretch and its oxygen-18 isotope shifts was shown by hydrogen ozonide (H_2O_3) and its oxygen-18 isotopomers.
6. The formation of H_2O_3 in this reaction is undoubtedly facilitated by the unusually strong intermolecular interactions of ozone with the Bz_2Cr hydrogen atoms, which have calculated interaction energies of −8.27 to −9.16 kcal/mol, rivaling those of traditional hydrogen bonding.
7. H_2O_3 can be formed in an energetically favorable process if the hydrogen-deficient benzene-benzyne-Cr (BzByCr) couples with another Bz_2Cr molecule and rearranges to benzene-Cr-biphenyl-Cr-benzene (BzCrBPCrBz), which was previously observed [21].

Author Contributions: Conceptualization, B.S.A. and R.W.K.; Methodology, R.W.K.; Software, R.W.K.; Validation, B.S.A. and R.W.K.; Formal Analysis, R.W.K.; Investigation, R.W.K.; Resources, B.S.A.; Data Curation, R.W.K.; Writing—Original Draft Preparation, R.W.K.; Writing—Review & Editing, B.S.A.; Visualization, R.W.K.; Supervision, B.S.A.; Project Administration, B.S.A.; Funding Acquisition, B.S.A. All authors have read and agreed to the published version of the manuscript.

Funding: This article was supported by Department of Chemistry, University of Cincinnati. There were no research grants involved.

Institutional Review Board Statement: Not applicable.

Informed Consent Statement: Not applicable.

Data Availability Statement: The original contributions presented in the study are included in the article, further inquiries can be directed to the corresponding author.

Acknowledgments: The authors are grateful to Hairong Guan and Allan Pinhas for useful discussions, and especially to Allan Pinhas for suggesting the possible formation of a biphenyl-linked BzCrBPCrBz product. The authors are also grateful to Joel Collett and Bedraj Pandey for loading the Bz_2Cr cell under inert atmosphere conditions.

Conflicts of Interest: The authors declare no conflicts of interests.

References

1. Tahir, A.A.; Wijayntha, K.G.U.; Saremi-Yarahmadi, S.; Mazhar, M.; McKee, V. Nanostructured α-Fe_2O_3 thin films for photoelectrochemical hydrogen generation. *Chem. Mater.* **2009**, *21*, 3763–3772. [CrossRef]
2. Martinson, A.B.F.; DeVries, M.J.; Libera, J.A.; Christensen, S.T.; Hupp, J.T.; Pellin, M.J.; Elam, J.W. Atomic layer deposition of Fe_2O_3 using ferrocene and ozone. *J. Phys. Chem. C* **2011**, *115*, 4333–4339. [CrossRef]
3. Aaltonen, T.; Ale'n, P.; Ritala, M.; Leskelä, M. Ruthenium thin films grown by atomic layer deposition. *Chem. Vap. Depos.* **2003**, *9*, 45–49. [CrossRef]
4. Nguyen, V.T.; Kim, Y.C.; Ahn, Y.H.; Lee, S.; Park, J.-Y. Large-area growth of high-quality graphene/MoS_2 vertical heterostructures by chemical vapor deposition with nucleation control. *Carbon* **2020**, *168*, 580–587. [CrossRef]
5. Topka, K.C.; Chliavoras, G.A.; Senocq, F.; Vergnes, H.; Samelor, D.; Sadowski, D.; Vahlas, C.; Caussat, B. Large temperature range model for the atmospheric pressure chemical vapor deposition of silicon dioxide films on thermosensitive substrates. *Chem. Eng. Res. Des.* **2020**, *161*, 146–158. [CrossRef]
6. Maury, F.; Vahlas, C.; Abisset, S.; Gueroudji, L. Low temperature metalloorganic chemical vapor deposition routes to chromium metal thin folms using bis(benzene)chromium. *J. Electrochem. Soc.* **1999**, *10*, 3716–3723. [CrossRef]
7. Kugel, R.W.; Pinelo, L.F.; Ault, B.S. Infrared matrix-isolation and theoretical studies of the reactions of ferrocene with ozone. *J. Phys. Chem. A* **2015**, *119*, 2371–2382. [CrossRef] [PubMed]
8. Kugel, R.W.; Ault, B.S. Infrared matrix-isolation and theoretical study of the reactions of ruthenocene with ozone. *J. Phys. Chem. A* **2019**, *123*, 5768–5780. [CrossRef] [PubMed]
9. Hoops, M.D.; Ault, B.S. Matrix isolation study of the early intermediates in the ozonolysis of cyclopentene and cyclopentadiene: Observation of two Criegee intermediates. *J. Am. Chem. Soc.* **2009**, *131*, 2853–2863. [CrossRef] [PubMed]

10. Ault, B.S. Matrix Isolation Spectroscopic Studies: Thermal and Soft Photochemical Bimolecular Reactions. In *Frontiers and Advances in Molecular Spectroscopy*; Laane, J., Ed.; Elsevier Press: Amsterdam, The Netherlands, 2017; pp. 667–712.
11. Boyd, J.W.; Lavoie, J.M.; Gruen, D.M. Direct synthesis and characterization of dibenzenechromium(0) in an argon matrix at 14°K. *J. Chem. Phys.* **1974**, *60*, 4088–4089. [CrossRef]
12. Ngai, L.H.; Stafford, F.E.; Schafer, L. The symmetry of gaseous dibenzenechromium. *J. Am. Chem. Soc.* **1969**, *91*, 48–49. [CrossRef]
13. Fritz, H.P.; Lüttke, W.; Stammreich, H.; Forneris, R. IR- und Raman-untersuchungen zur struktur des di-benzol-chroms, seines kations sowie verwandter verbindungen. *Spectrochim. Acta* **1961**, *17*, 1068–1091. [CrossRef]
14. Engdahl, A.; Nelander, B. The vibrational spectrum of H_2O_3. *Science* **2002**, *295*, 482–483. [CrossRef] [PubMed]
15. Pinelo, L.; Gudmundsdottir, A.D.; Ault, B.S. Matrix Isolation Study of the Ozonolysis of 1,3- and 1,4-Cyclohexadiene: Identification of Novel Reaction Pathways. *J. Phys. Chem. A* **2013**, *117*, 4174–4182. [CrossRef] [PubMed]
16. Braga, D.; Costa, A.L.; Grepioni, F.; Scaccianoce, L.; Tagliavini, E. OH–O and CH–O hydrogen bonding in hydrated crystals of paramagnetic $[(\eta^6\text{-}C_6H_6)_2Cr]^+$. *Organometallics* **1996**, *15*, 1084–1086. [CrossRef]
17. Miller, R.G.; Stiles, M. Reaction of Benzyne with Benzene and Naphthalene. *J. Am. Chem. Soc.* **1963**, *85*, 1798–1800. [CrossRef]
18. Stiles, M.; Burckhardt, U.; Freund, G. The Reaction of Benzyne with Benzene. *J. Org. Chem.* **1967**, *32*, 3718–3719. [CrossRef]
19. Seyferth, D. Bis(benzene)chromium. 1. Franz Hein at the University of Leipzig and Harold Zeiss and Minoru Tsutsui at Yale. *Organometallics* **2002**, *21*, 1520–1530. [CrossRef]
20. Hein, F.; Eissner, W. Über das Tetraphenylchrom $(C_6H_5)_4Cr$. (VI. Mitteilung über Chromorganische Verbindungen). *Ber. Dtsch. Chem. Ges.* **1926**, *59*, 362. [CrossRef]
21. Seyferth, D. Bis(benzene)chromium. 2. Its discovery by E. O. Fischer and W. Hafner and subsequent work by the research groups of E. O. Fischer, H. H. Zeiss, F. Hein, C. Elschenbroich, and others. *Organometallics* **2002**, *21*, 2800–2820. [CrossRef]
22. Frisch, M.J.; Trucks, G.W.; Schlegel, H.B.; Scuseria, G.E.; Robb, M.A.; Cheeseman, J.R.; Scalmani, G.; Barone, V.; Mennucci, B.; Petersson, G.A.; et al. *Gaussian 09, Revision C.01*; Gaussian, Inc.: Wallingford, CT, USA, 2010.

molecules

MDPI

Review

Beyond the Harmonic Oscillator; Highlights of Selected Studies of Vibrational Potential Energy Functions

Esther J. Ocola [1,2] and Jaan Laane [1,2,*]

[1] Department of Chemistry, Texas A&M University, College Station, TX 77843-3255, USA
[2] Institute for Quantum Science and Engineering, Texas A&M University, College Station, TX 77843-4242, USA
* Correspondence: laane@chem.tamu.edu

Abstract: Although the harmonic oscillator model has found wide use in physics and chemistry, there are more interesting potential energy functions (PEFs) which can tell us a great deal about molecular structure and energetics. In the present work, we show that for selected systems simple one- and two-dimensional potential functions can be used to very accurately fit detailed spectroscopic data and provide extensive additional information. Results for molecular inversion, ring puckering, the anomeric effect, pseudorotation, triplet-state puckering, internal rotation, and π-type hydrogen bonding in ground and excited electronic states are presented.

Keywords: potential energy functions; molecular inversion; ring puckering; anomeric effect; pseudorotation; internal rotation; π-type hydrogen bonding

1. Introduction

The harmonic oscillator (HO) potential energy function $V = \frac{1}{2}kx^2$ has found wide use in physics and chemistry, and, in particular, in molecular spectroscopy. This is in part due to the fact that the Schrödinger Equation can be solved exactly for the energy levels of the HO. However, there are other forms of potential energy functions (PEFs) and potential energy surfaces that are needed to describe particular molecular systems, and many of these are arguably more interesting. Laane et al. have been investigating these for six decades and periodically published reviews of the work [1–8]. Among these are double-minimum potential functions for cyclic ring inversions dominated by quartic terms. Another type of one-dimensional potential function is that for analyzing the internal rotations of molecules. These have periodic potential energy functions. In the present paper, we present a selection of experimental and theoretical results of work previously published and mostly from our own laboratory.

2. Experimental and Theoretical Results

2.1. Harmonic Oscillator

Schrödinger himself recognized that his equation could be solved exactly for the harmonic oscillator [9]. Figure 1 shows the Schrödinger Equation along with the solutions and a diagram of the HO potential function along with the energy levels and wave functions. This has been widely used to calculate force constants for the vibrations of diatomic molecules and for bond stretching vibrations of larger molecules when their frequencies are sufficiently higher than the other vibrations. However, it was recognized early on that, unlike for the harmonic oscillator, the energy spacing for the upper levels of real molecules progressively decreased as the dissociation limit was approached. This anharmonicity arises from the fact that the potential energy, V, does not approach infinity as the bond distance gets very large. Figure 2 compares the harmonic oscillator function to the Morse Potential, which was developed to more accurately calculate the energy levels for the upper states.

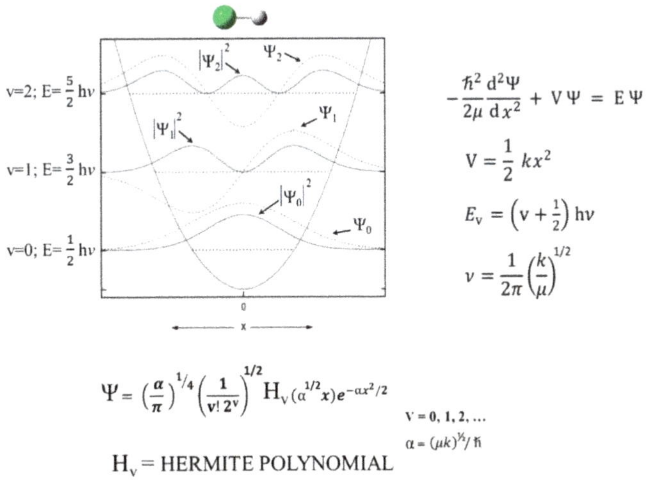

Figure 1. Harmonic oscillator potential energy function and solutions.

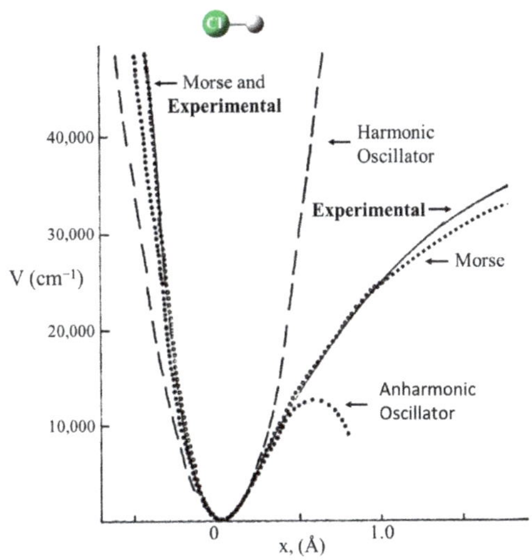

Figure 2. Harmonic oscillator corrected for anharmonicity.

2.2. *Ammonia Inversion*

The inversion vibration of ammonia can be studied as a one-dimensional problem since it has A_1 symmetry and has a frequency very much lower than the symmetric N-H_3 stretching vibration of the same symmetry. The potential function for the inversion has a double minimum for the two equivalent inverted structures and a barrier for the planar configuration. This results in inversion doubling. Figure 3 shows the infrared spectrum of gaseous ammonia, and the doubling can be seen for the inversion, v_2. Figure 3 also shows the observed energy spacings and the calculated potential function which best fits the experimental data [10]. This shows the barrier to planarity to be 2031 cm^{-1} or 5.8 kcal/mole. It also demonstrates how such potential energy functions are invaluable for providing energy and structural information on molecules. Coon and co-workers [11] used the following function to fit the data.

$$V = Ax^2 + B\exp\left(-Cx^2\right) \quad (1)$$

where x is the inversion coordinate and A, B, and C are adjusted potential energy constants. Laane [12] showed that the following potential energy function could also do a reasonably good job of fitting the experimental data.

$$V = ax^4 + bx^2 \quad (2)$$

where the constant b is negative.

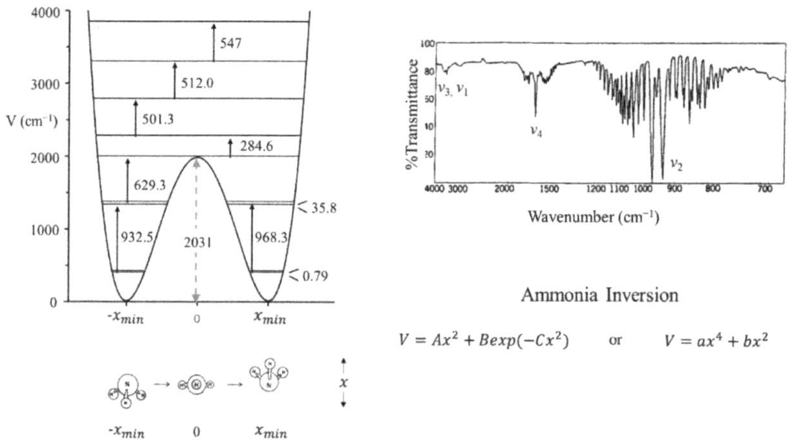

Figure 3. Vapor-phase infrared spectrum of ammonia and its vibrational potential energy function and energy levels for the inversion. The inversion doubling of v_2 can be seen.

2.3. Carbonyl Inversion in the Excited State

Molecules typically change structure after an electronic transition takes place, since electron distributions are changed. Figure 4 shows how the structures of formaldehyde and ethylene change upon such transitions. For formaldehyde, the lowest energy structure for the $S_1(n,\pi^*)$ excited state has the carbonyl oxygen out of the plane of the other three atoms. Hence, the carbonyl inversion (or carbonyl out-of-plane wagging) vibration will have a double-minimum PEF with a barrier at the planar conformation. Similarly, carbonyl groups in larger molecules will also have inversions governed by double-minimum PEFs. The Laane laboratory has investigated the fluorescence excitation spectra of a number of ring molecules containing carbonyl groups and determined their potential energy functions for the $S_1(n,\pi^*)$ states. Figure 5 shows the spectrum of 3-cyclopenten-1-one, and Figure 6 shows the carbonyl inversion PEF derived from that [13]. Table 1 presents the barriers to inversion in the excited state for this molecule as well as for several other cyclic molecules [14]. The barrier for 2-cyclopenten-1-one does not exist, since the conjugation is still present between the C=C and C=O groups to some extent in the $S_1(n,\pi^*)$ state. For the other molecules, the barrier increases with angle strain within the ring. As can be seen, the determinations of the PEFs again provide both structural and relative energy data for these molecules.

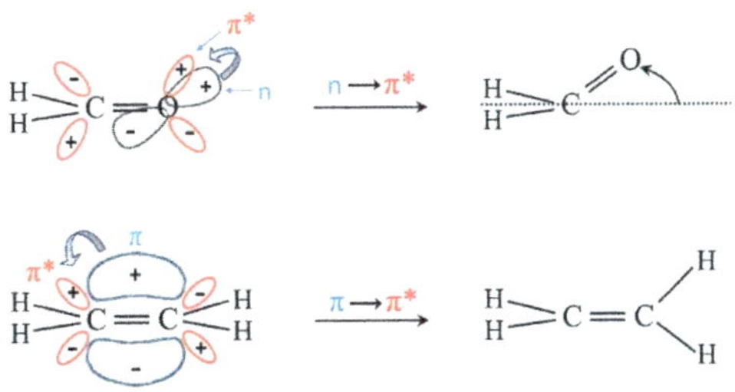

Figure 4. Structural changes for formaldehyde and ethylene following electronic transitions. The * means it is an antibonding orbital.

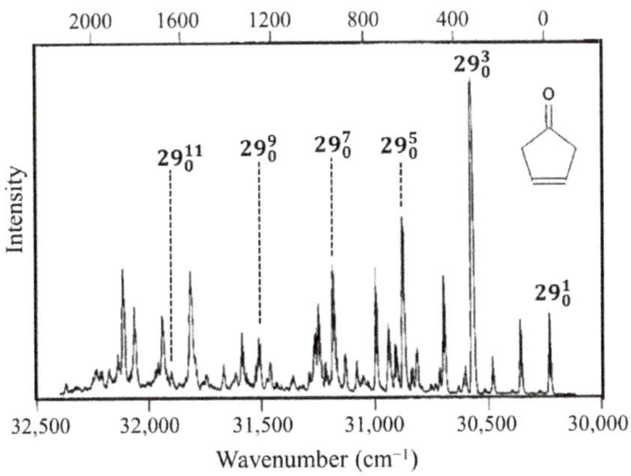

Figure 5. Fluorescence excitation spectrum of 3-cyclopentene-1one. υ_{29} is the carbonyl inversion vibration.

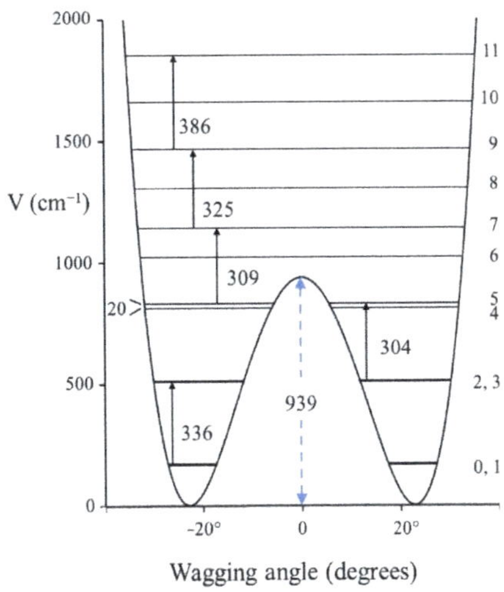

Figure 6. Carbonyl inversion potential energy function for 3-cyclopentene-1one.

Table 1. Inversion barriers for carbonyl groups in electronic excited states.

		C=O Inversion Barrier (cm^{-1})
	2-cyclopenten-1-one	0
	Cyclopentanone	672
	Cyclobutanone	939
	Cyclopentene-1-one	2149

2.4. Ring Puckering

In 1945, R. P. Bell [15] postulated that the ring-puckering vibration of cyclobutane should be governed by a quartic potential energy function:

$$V = ax^4 \quad (3)$$

where x is the ring-puckering coordinate reflecting the out-of-plane displacement. Bell did not present a mathematical explanation for this. Laane [16] later demonstrated that the function should have the form of Equation (2) if the angle-bending force constants are assumed to have quadratic dependence. He showed that angle strain resulted in the x^4 quartic term but also made a smaller positive contribution to the x^2 term. Negative contributions to the quadratic term generally arise for torsional forces such as CH_2-CH_2 interactions. Laane and Lord [17] also postulated that the ring puckering of so-called "pseudo-four-membered-rings", such as cyclopentene and 1,4-cyclohexadiene, could be investigated by one-dimensional PEFs. Figure 7 depicts these ring-puckering vibrations.

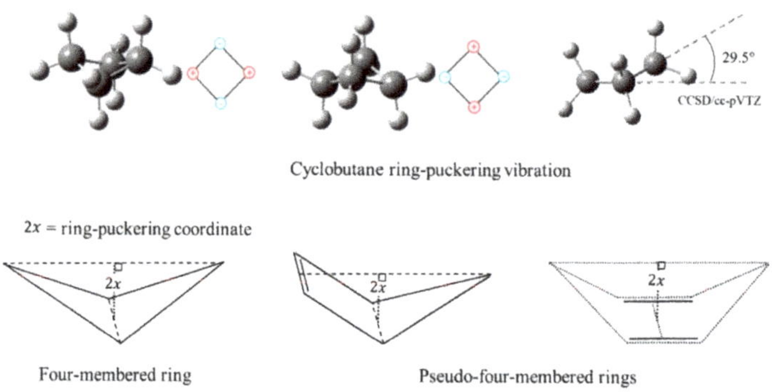

Figure 7. Ring-puckering vibrations for four-membered rings and pseudo-four-membered rings.

Bell's prediction of quartic oscillator functions excited the spectroscopic community in the 1960s when far-infrared equipment started to become available. References to the early work can be found in our previous reviews [1–8]. A dilemma at the time was that large-frame computers were slow and had memory capacities of only 512 K. Moreover, the wave equation for the potential function in Equation (2) could not be solved exactly and needed approximation methods. To facilitate the fitting of ring-puckering spectra, Laane [12] generated a table of energy levels for the reduced potential energy function:

$$V = A\left(Z^4 + BZ^2\right) \tag{4}$$

where the constants A and B can be related to a and b in Equation (2) if the reduced mass for the wave equation is known. Similarly, Z could be related to the puckering coordinate, x. At that time, the problem was that there were no methods for calculating the reduced mass, which, as it turns out, is also coordinate-dependent. Harthcock and Laane [18,19] later did develop methods for reduced mass calculations, but in the 1960s and 1970s these were not available.

Figure 8 presents a graph of how the energies of the quantum states vary as a function of the constant B in Equation (4). Once B is determined from relative energy spacings, the constant A can be used as a scaling factor to fit the spectrum. As can be seen, as the barrier $B^2/4$ increases, the energy levels begin to double up. Equation (4) was used to fit the far-infrared spectra of many molecules [12], and Figure 9 shows the potential functions and transitions for three nearly planar molecules for which the angle strain and torsional forces affecting the b coefficient in Equation (2) nearly balance out. Figure 10 shows the functions and transitions for molecules with greater torsional interactions giving rise to sizable barriers. The fitting of the observed spectroscopic data with Equation (2) is remarkably good, especially after the coordinate-dependent reduced mass is correctly calculated. To demonstrate this, Figure 11 presents the far-infrared spectrum of 2,3-dihydrofuran along with its puckering potential function and transitions [20]. The small barrier to planarity arises from the single CH_2-CH_2 interaction. Table 2 shows that the two-parameter potential function fits the observed data with a better than 1% accuracy.

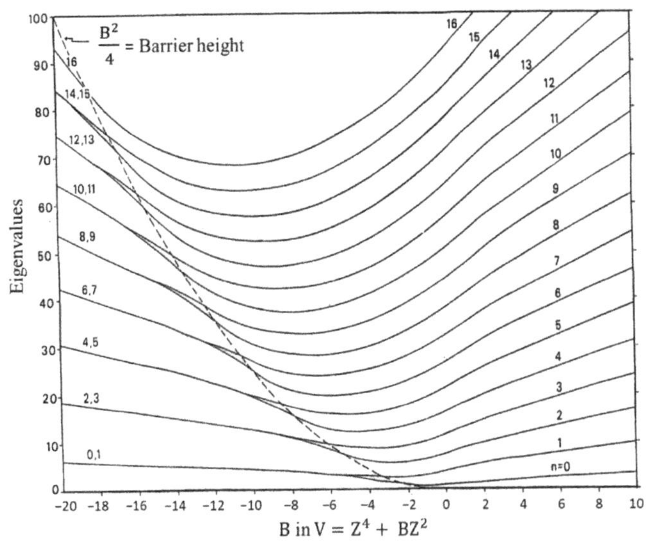

Figure 8. Energy level for the reduced potential energy function $V = Z^4 + BZ^2$ as a function of B.

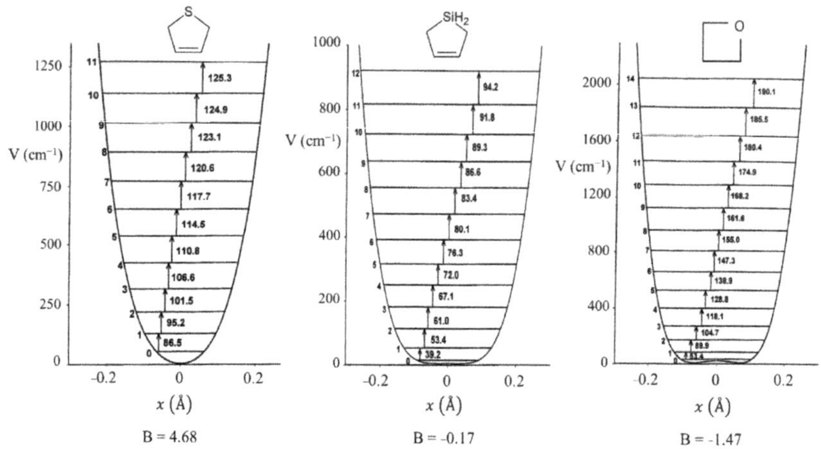

Figure 9. Ring-puckering potential energy functions for three nearly planar molecules.

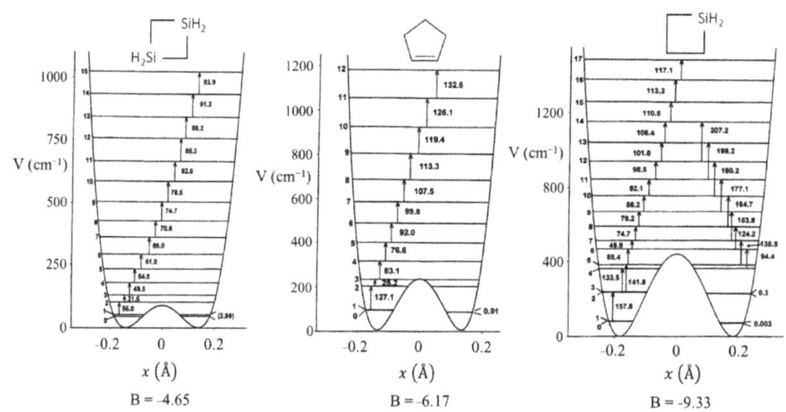

Figure 10. Ring-puckering potential energy functions for three puckered molecules.

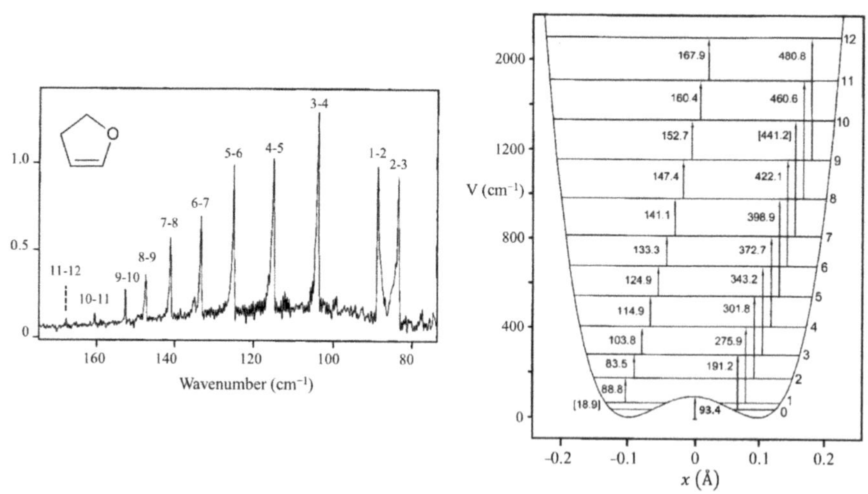

Figure 11. Far-infrared spectrum, potential energy function, and transitions for 2,3-dihydrofuran.

Table 2. Observed and calculated ring-puckering frequencies for 2,3-dihydrofuran.

Transition	Frequency (cm^{-1}) Observed	Calculated	Δ	Relative Intensity Observed	Calculated
$v_T = 0$					
1–2	88.8	88.6	0.2	0.80	0.90
2–3	83.5	83.5	0.0	0.76	0.87
3–4	103.8	103.6	0.2	(1.00)	(1.00)
4–5	114.9	114.8	0.1	0.77	0.83
5–6	124.9	124.8	0.1	0.75	0.61
6–7	133.3	133.4	−0.1	0.50	0.41
7–8	141.1	140.9	0.2	0.40	0.26
8–9	147.4	147.7	−0.3	0.22	0.15
9–10	152.7	153.9	−1.2	0.17	0.09
10–11	160.4	159.9	0.5	0.06	0.05
0–3	191.2	191.1	0.1	0.15	0.27
1–4	275.9	275.6	0.3	0.20	0.11
2–5	301.8	301.8	0.0	0.06	0.08
3–6	343.2	343.2	0.0	0.06	0.05
4–7	372.2	372.9	−0.7	0.05	0.03
5–8	398.9	398.1	0.8	0.04	0.02
6–9	422.1	422.0	0.1	0.02	0.01

$V (cm^{-1}) = 1.019 \times 10^6 x^4 - 1.946 \times 10^4 x^2$.

Since about the year 2000, ab initio calculations have progressively improved so that they can fairly accurately predict potential energy functions for the ring-puckering motions. Ocola and Laane [21], in 2020, presented a comparison of ab initio CCSD/cc-pVTZ and MP2/cc-pVTZ results with experimental ones. As can be seen in Table 3, in most cases the CCSD/cc-pVTZ calculation gives better agreement with the experimental data.

Table 3. Comparison of experimental barriers and x_{min} and θ_{min} values with those from CCSD/ccpVTZ and MP2/cc-pVTZ computations for cyclobutane and related molecules.

Molecule	Experimental Fit $V(x^6,x^4,x^2)$ [a]			Theoretical CCSD/cc-pVTZ			MP2/cc-pVTZ		
	Barrier cm^{-1}	x_{min} Å	θ_{min} Degrees	Barrier cm^{-1}	x_{min} Å	θ_{min} Degrees	Barrier cm^{-1}	x_{min} Å	θ_{min} Degrees
 C_4H_8	510	±0.142	±29.6°	586	±0.142	±29.5°	821	±0.153	31.9°
 C_4H_7O	15	±0.062	±13.4°	0	±0.001	±0.2°	21	±0.069	14.8°
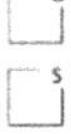 C_4H_7S	273	±0.141	±26.9	243	±0.133	±25.5°	412	±0.152	29.0°
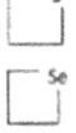 C_4H_7Se	378	±0.162	±30.4°	339	±0.148	±27.8°	512	±0.164	30.9°
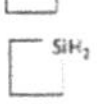 $C_4H_7SiH_2$	440	±0.163	±31.9°	472	±0.163	±31.7°	654	±0.176	34.5°
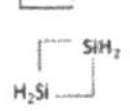 $C_4H_7(SiH_2)_2$	87	±0.130	±22.3°	89	±0.130	±22.4°	160	±0.150	25.8°
 $C_4H_7GeH_2$	---	---	---	409	±0.156	±30.1°	567	±0.169	32.6°
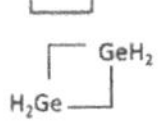 $C_4H_7(GeH_2)_2$	---	---	---	1	±0.047	±7.8°	24	±0.104	17.1°
 C_4H_4O	2	±0.043	±4.6°	36	±0.073	±15.5°	114	±0.097	20.8°

[a] Using $V = ax^4 + bx^2 + cx^6$.

2.5. Anomeric Effect

One of our ring-puckering investigations proved perhaps to be the best demonstration of the anomeric effect which occurs for molecules containing a XCH_2Y grouping, where X or Y = O, S, or Se. Figure 12 presents the infrared and Raman spectra of 1,3-dioxole [22], and Figure 13 shows the ring-puckering PEF determined from the data. The potential energy function demonstrates that the energy minima correspond to puckered structures brought on by the anomeric effect. Figure 13 shows how the puckering is brought about by the overlap of the non-bonded oxygen n orbital with the σ* orbital of the other O-C linkage.

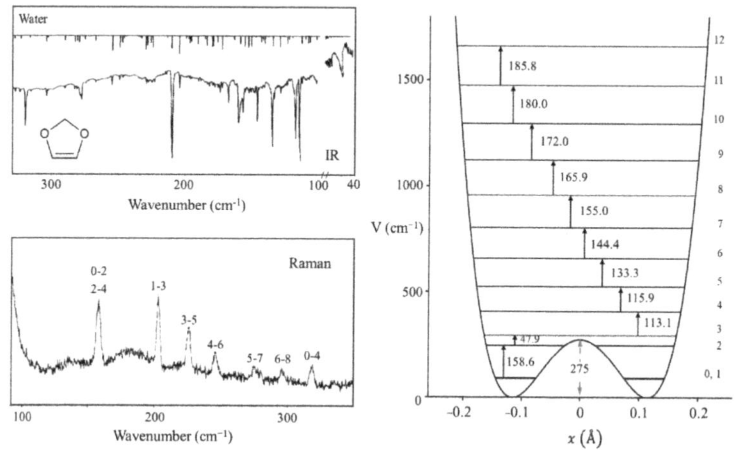

Figure 12. Far-infrared and Raman spectra and potential energy function of 1,3-dioxole.

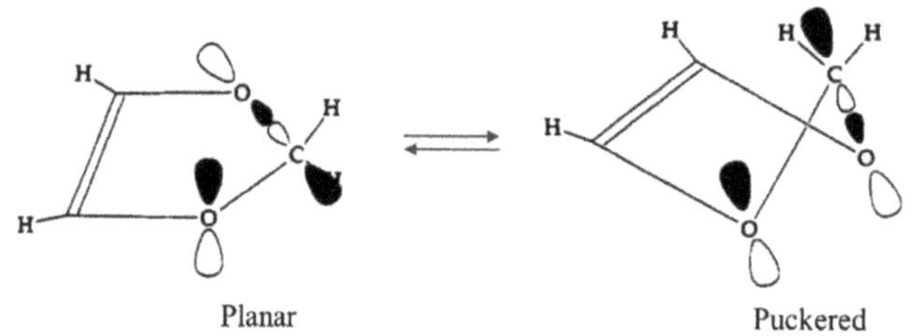

Figure 13. Orientation of the n and σ* orbitals involved in the anomeric effect for planar and puckered conformations of 1,3-dioxole.

Laane et al. have also reported the infrared and excited-state spectra for 1,3-benzodioxole [23,24] and showed that the anomeric effect again is responsible for the non-planarity of the five-membered ring. The data are actually best fit with a two-dimensional potential energy surface in terms of the ring-puckering and the ring-flapping vibrations. Figure 14 shows the spectra, and Figure 15 shows the one-dimensional potential energy along the puckering coordinate calculated for the two electronic states. The lower barrier to planarity for the excited state is attributed to competition for the oxygen non-bonded n orbital and the benzene ring π system.

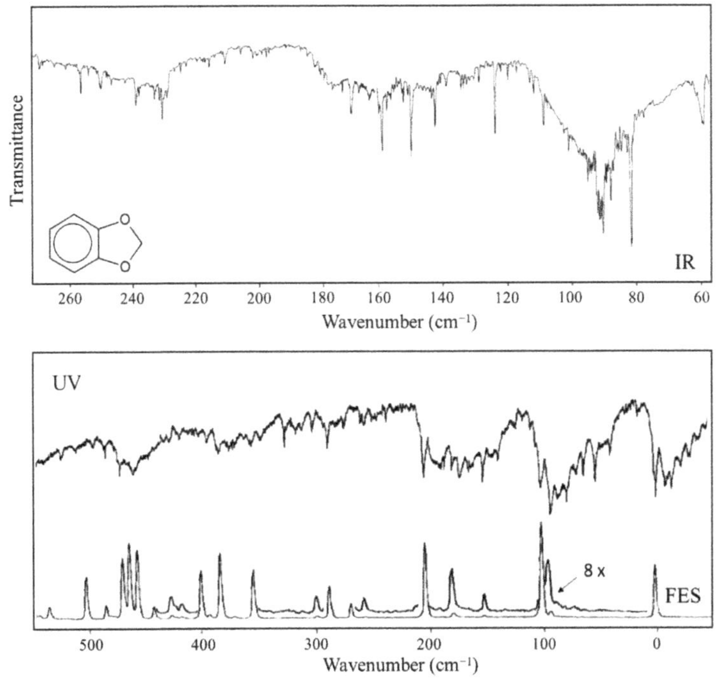

Figure 14. Far-infrared and excited-state spectra of 1,3-benzodioxole. The excited-state spectrum band origin is at 34,789.8 cm^{-1}.

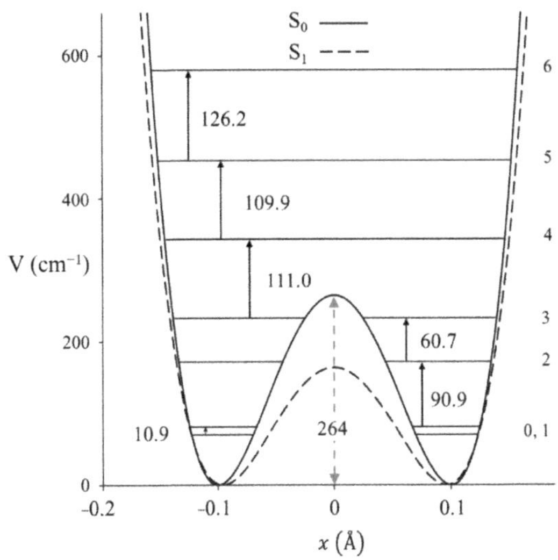

Figure 15. One-dimensional ring-puckering potential energy functions of 1,3-benzodioxole in its ground and excited $S_1(\pi,\pi^*)$ states.

The experimental results for 1,3-dioxole along with ab initio CCSD/cc-pVTZ and MP2/cc-pVTZ calculations allowed us to calculate the magnitude of the anomeric effect for different XCH_2Y arrangements, and the results are shown in Table 4 [25]. We believe that these results provide a greatly improved understanding of the anomeric effect for the different atoms in the XCH_2Y linkages.

Table 4. Relative magnitudes of the negative parameter b in Equation (2) reflecting the torsional contribution from the anomeric effect.

Molecule	Linkage	Relative b Value	$V_{anomeric}$ [a] (kcal/mole)
	OCO	1.00	5.97 [b]
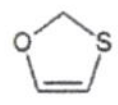	OCS	0.82	4.90
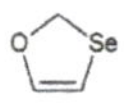	OCSe	0.66	3.94
	SCS	0.60	3.58
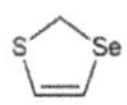	SCSe	0.45	2.69
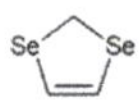	SeCSe	0.32	1.91

[a] Estimated torsional potential energy from the anomeric effect. [b] Magnitude from reference 22 based on MM3 molecular mechanics calculations.

2.6. Pseudorotation

Pitzer and co-workers [26–28], as early as 1947, postulated that cyclopentane should undergo free pseudorotation. The basic concept is that the out-of-plane bending and twisting vibrations, which would be degenerate for a planar cyclopentane molecule, can be transformed to a radial mode and a pseudorotational mode which "pseudorotates" from a bent structure to a twisted structure to another bent structure, and so on. Figure 16 shows this progression. The theory also predicted that the pseudorotation would be nearly free, with little or no energy differences between the bent and twisted conformations, and this would result in energy levels similar to pure rotation. Experimental evidence for this was first confirmed by Durig and Wertz [29], who reported low-resolution infrared combination band spectra in the CH_2 deformation region. High-resolution spectra were later published by Bauman and Laane [30], who also reported radial bands and the combination bands for several isotopic species. Figure 17 shows the higher-resolution spectrum.

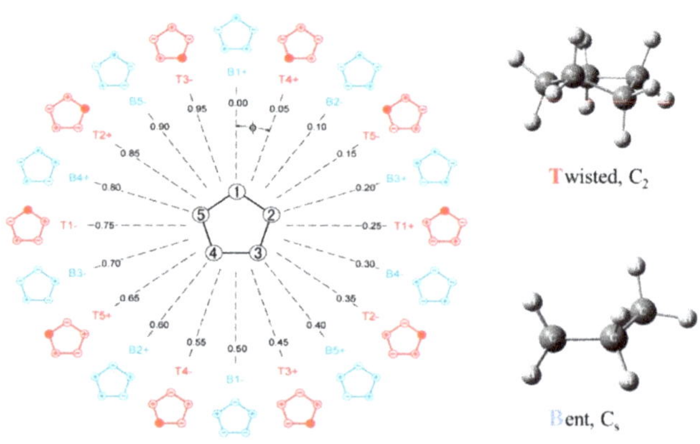

Figure 16. Pseudorotation of cyclopentane.

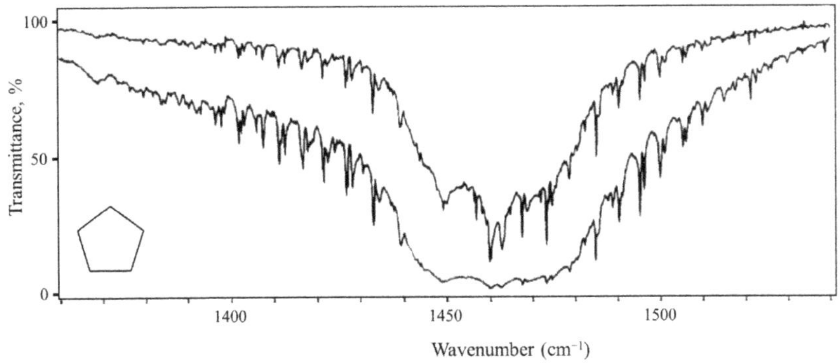

Figure 17. Pseudorotation combination band spectrum of cyclopentane.

Since what we have here is really a two-dimensional problem, Ocola, Bauman, and Laane also calculated the two-dimensional potential energy surface for cyclopentane [31]. This is shown in Figure 18 and can be viewed either in terms of the bending and twisting coordinate or in terms of the radial and pseudorotational modes. Complementary density function and ab initio calculations have shown that the pseudoration is free or nearly free, with energy differences between the ten bent and ten twisted conformations being no greater than about 5 cm^{-1} or 0.14 kcal/mole. This was consistent with our earlier study for the effect of a ten-fold barrier to pseudorotation [32].

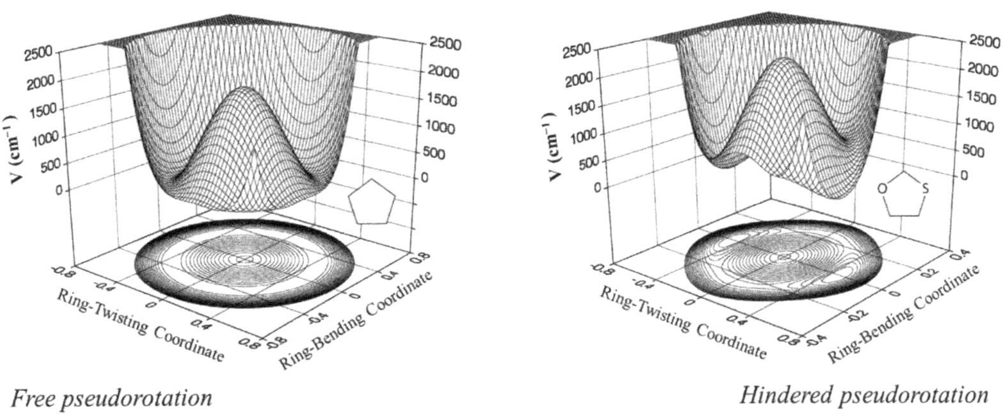

Figure 18. Two-dimensional potential energy surfaces for cyclopentane and 1,3-oxathiolane.

Laane et al. have also investigated the pseudorotation spectra of silacyclopentane [33] and 1,3-oxathiolane [34]. Both of these molecules hinder pseudorotation, since the bent structures are of higher energy than the twisted ones. The surface for 1,3-oxathiolane is also shown in Figure 18. A summary of early pseudorotational studies was published in 1972 [35]. In 1990, Rosas, Cooper, and Laane showed that molecular mechanics calculation did a fairly good job in predicting pseudorotational barriers [36].

2.7. Triplet-State Ring Puckering

In collaboration with the Drucker group, we investigated what the ring-puckering potential energy function looks like for the triplet state of 2-cyclopentene-1-one [37,38]. Figure 19 shows the cavity ringdown spectrum of this molecule. Vibration v_{30} is the puckering, with its lowest frequency for the triplet state at 37 cm^{-1}. Fitting the data results in the one-dimensional function in Figure 20, where the potential functions derived for the ground state and singlet excited S_1(n, π^*) state are also shown. It is remarkable that the unpaired electrons in the triplet state produce a small barrier and result in a non-planar molecule.

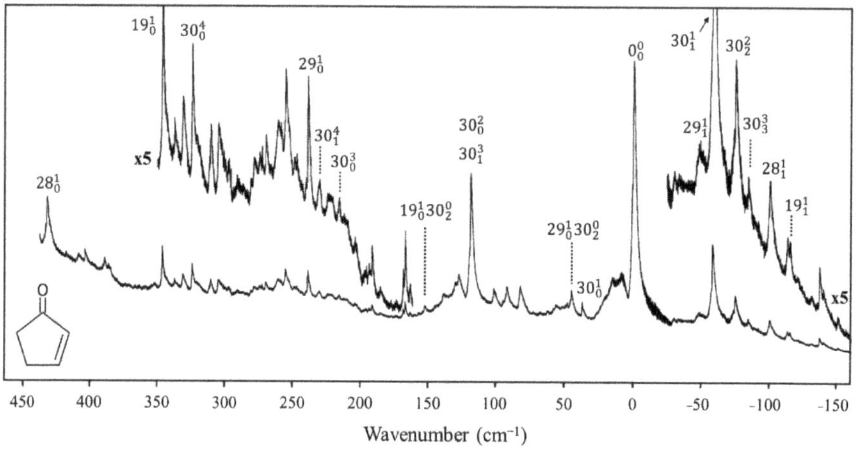

Figure 19. Cavity ringdown spectrum of 2-cyclopentenone.

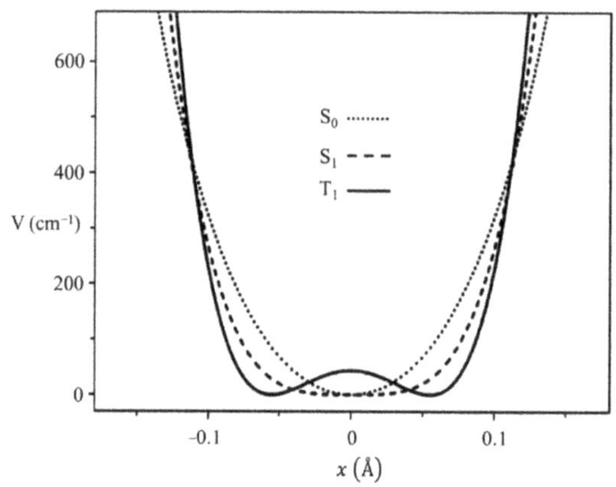

Figure 20. Ring-puckering potential energy functions for 2-cyclopentenone in three different electronic states.

2.8. Cyclohexane

We recorded the Raman spectrum for the bending vibration of cyclohexane [39], whose conformation has been investigated by many dozens of researchers. Figure 21 shows the spectrum, and Figure 22 shows the potential function based on Equation (2) calculated from the observed data for the out-of-plane vibration. The barrier to planarity for the calculated function is 8600 cm^{-1}, and this agrees well with the DFT B3LYP/cc-pVTZ computed value of 8804 cm^{-1}.

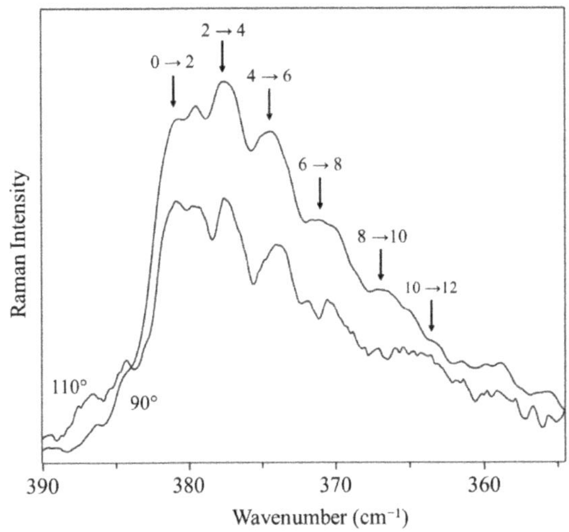

Figure 21. Vapor-phase Raman spectrum of heated cyclohexane vapor.

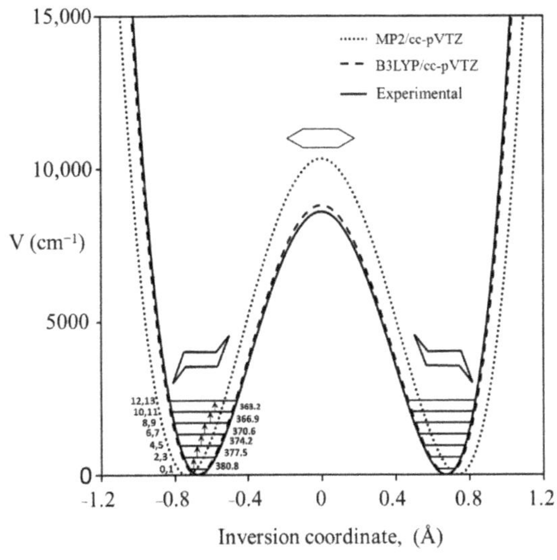

Figure 22. Cyclohexane potential energy function for the ring inversion.

2.9. Internal Rotation

Hundreds, if not thousands, of studies have been carried out on internal rotations. For a single internal rotation angle, Θ, the infrared and Raman spectra resulting from these can generally be reproduced using a function of the form

$$V = \sum V_n(1 - \cos(n\Theta) \tag{5}$$

The energy levels for simple three-fold or two-fold rotors were initially calculated using Mathieu tables. In 1972, Laane [40] developed the FORTRAN computer program VNCOSPX for calculating the energy levels for more complicated systems and fitting the experimental data. This program has been shared with dozens of labs around the world, and many more have written their own programs based on the matrix elements presented in our work. In 1977, a computer program was developed for even more complicated functions that also contained sine terms [41].

As an example of a system for a three-fold rotor, Figure 23 shows the mid-infrared combination band spectrum of cyclopropylgermane [42], and Figure 24 shows the experimental internal rotation PEF compared to those derived from theoretical calculations [43]. The torsional barrier can be seen to be about 450 cm^{-1}.

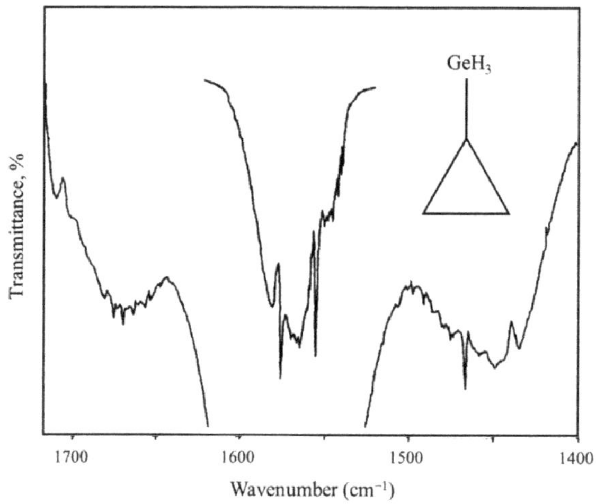

Figure 23. Internal rotation combination bands for cyclopropylgermane.

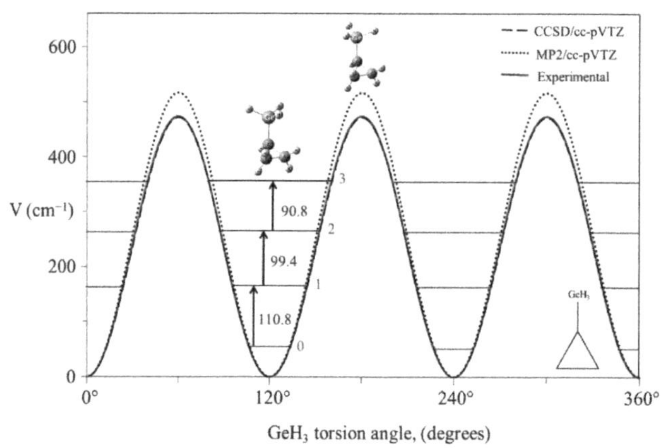

Figure 24. Experimental and theoretical potential energy functions for the internal rotation of cyclopropylgermane.

The potential energy function for cyclopropylamine shown in Figure 25 is quite different in that it has energy minima corresponding to two different conformations [43,44].

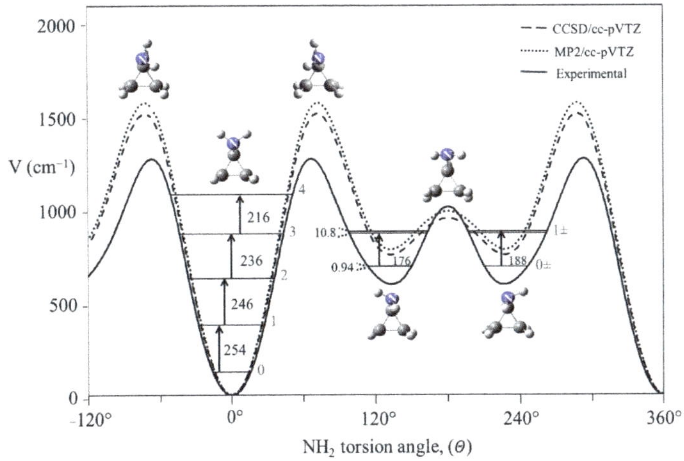

Figure 25. Internal rotation potential energy function for cyclopropylamine.

A more complex internal rotor case is that for the two internal rotations of *trans*-stilbene in its $S_1(\pi,\pi^*)$ electronic excited state [45]. Future Nobel laureate Ahmed Zewail visited Laane's research group in the 1990s at Texas A&M University, and after seeing the work on potential energy functions, he encouraged the study of this molecule, since his work at Caltech had raised some unanswered questions. Figure 26 shows the fluorescence excitation spectrum for this molecule and the two-dimensional internal rotation potential energy surface of Equation (6), which fits the data very well.

$$V_{(\phi_1,\phi_2)} = \frac{1}{2}V_2(2 + cos2\phi_1 + cos2\phi_2) + V_{12}cos2\phi_1 cos2\phi_2 + V'_{12}sin2\phi_1 sin2\phi_2 \quad (6)$$

The torsional modes v_{37} and v_{48} have their lowest vibrational levels at 9 and 118 cm^{-1}, respectively, for the excited state. This study helped clarify a long-standing uncertainty over the assignment of the electronic excitation spectra.

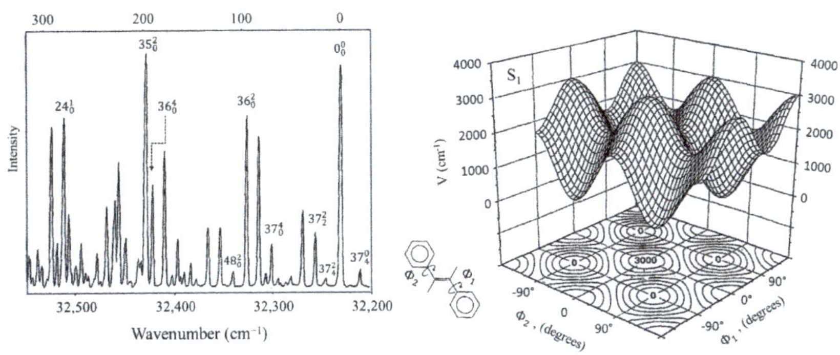

Figure 26. Fluorescence spectrum and two-dimensional internal rotation PES for *trans*-stilbene.

Another long-standing problem had been the internal rotation of 1,3-butadiene before we investigated the high-temperature Raman spectrum of this molecule and its isotopic species [46,47]. The spectrum of the normal species is shown in Figure 27. The data allowed the determination of its internal rotation PEF in Figure 28, and this showed that the long-sought second isomer of this molecule is not a *cis* structure but a *gauche* conformer corresponding to two shallow minima. There exists a small barrier at the *cis* conformation. This experimental result agrees very well with theoretical computations.

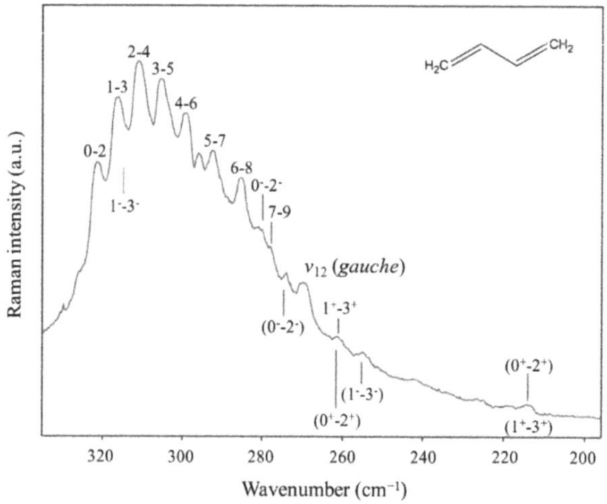

Figure 27. Low-frequency heated Raman spectrum of 1,3-butadiene.

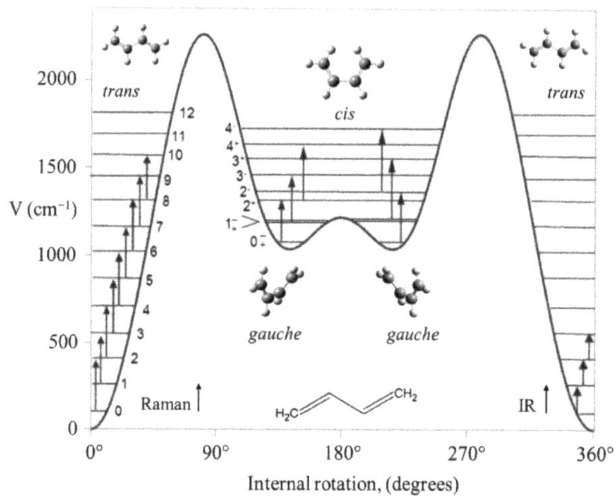

Figure 28. Potential energy function for the internal rotation of 1,3-butadiene.

2.10. π-Type Hydrogen Bonding

As shown in Figure 29, 3-cyclopenten-1-ol and 2-indanol are capable of π-type hydrogen bonding. Figure 30 shows four possible conformations of 3-cyclopenten-1-ol along with their relative energies derived from ab initio CCSD/6-311++G(d,p) computations [48,49]. As can be seen, conformation **A** has the lowest energy, and this is due to π-type hydrogen bonding between the hydrogen atom of the OH group and the C=C double bond. This can only be achieved when both the ring-puckering coordinate and the internal rotation coordinate for the OH group allow the hydrogen atom to be close enough to the double bond to allow the π bonding. Figure 31 shows the infrared and Raman spectra in the O-H stretching region. Bands from all conformers can be seen, and that from the one with the π bonding is at the lowest frequency, as expected. Figure 32 shows the two-dimensional potential energy surface calculated for this molecule from ab initio MP2/6-31+G(d,p) computations. The minima occur where the puckering and internal rotation coordinates allow the π bonding.

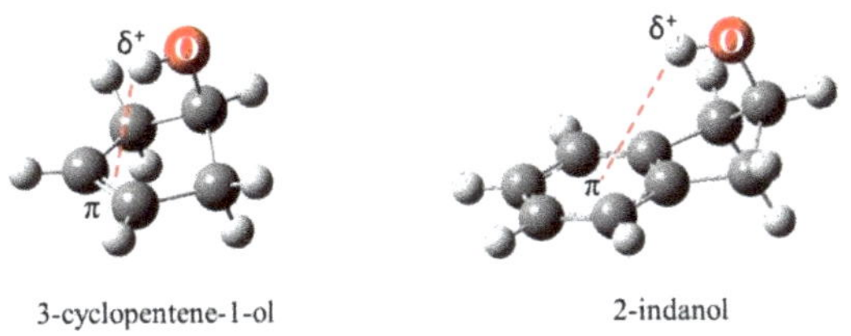

Figure 29. π-type hydrogen bonding of 3-cyclopenten-1-ol and 2-indanol.

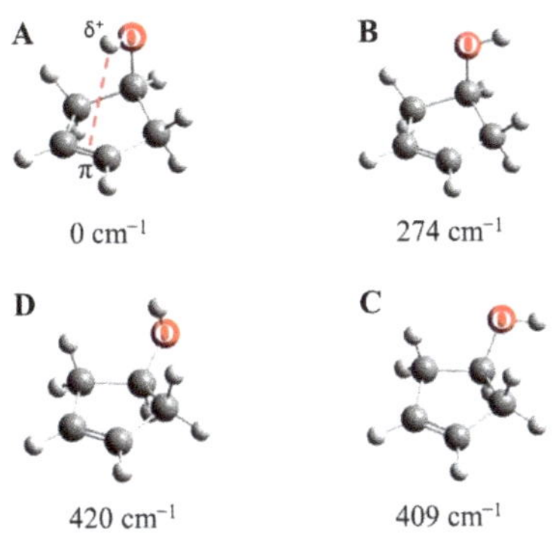

Figure 30. Conformers **A**, **B**, **C** and **D** of 3-cyclopenten-1-ol. The energies from CCSD/6-311++G(d,p) computations are shown.

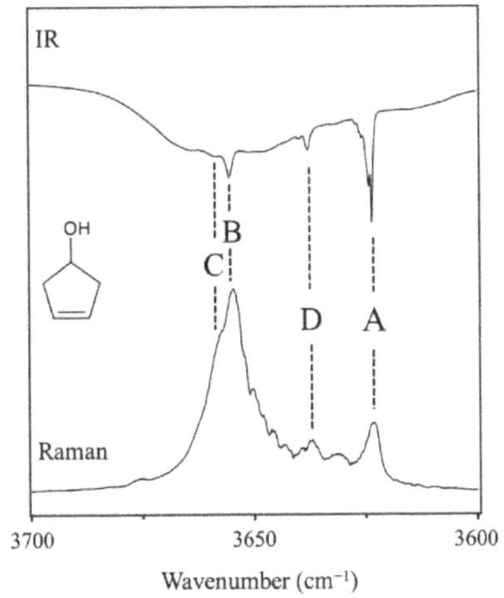

Figure 31. Infrared and Raman spectra of 3-cyclopenten-1-ol in the O-H stretching region showing bands from the four conformers **A**, **B**, **C** and **D**.

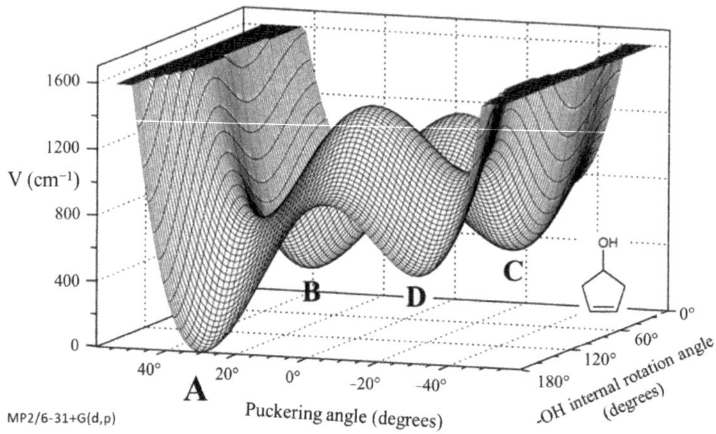

Figure 32. Two-dimensional potential energy surface for 3-cyclopenten-1-ol from ab initio MP2/6-31+G(d,p) computations. The minima correspond to the four conformations **A**, **B**, **C** and **D**.

The two-dimensional surface calculated for 2-indanol is similar [50]. Al-Saadi, Wagner, and Laane have recorded the fluorescence excitation spectrum for this molecule, as shown in Figure 33. The spectra of all four conformers, **A**, **B**, **C**, and **D**, can be seen. Their relative energies are similar to those of the 3-cyclopenten-1-ol.

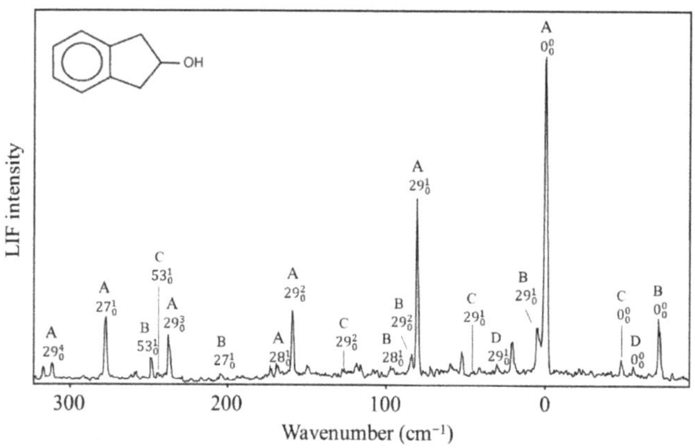

Figure 33. Fluorescence excitation spectrum of 2-indanol showing bands from its four conformers **A**, **B**, **C** and **D**.

We have also investigated the spectra of several other molecules with the π-type hydrogen bonding [51–53], which in each case lowers their conformational energy.

2.11. Pyridine and Sir Harry Kroto

Nobel laureate Sir Harry Kroto visited us in the early 2000s and told us about his work on pyridine [54] when he was a post-doc in Ottawa in the 1970s. He had recorded the ultraviolet absorption spectrum of this molecule but had not fully assigned the spectrum. Since we had a high-resolution Bomem DA8.02 FT spectrometer available, this molecule along with its d_5 isotopic species was reinvestigated by our research group [55]. The spectrum for the d_0 molecule is shown in Figure 34. The PEF derived for the out-of-plane ring-bending vibration in its $S_1(n, \pi^*)$ state is shown in Figure 35, which also compares it to the function for the ground state. As can be seen, the electronic transition to the antibonding π orbital greatly reduces the rigidity of the pyridine ring. In fact, there is a tiny 4 cm^{-1} barrier to planarity for the excited state.

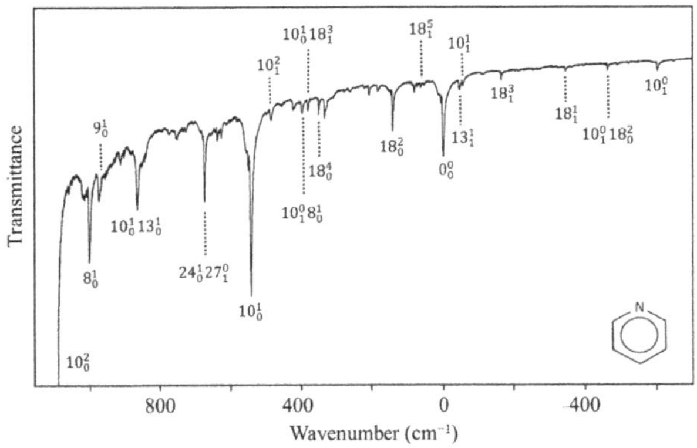

Figure 34. Ultraviolet absorption spectrum of pyridine.

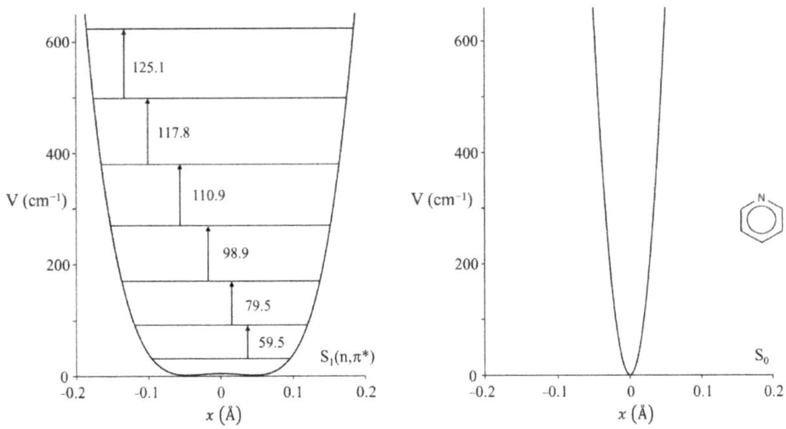

Figure 35. Ring-bending potential energy function for pyridine in its $S_1(n,\pi^*)$ state (**left**) and compared to that in the ground state (**right**).

Laane's research group has also investigated the ultraviolet absorption spectra of several fluoropyridines [56,57]. Some have small barriers to planarity for the excited states, while others are planar but are still not very rigid.

3. Conclusions

Compared to the potential energy functions described in this paper, the harmonic oscillator is indeed boring. We have demonstrated in the present paper how one- and two-dimensional potential energy functions for selected systems can be determined very accurately to fit experimental data and therefore lead to detailed information on molecular structures and energetics. As an example, in our Table 2 we show the remarkably good fit between experimental far-infrared frequencies and the values calculated from the simple potential function of Equation (2). Similarly excellent fits were achieved for the other potential functions discussed in the present work, and these can be seen in the references provided. Physical chemistry textbooks generally show solutions of the wave equation for the harmonic oscillator, but these are not terribly convincing since only one parameter (k in kx^2) is used to fit a single vibrational frequency. In our studies described in this paper, the two-parameter potential energy function in Equation (2) typically fits more than a dozen observed frequencies extremely well.

Author Contributions: E.J.O. and J.L. contributed equally. All authors have read and agreed to the published version of the manuscript.

Funding: This research received no external funding.

Institutional Review Board Statement: Not applicable.

Informed Consent Statement: Not applicable.

Data Availability Statement: The data are available in this article.

Acknowledgments: J.L. wishes to thank co-author E.J.O. for her work on this manuscript even after we have both retired. He also thanks her for her dedicated work in our laboratory for almost twenty years as a student and post-doc. More than a hundred graduate students, undergraduate students, post-docs, visiting professors, and colleagues at other institutions have contribute extensively to the work described in the present paper. J.L. thanks the National Science Foundation for almost forty years of financial support, and he also thanks the Welch Foundation for five decades of support. J.L. also thanks Rui Fausto and Gulce Ogruc Ildiz for overseeing and editing this special honor issue.

Conflicts of Interest: The authors declare no conflicts of interest.

References

1. Laane, J.; Ocola, E.J.; Chun, H.J. Vibrational Potential Energy Surfaces in Ground and Electronic Excited States. In *Frontiers and Advances in Molecular Spectroscopy*; Laane, J., Ed.; Elsevier: Amsterdam, The Netherlands, 2017; pp. 101–142. [CrossRef]
2. Laane, J. Vibrational Potential Energy Surfaces in Electronic Excited States. In *Frontiers of Molecular Spectroscopy*; Laane, J., Ed.; Elsevier: Amsterdam, The Netherlands, 2009; pp. 63–132. [CrossRef]
3. Laane, J. Experimental Determination of Vibrational Potential Energy Surfaces and Molecular Structures in Electronic Excited States. *J. Chem. Phys. A* **2000**, *104*, 7715–7733. [CrossRef]
4. Laane, J. Spectroscopic Determination of Ground and Excited State Vibrational Potential Energy Surfaces. *Int. Rev. Phys. Chem.* **1999**, *18*, 301–341. [CrossRef]
5. Laane, J. Vibrational Potential Energy Surfaces and Conformations of Molecules in Ground and Excited Electronic States. *Annu. Rev. Phys. Chem.* **1994**, *45*, 179–211. [CrossRef]
6. Laane, J. Vibrational Potential Energy Surfaces of Non-Rigid Molecules in Ground and Excited Electronic States. In *Structures and Conformations of Non-Rigid Molecules*; Laane, J., Dakkouri, M., Eds.; Kluwer Publishing: Amsterdam, The Netherlands, 1993; pp. 65–98. [CrossRef]
7. Laane, J. Determination of Vibrational Potential Energy Surfaces from Raman and Infrared Spectra. *J. Pure Appl. Chem.* **1987**, *59*, 1307–1326. [CrossRef]
8. Laane, J. One-dimensional Potential Energy Functions in Vibrational Spectroscopy. *Q. Rev. Chem. Soc.* **1971**, *25*, 533–552. [CrossRef]
9. Schrödinger, E. Der stetige Übergang von der Mikro- zur Makromechanik. *Naturwissenschaften* **1926**, *14*, 664–666. [CrossRef]
10. Benedict, W.S.; Player, E.L. Vibration–Rotation Bands of Ammonia: II. The Molecular Dimensions and Harmonic Frequencies of Ammonia and Deuterated Ammonia. *Canad. J. Phys.* **1957**, *35*, 1235–1241. [CrossRef]
11. Coon, J.B.; Naugle, N.W.; McKenzie, R.D. The Investigation of Double-Minimum Potentials in Molecules. *J. Mol. Spectrosc.* **1966**, *20*, 107–129. [CrossRef]
12. Laane, J. Eigenvalues of the Potential Function $V = Z^4 \pm BZ^2$ and the Effect of Sixth Power Terms. *Appl. Spectrosc.* **1970**, *24*, 73–80. [CrossRef]
13. Sagear, P.; Laane, J. Jet-Cooled Fluorescence Excitation Spectrum, Carbonyl Wagging, and Ring-Puckering Potential Energy Functions of 3-Cyclopenten-1-one in its $S_1(n,\pi^*)$ Electronic Excited State. *J. Chem. Phys.* **1995**, *102*, 7789–7797. [CrossRef]
14. Laane, J.; Zhang, J.; Chiang, W.-Y.; Sagear, P.; Cheatham, C.M. Jet-Cooled Fluorescence Excitation Spectra and Carbonyl Wagging Potential Energy Functions of Cyclic Ketones in Their Electronic Excited States. In *Structure and Dynamics of Non-Rigid Systems*; Smeyers, Y.G., Ed.; Kluwer Publishing: Amsterdam, The Netherlands, 1994; pp. 181–201. [CrossRef]
15. Bell, R.P. The Occurrence and Properties of Molecular vibrations with $V(x)=ax^4$. *Proc. R. Soc. A* **1945**, *183*, 328–337. [CrossRef]

16. Laane, J. The Origin of the Ring-Puckering Potential Energy Function for Four-Membered Rings and Spiro Compounds. A Possibility for Pseudorotation. *J. Phys. Chem.* **1991**, *95*, 9246–9249. [CrossRef]
17. Laane, J.; Lord, R.C. Far-Infrared Spectra of Ring Compounds. II. The Spectrum and Ring-puckering Potential Function of Cyclopentene. *J. Chem. Phys.* **1967**, *47*, 4941–4945. [CrossRef]
18. Laane, J.; Harthcock, M.A.; Killough, P.M.; Bauman, L.E.; Cooke, J.M. Vector Representation of Large Amplitude Vibrations for the Determination of Kinetic Energy Functions. *J. Mol. Spectrosc.* **1982**, *91*, 286–299. [CrossRef]
19. Harthcock, M.A.; Laane, J. Calculation of Kinetic Energy Terms for the Vibrational Hamiltonian: Application to Large Amplitude Vibrations Using One-, Two-, and Three-Dimensional Models. *J. Mol. Spectrosc.* **1982**, *91*, 300–324. [CrossRef]
20. Autrey, D.; Laane, J. Far-Infrared Spectra, ab initio Calculations, and the Ring-Puckering Potential Energy Function of 2,3-Dihydrofuran. *J. Phys. Chem. A* **2001**, *105*, 6894–6899. [CrossRef]
21. Ocola, E.J.; Laane, J. Ring-puckering Potential Energy Functions for Cyclobutane and Related Molecules Based on Refined Kinetic Energy Expansions and Theoretical Calculations. *Chem. Phys.* **2020**, *532*, 110647. [CrossRef]
22. Cortez, E.; Verastegui, R.; Villarreal, J.R.; Laane, J. Low-Frequency Vibrational Spectra and Ring-Puckering Potential Energy Function of 1,3-Dioxole. A Convincing Demonstration of the Anomeric Effect. *J. Amer. Chem. Soc.* **1993**, *115*, 12132–12136. [CrossRef]
23. Sakurai, S.; Meinander, N.; Morris, K.; Laane, J. Far-Infrared, Raman and Dispersed Fluorescence Spectra, Vibrational Potential Energy Surface, and the Anomeric Effect of 1,3-Benzodioxole. *J. Amer. Chem. Soc.* **1999**, *121*, 5056–5062. [CrossRef]
24. Laane, J.; Bondoc, E.; Sakurai, S.; Morris, K.; Meinander, N.; Choo, J. Spectroscopic Determination of the Vibrational Potential Energy Surface and Conformation of 1,3-Benzodioxole in its $S_1(\pi,\pi^*)$ Excited State. The Effect of the Electronic Excitation on the Anomeric Effect. *J. Amer. Chem. Soc.* **2000**, *122*, 2628–2634. [CrossRef]
25. Ocola, E.J.; Laane, J. The Anomeric Effect in Five-Membered Ring Molecules. Comparison of Theoretical Computations and Experimental Spectroscopic Results. *J. Phys. Chem. A* **2021**, *125*, 327–337. [CrossRef]
26. Kilpatrick, J.E.; Pitzer, K.S.; Spitzer, R. The Thermodynamics and Molecular Structure of Cyclopentane. *J. Amer. Chem. Soc.* **1947**, *69*, 2483.
27. Kilpatrick, J.E.; Pitzer, K.S.; Spitzer, R. Additions and Corrections: The Thermodynamics and Molecular Structure of Cyclopentane. *J. Amer. Chem. Soc.* **1958**, *80*, 6697–6703.
28. Pitzer, K.S.; Donath, W.E. Conformations and Strain Energy of Cyclopentane and its Derivatives. *J. Amer. Chem. Soc.* **1959**, *81*, 3213–3218. [CrossRef]
29. Durig, J.R.; Wertz, D.W. Vibrational Spectra and Structure of Small-Ring Compounds. X. Spectroscopic Evidence for Pseudorotation in Cyclopentane. *J. Chem. Phys.* **1968**, *49*, 2118–2121. [CrossRef]
30. Bauman, L.E.; Laane, J. Pseudorotation of Cyclopentane and Its Deuterated Derivatives. *J. Phys. Chem.* **1988**, *92*, 1040–1051. [CrossRef]
31. Ocola, E.J.; Bauman, L.E.; Laane, J. Vibrational Spectra and Structure of Cyclopentane and its Isotopomers. *J. Phys. Chem. A* **2011**, *115*, 6531–6542. [CrossRef] [PubMed]
32. Chao, T.H.; Laane, J. Vibrational Studies of Cyclopentane: Effect of 10-Fold Barrier to Pseudorotation. *J. Mol. Spectrosc.* **1978**, *70*, 357–360. [CrossRef]
33. Laane, J. Far-Infrared Spectrum and the Barrier to Pseudorotation of Silacyclopentane. *J. Chem. Phys.* **1969**, *50*, 1946–1951. [CrossRef]
34. Leibowitz, S.J.; Laane, J.; Verastegui, R., Jr.; Villarreal, J.R. Far-Infrared Spectra and Hindered Pseudorotation of 1,3-Oxathiolane. *J. Chem. Phys.* **1992**, *96*, 7298–7305. [CrossRef]
35. Laane, J. Pseudorotation of Five-Membered Rings. In *Vibrational Spectra and Structure*; Durig, J.R., Ed.; Marcell Dekker, Inc.: New York, NY, USA, 1972; Volume 1, pp. 25–50. ISBN 0-8247-1149-1.
36. Rosas, R.L.; Cooper, C.; Laane, J. Evaluation of Molecular Mechanics Methods for the Calculation of the Barriers to Planarity and Pseudorotation of Small Ring Molecules. *J. Phys. Chem.* **1990**, *94*, 1830–1836. [CrossRef]
37. Pillsbury, N.; Choo, J.; Laane, J.; Drucker, S. Lowest n,π^* Triplet State of 2-Cyclopenten-1-one: Cavity Ringdown Absorption Spectrum and Ring-Bending Potential-Energy Function. *J. Phys. Chem. A* **2003**, *107*, 10648–10654. [CrossRef]
38. Choo, J.; Kim, S.; Drucker, S.; Laane, J. Density Functional Calculations, Structure, and Vibrational Frequencies of 2-Cyclopenten-1-one in its S_0, $S_1(n,\pi^*)$, $T_1(n,\pi^*)$ and $T_2(\pi,\pi^*)$ States. *J. Phys. Chem. A* **2003**, *107*, 10655–10659. [CrossRef]
39. Chun, H.J.; Ocola, E.J.; Laane, J. Vapor-Phase Raman Spectra and the Barrier to Planarity of Cyclohexane. *J. Phys. Chem. A* **2016**, *120*, 7077–7680. [CrossRef]
40. Lewis, J.D.; Malloy, T.B., Jr.; Chao, T.H.; Laane, J. Periodic Potential Functions for Pseudorotation and Internal Rotation. *J. Mol. Struct.* **1972**, *12*, 427–449. [CrossRef]
41. Lewis, J.D.; Laane, J. Periodic Potential Energy Functions with Sine and Cosine Terms. *Mol. Spectrosc.* **1977**, *65*, 147–154. [CrossRef]
42. Kelly, M.B.; Laane, J.; Dakkouri, M. Barriers to Internal Rotation of Cyclopropylsilane-d_3 and Cyclopropylgermane from Combination Band Spectra. *J. Mol. Spectrosc.* **1989**, *137*, 82–86. [CrossRef]

43. Ocola, E.J.; Laane, J. Internal Rotation of Methylcyclopropane and Related Molecules; Comparison of Experimental Results and Theory. *J. Phys. Chem. A* **2016**, *120*, 7269–7278. [CrossRef]
44. Kalasinsky, V.F.; Powers, D.E.; Harris, W.C. Vibrational Spectra and Conformations of Cyclopropylamine. *J. Phys. Chem.* **1979**, *83*, 506–510. [CrossRef]
45. Chiang, W.-Y.; Laane, J. Fluorescence Spectra and Torsional Potential Functions for *trans*-Stilbene in Its S_0 and $S_1(\pi,\pi^*)$ Electronic States. *J. Chem. Phys.* **1994**, *101*, 8755–8767. [CrossRef]
46. Boopalachandran, P.; Craig, N.; Groner, P.; Laane, J. Gas-Phase Raman Spectra and the Potential Energy Function for the Internal Rotation of 1,3-Butadiene and its Isotopologues. *J. Phys. Chem. A* **2011**, *115*, 8920–8927. [CrossRef]
47. Boopalachandran, P.; Craig, N.; Laane, J. Gas-Phase Raman Spectra of Hot Bands of Fundamentals and Combinations Associated with the Torsional Vibration of *s-trans*-1,3-Butadiene and its Deuterated Isotopologues. *J. Mol. Spectrosc.* **2011**, *269*, 236–241. [CrossRef]
48. Al-Saadi, A.A.; Ocola, E.J.; Laane, J. Intramolecular π-Type Hydrogen Bonding and Conformations of 3-Cyclopenten-1-ol. 1. Theoretical Calculations. *J. Phys. Chem. A* **2010**, *114*, 7453–7456. [CrossRef]
49. Ocola, E.J.; Al-Saadi, A.A.; Mlynek, C.; Hopf, H.; Laane, J. Intramolecular π-Type Hydrogen Bonding and Conformations of 3-Cyclopenten-1-ol. 2. Infrared and Raman Spectral Studies at High Temperatures. *J. Phys. Chem. A* **2010**, *114*, 7457–7461. [CrossRef] [PubMed]
50. Al-Saadi, A.A.; Wagner, M.; Laane, J. Spectroscopic and Computational Studies of the Intramolecular Hydrogen Bonding of 2-Indanol. *J. Phys. Chem. A* **2006**, *110*, 12292–12297. [CrossRef] [PubMed]
51. Ocola, E.J.; Laane, J. Spectroscopic and Theoretical Study of the Intramolecular π-Type Hydrogen Bonding and Conformations of 2-Cyclohexen-1-ol. *J. Phys Chem. A* **2016**, *120*, 74–80. [CrossRef]
52. Ocola, E.J.; Laane, J. Theoretical Investigation of Intramolecular π-Type Hydrogen Bonding and Internal Rotation of 2-Cyclopropen-1-ol, 2-Cyclopropen-1-thiol and 2-Cyclopropen-1-amine. *Mol. Phys.* **2019**, *17*, 1404–1412. [CrossRef]
53. Ocola, E.J.; Laane, J. Spectroscopic and Theoretical Study of the Intramolecular π-type Hydrogen Bonding and Conformations of 3-Cyclopentene-1-amine. *J. Phys. Chem.* **2020**, *124*, 5907–5916. [CrossRef]
54. Jesson, J.P.; Kroto, H.W.; Ramsay, D.A. Quasiplanarity of Pyridine in Its First Excited Singlet State. *J. Chem. Phys.* **1972**, *56*, 6257–6258. [CrossRef]
55. Boopalachandran, P.; Jaan, J. Ultraviolet Absorption Spectra of Pyridine-d_0 and -d_5 and their Ring-Bending Potential Energy Function in the $S_1(n,\pi^*)$ State. *Chem. Phys. Lett.* **2008**, *462*, 178–182. [CrossRef]
56. Sheu, H.-L.; Boopalachandran, P.; Kim, S.; Laane, J. Infrared, Raman, and Ultraviolet Absorption Spectra and Theoretical Calculations and Structure of 2,3,5,6-Tetrafluoropyridine in its Ground and Excited Electronic States. *Chem. Phys.* **2015**, *456*, 28–33. [CrossRef] [PubMed]
57. Boopalachandran, P.; Kim, S.; Choo, J.; Laane, J. Ultraviolet Absorption Spectra, Structure, Vibrations, and Theoretical Calculations of 2-Fluoro- and 3-Fluoropyridine in their Electronic Excited States. *Chem. Phys. Lett.* **2011**, *514*, 214–219. [CrossRef]

molecules

MDPI

Article

Vibrational Properties of Doped P3HT Chains in Solution: Insight into the Doping Mechanism from Infrared IRAV and Raman RaAV Bands

Kaiyue Hu, Sara Doti, Luigi Brambilla, Mirella Del Zoppo, Chiara Castiglioni * and Giuseppe Zerbi

Dipartimento di Chimica, Materiali e Ingegneria Chimica Giulio Natta, Politecnico di Milano, 20133 Milano, Italy; kaiyue.hu@polimi.it (K.H.); sara.doti@mail.polimi.it (S.D.); luigi.brambilla@polimi.it (L.B.); mirella.delzoppo@polimi.it (M.D.Z.); giuseppe.zerbi@polimi.it (G.Z.)

* Correspondence: chiara.castiglioni@polimi.it

Abstract: Chemical doping is a well-established technique for increasing the electrical conductivity of polyconjugated polymers, and its effectiveness can be assessed through IR spectroscopy, thanks to the rise of the so-called IRAVs (infrared activated vibrations), which prove the formation of polarons on the polymer chain. While the mechanism of the IRAVs activation has been widely explored in the past, several peculiar features remain unclear. Changes in the Raman spectrum of doped polymers (RaAV, Raman activated vibrations) are widely used as well for monitoring the doping process, but the interpretation is often limited to purely empirical correlations. By means of an experimental campaign on doped regio-regular poly(3-hexylthiophene-2,5-diyl) (P3HT) samples in chloroform solution and on the solid samples cast from the same solutions, this paper presents for the first time a thorough comparative analysis of IRAVs and RaAVs, aiming at a unified description of the structure of doped P3HT. In particular, we will discuss the effect of the doping level on the vibrational features of the polymer and the dopant so that spectroscopic markers can be found to be used in the identification of the presence of ICT (integer charge transfer) complexes in different doping regimes. This study demonstrates that combining IR, Raman, and UV-Vis-NIR spectroscopies provides a powerful, complementary set of tools to diagnose not only the doping level but also the detailed molecular and supramolecular structure of the doped P3HT, useful for the development of structure/properties relationships in the perspective of the optimization of the charge transport performances.

Keywords: polaron; F4TCNQ; conjugation length; UV-vis-NIR absorption; conducting polymers

1. Introduction

1.1. Spectroscopy of Doped Polymers

Since the early discovery of the peculiar chemical-physical properties of polyacetylene [1–4] the design and synthesis of organic semiconducting polymers has been and is still a flourishing research field [5–8]. It boasts successful applications in the development of the so-called plastic electronics—e.g., flexible opto-electronic devices, light emitting diodes—and active layers for energy storage (e.g., photovoltaics, thermoelectric devices) and sensors [8–12].

In the last decade we have witnessed a renewed interest in developing polymeric materials reaching high conductivities when suitably doped [10,13]. Chemical or electrochemical doping involves a redox reaction that results in the injection of an electron or, more frequently, a hole into a polymer chain characterized by a conjugated network of π-electrons. The so-called "conducting polymers" show an intriguing interplay between electronic and molecular structure, and chemists and physicists devoted great efforts to the study of doping-induced charge carriers (polarons). Theoretical models and extended experimental studies contributed to the rationalization of the fundamental phenomena leading to charge transport and the optical properties of doped polymers, specifically of polarons [14–18].

On the other hand, organic materials can be tailored through a top-down approach, which exploits the potential in manipulating and finely tuning the morphology of the polymer materials. Properties can be modified by means of a variety of methods of preparation—e.g., film deposition from different solvents, production of fibers, different dopant and doping procedures, and post-processing, such as thermal or mechanical treatments [11,19–27].

Molecular spectroscopy is widely exploited for the characterization of novel materials, and well-consolidated correlations allow the diagnosis and the comparison of different samples, offering guidelines for the synthesis/preparation of improved materials [6,17]. Despite experimental data being available on several different materials, a unified description of the structure of polarons and precise knowledge of how polarons modify the hosting polymer chains and the doped crystal domains is still lacking.

This paper deals with a very renowned polymer, which can be regarded as a record in terms of successful applications and number of scientific studies, namely regio-regular poly(3-hexylthiophene-2,5-diyl) (P3HT). Our aim is to provide a critical and comprehensive analysis of the spectroscopic features associated with the doping of P3HT, offering a guided tour through several experimental evidences obtained from different samples by means of several techniques in the effort to fit them in a unified picture of polaron structure and morphology of the doped material. The study focuses on the UV-vis, IR, and Raman response of doped P3HT by means of a systematic investigation on doped P3HT in chloroform solutions and on the corresponding solid samples obtained as thin films by casting the same solutions on KBr or on glass windows. These experiments highlight phenomena that cannot be disclosed by analyzing only solid-state samples.

The interpretation of the vibrational spectra of P3HT in its neutral (pristine) and doped state cannot overlook the very peculiar structural/morphological characteristics of polymeric materials featuring the simultaneous presence of different phases and molecular structures in the same sample. Each individual polymer chain is a molecule, which can assume different geometries depending on its conformation: the polymer shows a flat backbone with co-planar (or quasi-coplanar) thiophene rings in the crystalline domains, while in the amorphous phase and in solution the polymer backbone takes on a more coiled, disordered conformation. Moreover, the lateral hexyl chains can assume straight, trans-planar geometries or disordered conformations, the last being typical of the amorphous phases and solutions. Interestingly, the structure of the lateral chains rules the so-called inter-lamellar spacing in the crystals. Indeed, P3HT crystals can host polymers with some amount of conformational disorder of the lateral hexyl chains, giving rise to "hairy" structures [28,29], characterized by a different three dimensional (3D) packing.

The description of the chemical doping, namely the formation of a complex between P3HT and a dopant molecule, followed by a charge transfer between partners, should consider this complicated scenario. Moreover, doping can lead to the formation of complexes of different kinds, namely Charge Transfer Complexes (CTC) with a fractional charge transfer, or Integer Charge Transfer (ICT) complexes [24,25,30–32]. In case of p-doping, ICT entails a hole injection into the polymer chain and the formation of a polaron: the hole is accompanied by a remarkable relaxation of the polymer structure, involving several thiophene units; in addition, the presence of charged P3HT segments and of the counterions affects the structure of the crystalline domains which host the polarons [20,26,27,33–37].

Doping has remarkable effects on the spectroscopic response of P3HT, and clear evidence of doping can be collected by means of UV-visible absorption spectroscopy and vibrational (IR and Raman) spectroscopies. In addition to the recognition of the spectroscopic signature of the doping, the spectra provide a rich set of data that can shed light on the structure and morphology of doped P3HT. However, the interpretation of these data is not straightforward because the response of the sample is often the superposition of different contributions, which come from the different phases/polymer structures simultaneously probed by the experiment. A comprehensive picture can be obtained only through the joint use of several techniques and the comparison of different samples.

In this work, we will analyze in a systematic and comparative way the UV-vis, IR, and Raman spectra of p-doped P3HT samples in solution and in the solid state, focusing on the spectral changes observed varying the dopant percentage. For each spectroscopic technique, we will bring to light which phases are preferentially probed (e.g., doped/charged domains are preferentially probed by IR) and the relationship between given spectroscopic markers and specific structural features. The analysis of the vibrational spectra will be carried out considering the peculiarity of the vibrational dynamics and vibrational spectra of polymers and, in particular, π-conjugated polymers [17,38,39]. We will refer to the Effective Conjugation Coordinate (ECC) theory [17,18] for the interpretation of the Raman spectrum of P3HT and of the doping-induced IR features (IRAVs) of the doped samples [39–41].

There are several different ways to dope P3HT, and a possible strategy is to perform doping in solution, which can affect the morphology and properties of the film deposited because of pre-doping and possibly pre-aggregation of doped polymer molecules [25,27,33,34,36,42]. This paper will focus on P3HT samples doped in chloroform solution. 2,3,5,6-tetrafluoro-7,7,8,8-tetracyanoquinodimethane (F4TCNQ)—a molecule widely employed as a very effective p-dopant—was used to prepare solutions at different relative polymer/dopant concentrations. The spectroscopic response of these solutions and that of the films obtained by casting the corresponding solutions will be presented and analyzed in a comparative way.

1.2. Ideal Structural Models of Doped P3HT Chains

Given the complex nature of the systems considered, we will need to take into account the coexistence of different phases/structures. For this reason, to describe the real samples, it is useful to introduce some simple and schematic, ideal models that have been devised considering the spectroscopic evidence, the results of quantum chemical calculations, and data from the literature and will be of help in the analysis of the spectra that follows. These models reflect different scenarios, consistent with our choice of dopant concentration for the preparation of the samples. We selected the dopant polymer molar ratio of 11.1% as representative of the doping regime suitable to maximize the conductivity of the material [33], while moderate (5.6%) and low (2.7%) doping ratios have been chosen in order to grasp trends in the spectra. The use of low dopant concentration is especially relevant for the study of doped samples in solutions, where we were interested in investigating the possible presence of isolated doped chains. Moreover, some experiments performed on P3HT doped in large excess of F4TCNQ will help to highlight the peculiarities of the Raman spectrum of a very dense polaron network. It will be shown that at every doping level, the dominant mechanism is that of ICT, confirming what is already known from the literature [25,30]. Polarons are thus the main kind of charge defect: the dopant, in this case F4TCNQ, becomes an anion, and a hole is injected into the polymer chain.

A delicate issue when modeling doped polymers is that of giving a reliable estimate of the extent of the polaron domain. Theoretical and experimental studies [41,43–47] suggest that the chain segment perturbed by the polaron defect extends from 5 to 11 rings. The structural rearrangement following the charge injection involves a change in the bond lengths of the backbone rings (quinoidization of the structure) and a planarization of the backbone, which might extend well beyond the electronically perturbed region. Theoretical indications that this is the case can be found in [41]. This is deemed to be true also for doped chains in solution [34]. In the following we will describe two different ideal, limiting situations that can be found in our samples: namely, A. isolated chains such as can be found in solutions with a good solvent; B. aggregated chains as found in solid-state samples, especially in crystalline domains.

Case A. Polarons in isolated P3HT chains. Depending on the concentration of charge defects, different structures are possible:

1. Neutral chains, which have a prevalently disordered conformation (both for the backbone and the alkyl chains). Neutral coils in Scheme 1a.
2. Slightly doped chains where few backbone segments are planar due to the presence of the polaron defect (even if more than one polaron defect is present, they are far apart) while the remaining part of the polymer chain is conformationally distorted. Isolated polarons in Scheme 1b.
3. Heavily doped chains where long planar segments of the backbone are present (Scheme 1c). In an ideal limiting case, the entire polymer backbone would be planar, and we might even think of an ideal 1D crystal with equally spaced polaron defects.

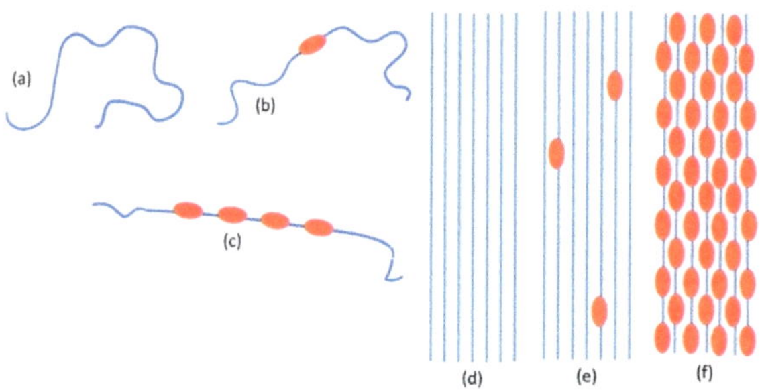

Scheme 1. Sketches illustrating the ideal structural models of undoped (**a**) and doped (**b**,**c**) P3HT chains and of P3HT crystalline domains in the undoped (**d**) and doped states (**e**,**f**).

Case B. Polarons in interacting P3HT chains. In the doped solid-state samples, it is possible to find:

1. Neutral, crystalline P3HT: the structure is the same as that of the pristine polymer with planar backbone chains regularly packed in a 3D crystalline structure (Scheme 1d).
2. P3HT crystals with isolated polarons. It is reasonable to think that the structure is similar to that of the pristine polymer perturbed just in proximity to the isolated polaron defects (Scheme 1e).
3. P3HT crystals with a high polaron density. They can be thought of as P3HT-F4TCNQ co-crystals [20,30,37] (Scheme 1f).
4. Chains with distorted conformation in the amorphous phase. Doping in the disordered amorphous phase seems unlikely since ICT would imply planarization of the backbone. Probably only with an excess of F4TCNQ might some doping occur. On the other hand, if ICT does occur, the planarization of the chain would favor clustering of the chains and hence the obtainment of an ordered 3D domain.

It must be noticed that two possible 3D crystalline structures have been identified in the literature [25,30] namely, Π or IL (inter-lamellar) crystals. In the former case, the dopant (either forming CTC or less frequently ICT complexes) is in between π-stacked thiophene rings, while in the latter (forming only ICT complexes), it is in between the layers of the side alkyl chains.

In real samples, the conformation of the hexyl chains also plays a role in driving the doping mechanism. In isolated P3HT chains or in very small clusters, the side chains will be free to assume a highly disordered conformation. Only in the presence of larger clusters or in crystalline domains can the side chains arrange themselves in a more ordered phase. It seems then reasonable to hypothesize that crystals with isolated polarons can host ICT Π complexes, while to obtain the IL architecture a co-crystal is required.

In order to analyze and rationalize the vibrational spectra, it is necessary to consider how the thiophene features (in terms of geometry, force constants, and polarization) are modified near the polaron defect. The following schemes can be introduced:

- Isolated polaron (Scheme 2a). We can distinguish three regions of the polymer chain: (1) a charge defect of about 4–5 thiophene rings characterized by large spin density and polarization of the CC bonds (red region); (2) a perturbed region which extends to a certain number X of rings with perturbed geometry and force constants (orange region); and (3) a planar and neutral region which extends for Y rings (yellow region). In the case of crystalline samples, Y can be very large because of the planarization induced by the crystalline field.
- Region with high polaron concentration (Scheme 2b). The charge defects are similar to the previous case and are separated by a perturbed region (K rings). We can assume that in this case the third region characterized by planar, neutral rings is lacking.
- Region with very high polaron concentration (Scheme 2c) can be foreseen, e.g., in the case of excess doping. In this case, the perturbed regions, which separate the individual charge defects, are vanishingly small. The discussion of the Raman spectra (Section 2.3) will show that postulating the existence of such a limiting structure is needed to explain experiments on samples prepared with a large excess of dopants.

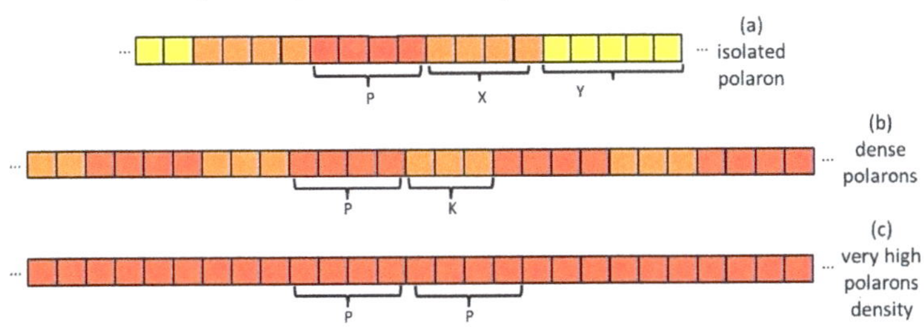

Scheme 2. Schematic illustration of the effect of doping on a P3HT chain. Each square represents a thiophene ring: red domains are characterized by large positive charge and highly polar CC bonds, orange domains are slightly perturbed (transition region), and yellow domains maintain the same structure as in the pristine polymer.

As noticed before, the determination of the exact extent of the various regions is a very subtle issue that depends first of all on the nature of the sample analyzed and on the parameters chosen to describe the perturbation decay length.

1.3. *Vibrational Spectra of P3HT and the ECC Theory*

A suitable tool to rationalize the most characteristic features of the vibrational spectra of conjugated polymers is given by the ECC theory [17,38–41]. The ECC describes a collective CC stretching coordinate represented in Scheme 3 in the case of polythiophene.

ECC is the key vibrational coordinate ruling the spectroscopic response of polyconjugated polymers because it describes the nuclei trajectory with the largest coupling with electrons (in the case of crystals, the phenomenon is referred to as electron-phonon coupling). In the case of P3HT, ECC describes the CC Bond Length Alternation (BLA) oscillation between two ideal structures, namely the aromatic and the quinoid structures (Scheme 3, top). In the aromatic structure the inter-ring CC bonds are single bonds, and the quasi-double and quasi-single CC bonds of the ring are arranged as shown in Scheme 3, top-left sketch, while the quinoid structure shows a double inter-ring bond and a reversal of the quasi-single and quasi-double bonds in the rings. Because of the strong coupling with the electron cloud, modes with an ECC character—namely, vibrational trajectories with large projection along ECC—exhibit large fluctuations of the molecular polarizability and thus have large Raman activity.

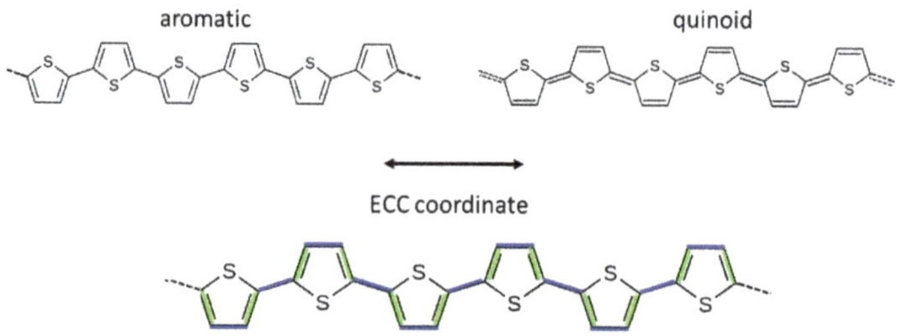

Scheme 3. Sketch of the two—aromatic and quinoid—ideal structures of polythiophene and vibrational displacements associated with the collective ECC for polythiophene, which describes the vibrational trajectory linking the two structures: green CC bonds stretch, while blue CC bonds shrink.

The ECC theory is capable of explaining some of the peculiar features of the Raman spectrum of pristine polymers. In Figure S1 we present the Raman spectra of pristine P3HT in the solid state and in $CHCl_3$ solution. Especially in solid samples, the spectrum is highly selective, showing only a few, very strong bands assigned to collective ECC modes; one can observe an extremely strong ECC band and a satellite peak [28,41]. In the solution, the ECC mode shifts to higher wavenumbers because of conformational disorder, the intensity decreases [48], and the spectrum is less selective. It can also be noticed that the strongest peak is broader, indicating that it results from the convolution of many components, which can be ascribed to ECC-like modes localized on small co-planar sequences of rings or on individual rings. Within the ECC theory it is also possible to understand, at a molecular level, the activation in the infrared spectrum of the IRAV modes upon doping [17,18,38,39]. In Figure 1, one can see the IRAV modes of doped polythiophene, which form a series of strong bands in the IR region below 1400 cm^{-1}. As demonstrated by means of theoretical modeling [41], the activation mechanism of IRAVs and their significant intrinsic IR intensity is due to ECC-like modes mainly involving the rings in the charged region. In P3HT the band multiplicity is due to the dynamical coupling of the ECC with vibrations of the hexyl side chains. This coupling makes the IRAV bands sensitive to the hexyl chain conformation, both in frequency and intensity. We can summarize the spectroscopic features of doped P3HT as follows.

ECC modes are collective CC stretching vibrations that involve a whole planar backbone segment (yellow, red, and orange regions of Scheme 2). The presence of structural defects can localize these modes, and the coupling with the alkyl chains through the linking C-C bond can activate IR and Raman modes that could not be seen in the spectrum of the highly crystalline pristine polymer. The frequency of ECC-like modes is sensitive to conjugation: it decreases for increasing conjugation length. Indeed, the comparison between the solution and solid-state spectra of pristine P3HT shows a remarkable decrease in the

ECC band frequency due to the planarization of the backbone. Moreover, upon doping, also the hole, delocalized along the chain, contributes to the lowering of the frequency through the change in the CC force constants associated with changes in BLA towards a more quinoid structure.

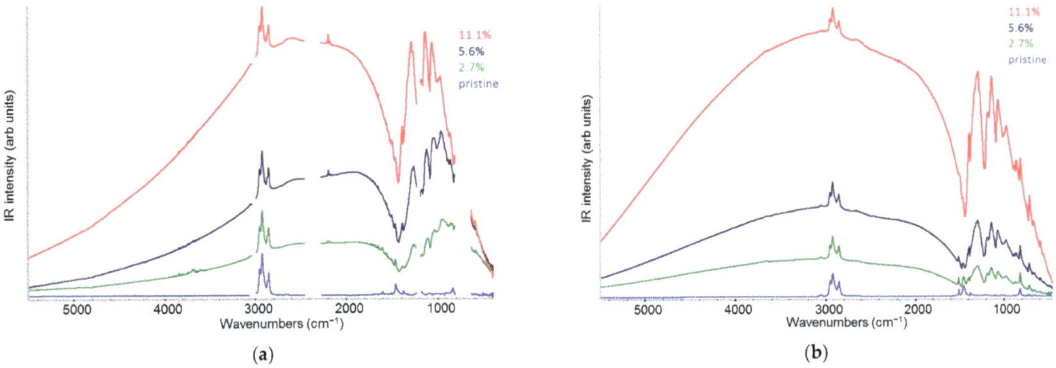

Figure 1. IR absorption spectra of pristine P3HT (light blue line) and doped P3HT at 2.7% (green line), 5.6% (dark blue line), and 11.1% (red line) F4TCNQ/P3HT relative molar concentration. (**a**) Samples in chloroform solution. (**b**) Solid thin films obtained from the same solutions of panel (**a**). Spectra have been normalized to the IR intensity of the CH stretching band of the hexyl chains (3000–2800 cm^{-1}), which is not affected by the doping.

In general, the polarizability and hence the Raman intensity associated with the stretching of a polar bond is lower than that of an apolar bond. Therefore, it is reasonable to assume that the modes with the highest Raman intensities are those associated with co-planar and neutral ring sequences. It is also well known that the intensity of collective modes is stronger than that of localized modes. From these observations one can conclude that the Raman spectrum is dominated by the vibrational modes of the planar, neutral thiophene segments. Thus, for example, in the Raman spectrum of pristine polythiophene, the dominant contribution is that of the crystalline phase, while the amorphous fraction gives an almost negligible contribution. A further remarkable factor affecting the Raman activity is the resonance enhancement, which occurs when the laser energy matches the allowed electronic transition of the material. In the following experimental investigation, we performed Raman measurements with an exciting laser wavelength of 1064 nm, well below the electronic transitions of pristine P3HT but in the spectral region of the infrared (P1) polaron band. Notwithstanding the potential resonance enhancement of the Raman modes of doped chains, because of the presence of highly polar bonds, the Raman spectrum of heavily doped P3HT is weak, and clear evidence of this will be given in Section 2.3.

Unlike the Raman response, polar bonds give rise to a large infrared intensity. This explains the significant intensity of the ECC modes responsible for the IRAV bands, which represent the vibrations of the polar CC bonds in the polaron defect and in the highly perturbed adjacent units.

2. Results and Discussion

2.1. IR Features of F4TCNQ: Evidence of ICT Doping in Solution and in the Solid-State

The absorption intensity of each IR spectrum in Figure 1 has been normalized to the group of bands (2800–3000 cm^{-1}) associated with CH stretching modes of the hexyl P3HT chains, whose absorption intensity, according to theoretical predictions [41] and chemical-physical intuition, is almost unaffected by the charge transfer from the dopant.

Figure 1 clearly shows the effectiveness of the doping process through the activation of the IRAV bands both in solutions and in the films. Moreover, the intensity of the IRAVs and of the lower energy electronic transition of the polarons monotonically grows with increasing dopant concentration. In addition, the comparison of the spectra of the solutions (Figure 1a) and of the solids (Figure 1b) suggests that, at a given dopant concentration, the doping yield is higher in the solid state than in solution. Indeed, the spectra of panel (a) displays similar IRAVs intensities as those of panel (b), while the CH stretching band appears to be significantly weaker in solid-state spectra. Figure S2 (Supporting Information) illustrates this feature better by means of the comparison of the spectra of solution and film at 11.1% dopant ratio after normalization to CH stretching absorptions. Since the CH stretching intensity is proportional to the P3HT amount—irrespective of its doped or pristine state—it is immediately clear that both the polaron band (P1) and the IRAV bands are stronger in the solid sample, implying that the amount of doped material is larger in the solid. Such behavior is due to the larger doping in the solid, which is expected since during solvent evaporation further ICT complexes can form, involving the neutral F4TCNQ species still present in the doped solution.

Doping with F4TCNQ gives further proof of the formation of ICT complexes, thanks to the peculiar IR features of the F4TCNQ anion, which has CN stretching frequencies markedly different from those of the neutral molecule [31,32,43,49]. The anion shows an intense B_{1u} symmetric stretching band at about 2190 cm^{-1} and a weaker B_{2u} antisymmetric stretching band at about 2170 cm^{-1}, while the corresponding wavenumbers of the neutral species in chloroform are observed at 2227 cm^{-1} (B_{1u}, very weak) and 2213 cm^{-1} (B_{2u}). The B_{1u} band of the anion is highly sensitive (both in frequency and IR intensity) to the environment [43,49]. Furthermore, when F4TCNQ is involved in CTC, characterized by a partial charge transfer, its peak shifts to a wavenumber between that of the neutral molecule and that of the anion, showing a pretty linear trend with the amount of charge transferred [25,32]. In ref. [43], the Coulomb interaction between the anion and the hole delocalized on the polaron has been modeled to explain the small wavenumber shifts of the B_{1u} band in doped solid P3HT samples, characterized by a different degree of order. In the following, thanks to the comparison between the CN stretching features of solid samples and of doped P3HT in solution, we will obtain additional information on the ICT complexes involving isolated chains and small clusters or crystalline domains in the solid.

Figure 2 shows the IR spectra of the doped samples in the CN stretching region. The spectra of doped P3HT in solution display only the absorption features of the anion (the B_{1u} and B_{2u} bands occur at 2195 and 2173 cm^{-1} respectively). The concentration of the dopant does not affect the band shape and the position: the spectra of all the samples are almost superimposable when normalized to the height of the B_{1u} band (spectra in "full scale", Figure 2, top). The B_{1u} wavenumber observed for samples in solution coincides with that reported in [43] for solid P3HT samples characterized by a remarkable disorder and is ascribed to the formation of more localized polarons. Such behavior in solution could be ascribed to the relative geometry of the F4TCNQ anion and the polymer segment involved in the charge transfer. In solution, we expect that the anion is placed on top of a thiophene ring, with the phenyl unit π-stacked above P3HT (Π-ICT complex), while it is hard to imagine a geometry like that of the IL ICT crystals, since in solution the hexyl chains are very mobile and highly distorted.

Also, solid samples (Figure 2, bottom) show the characteristic doublet of the anion, but in this case the maximum of the B_{1u} band systematically shifts toward lower wavenumbers going from the sample at lower dopant concentration (2.7%, $\nu(B_{1u)}$ = 2190 cm^{-1}) to the sample at higher doping (11.1%, $\nu(B_{1u)}$ = 2186 cm^{-1}). The B_{2u} peak shows a similar shift. It is clear that, on average, in the solid samples at different dopant concentrations, the anion feels different environments, and this suggests that a variety of crystalline domains exist. Some P3HT crystals are expected to accommodate few polarons as diluted defects into the lattice of pristine P3HT (see Scheme 1e), while co-crystals with a high concentration of polarons (Scheme 1f) will host the anions quite far from the polymer chain, in the inter-lamellar regions between hexyl chains (IL crystals) [20,25,43]. We could argue that in the case of diluted polarons, the anion chooses a disposition similar to that in solution and intercalates between π-stacked P3HT chains. This picture seems to be acceptable in light of the peculiar crystal modifications of doped P3HT observed in [25].

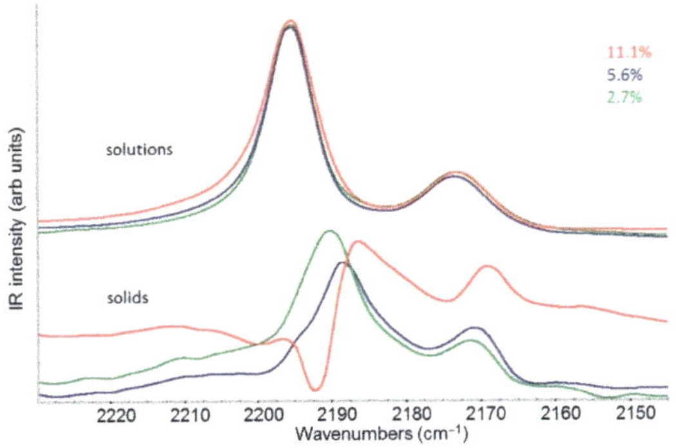

Figure 2. CN stretching region of the IR spectra of P3HT samples doped with F4TCNQ at 2.7% (green line), 5.6% (dark blue line), and 11.1% (red line) F4TCNQ/P3HT relative molar concentration. Stacked on top: spectra from chloroform solutions; bottom: spectra from solid thin films. The spectra are displayed at full scale.

The shoulder on the high-frequency side of the B_{1u} band of the 5.6% solid sample, at the same wavenumber as the samples in solution, proves that the observed CN stretching peak frequency in solids results from different contributions from possibly different phases. Unfortunately, it is impossible to analyze the shape of the absorption band of the 11.1% solid sample because of the distortion of its profile, probably due to the occurrence of Fano resonance in the presence of highly doped P3HT, characterized by metal-like features [50,51].

Another interesting phenomenon is the systematic decrease in the intensity of the CN stretching bands in going from samples in solution to the corresponding solid-state samples. Even if IRAVs are stronger in the solid, suggesting that doping proceeds during the film deposition, the absorption features of the anion in solid-state samples weaken. This apparent contradiction is due to the exceptional sensitivity of the CN stretching IR intensities to the environment, which makes it difficult to compare band intensities in solution and in the solid state [49]. Moreover, it has been proven that the CN stretching absorption intensities dramatically increase in going from the neutral molecule to the anion (according to [32,49], the increase in the intensity of the whole doublet is by a factor of about 60). This fact almost prevents the possibility of evaluating the doping yield—both in solution and in the solid state—looking at the ratio of the CN stretching band intensities of the neutral species to that of the anions. In fact, low concentrations of neutral—unreacted—F4TCNQ molecules cannot be detected.

2.2. UV-Vis-NIR Absorption in Solution and in Solid-State

Figure 3 shows the UV-vis-NIR spectra of solutions—panel (a)—and solid samples—panel (b)—of doped P3HT. All the spectra demonstrate the effectiveness of the doping at any dopant concentration: the characteristic absorption features of the F4TCNQ anions [30] are evident, and the rise of the polaron absorption (the so-called P2 band) in the region

800–1100 nm is appreciable. The last feature is accompanied by the rise of the P1 polaron band, already observed in the IR spectra of Figure 1.

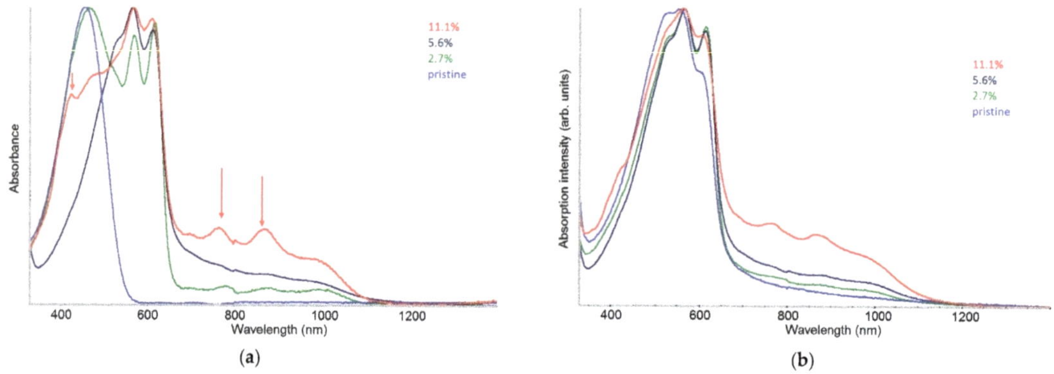

Figure 3. UV-vis-NIR absorption spectra of pristine P3HT (light blue line) and P3HT doped at 2.7% (green line), 5.6% (dark blue line), and 11.1% (red line) F4TCNQ/P3HT relative molar concentration. (**a**) Samples in chloroform solution. (**b**) Thin films from the same solutions of panel (**a**). The red arrows indicate the absorption features of the F4TCNQ anion.

The main P3HT absorption peak of the polymer shows a remarkable evolution upon doping in solution. Beside the strong band peaking at 460 nm, due to P3HT chains with disordered conformations and typical of amorphous phases and P3HT solutions, two strong peaks at longer wavelength appear at any doping level. According to the literature [52], they correspond to the 0–0 and to the 0–1 transition of the vibronic sequence of an electronic transition at remarkably lower energy (around 600 nm) compared to the absorption band of the pristine polymer in solution. The appearance and the characteristic shape of the P3HT vibronic peaks have been rationalized by Spano et al. [52] by modeling the spectroscopic response of J and H aggregates. J aggregates feel the intra-chain interactions between adjacent monomer units of P3HT, covalently bonded in the polymer chain with a regular, planar conformation. H interactions affect adjacent polymer chains π-stacked in the crystal. The interplay between J and H interactions leads to a modulation of the relative intensity of the vibronic transitions and explains why, in pristine P3HT solids, the 0–1 line is the dominant one.

The characteristic shape of the two vibronic peaks in the spectra of solutions (Figure 3a) fits the predicted behavior of J aggregates and suggests that doped chains in solution behave as isolated chains that contain planar P3HT domains (see Scheme 1), namely that chain aggregation is modest. Moreover, there is an evolution of the relative intensity of the vibronic peaks, and the spectrum of the sample obtained at the highest dopant concentration (11.1%) approaches the band-shape of the solid-state samples, showing a remarkable weakening of the 0–0 line. This can be explained considering that at the highest dopant concentration chains densely populated by polarons are present, showing large planar domains that are prone to aggregate already in solution.

Interestingly, the solid doped samples (Figure 3b) have a band shape more similar to that of the pristine solid, but a sharp 0–0 feature, reminiscent of the characteristic absorption of the corresponding solution, is still present. It seems that for these samples the model of J-H aggregates, which describes well the pristine material, is no longer fully adequate. A possible explanation, which allows us to justify the observed J-like character, is the presence of doped crystalline domains with different structures. For instance, the presence of co-crystals where the anions are sandwiched between π-stacked chains should drastically reduce the H interactions. It is nice to see that at the highest doping (11.1%), the band shape of the solid sample is very close to that of the pristine polymer, strongly suggesting that IL crystalline domains dominate in this case; indeed, in IL crystals the anions do not interfere with the close π-stacking of the P3HT chains. This feature could be related to the conductivity measurement reported in [33], showing that dopant concentration close to 10% results in samples approaching a maximum value of conductivity. The phenomenon could be associated with the prevalence of IL crystals where hole hopping between closest neighboring P3HT chains is not hampered by the intercalated anions. Figure S3 shows the direct comparison between solids and solutions spectra, which underlines the differences and similarity of the absorption band shapes.

2.3. Raman Spectra of Doped P3HT: Evidence of Chain Planarization in Solution and Interplay of Different Phases

Raman spectroscopy is frequently used for the characterization of doped polymers because, unlike IR, it does not require any special sample preparation and can also be exploited for in situ analyses, e.g., to follow electrochemical doping. While doping-induced changes in Raman band frequencies and band shapes are often analyzed, few studies focus on the remarkable changes in the Raman cross section of doped P3HT samples. On the contrary, one of the most intriguing pieces of evidence is that Raman Active Vibrations (RaAVs) of doped P3HT in the typical region of the ECC modes (1450–1350 cm^{-1}) have lower absolute intensity than the ECC Raman bands of the pristine material in the solid state. Figure 4 clearly illustrates this feature for solid-state samples: the spectra are displayed after normalization of the Raman intensity to the CS stretching band at 720 cm^{-1}, which is very stable in peak position and shape in all the samples. The very strong Raman doublet (1443 and 1380 cm^{-1}) of the pristine, highly crystalline P3HT polymer, assigned to collective ECC modes (see Section 1), dramatically weakens already in the sample at 2.7% dopant concentration, then slightly weakens monotonically with the dopant amount. The intensity of the Raman transitions is very low in the case of a sample doped with a large excess of F4TCNQ, which shows a weak and very broad band with the maximum shifted to 1410 cm^{-1}. The downward shift in the RaAV main band of doped samples is well known and is related to the changes in the vibrational dynamics of P3HT chains affected by doping. Furthermore, Figure 4 provides the experimental evidence that ECC-like modes have a rather low Raman polarizability when the vibrating CC bonds belong to charged/polarized domains, thus resulting in a smaller Raman activity compared to that of the un-doped species.

The same decreasing intensity trend with dopant concentration is observed in the case of the Raman spectra of doped P3HT in solution—see Figure 5a, which shows Raman spectra normalized to the CS stretching band. Figure 5b compares the same Raman spectra of doped P3HT solutions with the spectrum of a solution of pristine P3HT, which does not display the CS stretching band at 720 cm^{-1}, typical of planar chains. The P3HT concentration in chloroform is the same for all the samples; thus, in Figure 5b, the normalization of the intensities to the chloroform band at 667 cm^{-1} enables a comparison among the doped and the neutral polymer. The ECC band of pristine P3HT chains with disordered conformation appears very weak and broad if compared with RaAVs of doped P3HT solutions: this feature is perfectly explained considering that only planar chains can develop strong Raman-active collective ECC vibrations. In other words, the broad band of the pristine polymer results from the convolution of weakly active Raman transitions associated with localized ECC-like vibrations. Very interestingly, in the case of the solutions of doped P3HT, the effect of planarization upon doping causes a meaningful intensification of the RaAV bands, despite the quite small Raman polarizability of charged/polarized CC bonds belonging to the doped domains. However, it is worth stressing that the RaAVs of doped samples in solution have appreciably lower Raman activity than the ECC band of solid pristine P3HT, as it can be immediately realized by comparison of Figure 4 with Figure 5a, both displaying Raman spectra normalized to the CS stretching band.

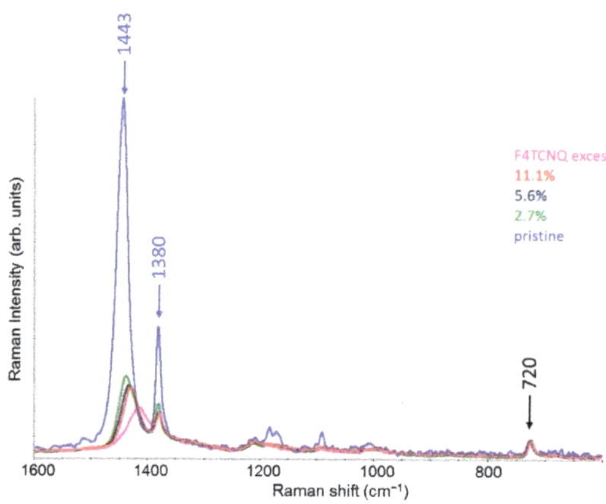

Figure 4. FT-Raman spectra of pristine solid P3HT (blue line) and of solid P3HT samples doped at 2.7% (green line), 5.6% (dark blue line), and 11.1% (red line) F4TCNQ/P3HT relative molar concentration. Raman intensities are normalized to the CS stretching band at 720 cm^{-1} (black arrow). ECC Raman bands of pristine P3HT are indicated by blue arrows.

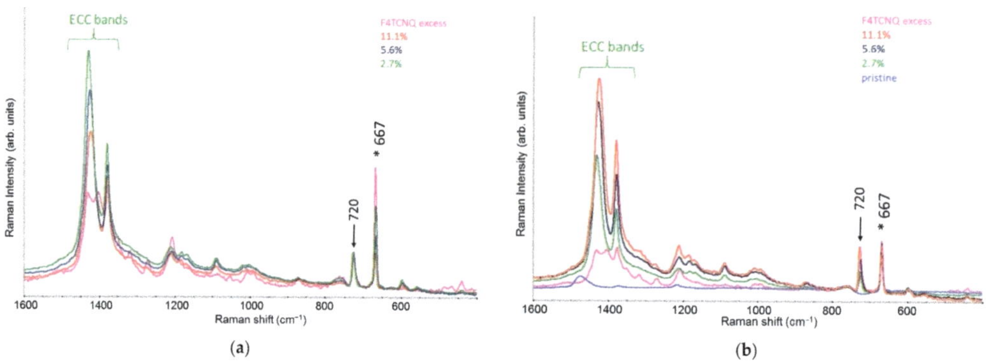

Figure 5. (**a**) FT-Raman spectra of doped P3HT samples in chloroform solutions: 2.7% (green line), 5.6% (dark blue line), and 11.1% (red line) F4TCNQ/P3HT relative molar concentration. The pink line is the Raman spectrum of a sample doped with a large excess of F4TCNQ, in solution. Raman intensities are normalized to the CS stretching band at 720 cm^{-1} (black arrow). The Raman peak of the solvent is indicated by an asterisk. (**b**) Same FT-Raman spectra of panel (**a**) normalized to the chloroform Raman line at 680 cm^{-1} and compared with the spectrum of pristine P3HT in solution (light blue line). The polymer concentration is the same for doped and pristine samples.

As a last comment on Figure 5b, we observe that, when they are normalized to the intensity of the solvent band, the RaAV bands show an increasing intensity trend with dopant concentration, as expected because of the increasing amount of doped chains in solution. The only exception is the spectrum of the sample prepared with a large excess of dopant. As it will be clear at the end of our Raman analysis, all the rings of a highly doped chain are strongly polarized (see Scheme 2c), resulting in a dramatic weakening of the Raman polarizability and hence of the RaAV bands, in spite of the increased number of doped chains.

The above analysis and discussion allowed us to demonstrate that doping in solution determines the conformational change in a segment of the polymer chain that is affected by the hole injection. We can also presume that the planarization of the sequence of thiophene rings in the rather small domain hosting the polaron can promote a domino effect, leading to a planarization of the rings adjacent to the polaron, which in turn can favor further doping of the same chain and possibly clustering of the doped chains. We have also proven experimentally that the doped domains develop ECC-like modes with a slightly lower frequency than the ECC of the pristine solid polymer and with a remarkably lower Raman activity.

A more detailed analysis of the characteristics of the main RaAV in our samples allows a further insight into the origin of such behavior. In Figure 6 we report the Raman spectra in the region of the ECC modes: Raman intensities are displayed at full scale, which clearly highlights the shifts in the peaks and change in the band shape of the main RaAV band. The spectrum of pristine P3HT is also reported, for the sake of comparison. As already observed, the main RaAV band of solutions (Figure 6a) is displaced by about 40 cm^{-1} from the ECC band of the pristine polymer in chloroform. Moreover, it further shifts 8 cm^{-1} going from the lowest (2.7%) to the highest (11.1%) dopant concentration. In the case of solid samples (Figure 6b), the ECC band of the pristine material is close to the main RaAV of the doped material, but also in this case there is an appreciable shift in the band maximum ($\Delta\nu$ = 14 cm^{-1}) from the pristine to the 11.1% doped sample. A remarkable downward shift at 1410 cm^{-1} is observed for the sample with a large excess of dopant.

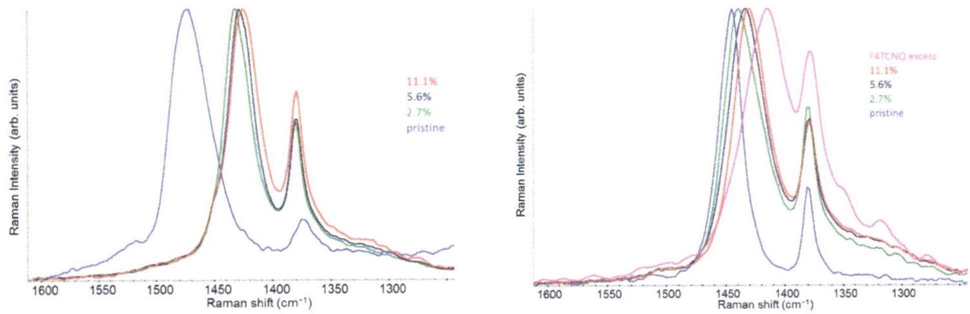

(a) (b)

Figure 6. FT-Raman spectra of pristine P3HT (light blue line) and P3HT doped at 2.7% (green line), 5.6% (dark blue line), and 11.1% (red line) F4TCNQ/P3HT relative molar concentration. (**a**) Doped samples in CHCl3 solution. (**b**) Solid samples from the same solutions of Panel (**a**). In Panel (**b**) the spectrum of a solid P3HT sample—doped in solution with a large excess of F4TCNQ—is also reported (pink line). All the spectra are displayed at full scale.

The main RaAV band of solid doped P3HT is remarkably broader than the ECC band of the pristine polymer: its FWHM goes from 25 cm^{-1} in the pristine case to 40 cm^{-1} for the doped sample—thus suggesting that several components (with different frequencies and Raman cross sections) could contribute to the band. Different components of the main RaAV band might arise because of the following reasons:

1. A variety of different phases present in the material: for example, undoped polymer chains and chains/crystal domains doped to a different extent, namely chains belonging to crystals that host diluted polaron defects (Scheme 1e) or chains belonging to co-crystals with a dense polaron population (Scheme 1f).
2. Different ECC-like normal modes, which mainly involve different domains—chain segments—of each doped chain. Scheme 2 helps to illustrate this feature, which can be corroborated considering that the chain segments affected by the doping (the red and orange boxes in Scheme 2b) show a more quinoid geometry of the thiophene rings as

well as CC stretching force constants different from those of the pristine case. This kind of phenomenon often causes localization of the vibrational modes in the regions featuring different structural characteristics and a different vibrational dynamic.

Hypotheses 1 and 2 correspond to two simplified and extreme viewpoints, and hypothesis 1 is the assumption commonly reported in the literature [53]. We will show that a reasoning grounded on hypothesis 2 can coherently explain the many different Raman features observed so far.

Figure 7 has been fabricated based on hypothesis 2, considering that experiments that probe the two extreme cases, namely the Raman spectrum of a pristine sample and that of a heavily doped sample prepared in large excess of dopant, are available. For the pristine crystalline polymer, we expect a strongly active Raman ECC mode of planar (extended) neutral domains (yellow boxes in Scheme 2a), while the heavily doped P3HT crystals, corresponding to Scheme 2c, should have CC stretching modes typical of polymer chains with highly perturbed/highly polar thiophene rings, characterized by low Raman intensities. Modes like those illustrated above should be present, to a different extent, for any doping regime. Furthermore, each doped chain can show domains similar to those of the pristine material (yellow boxes in Scheme 2a) and domains similar to segments of heavily doped chains, i.e., the inner region of the polarons (red boxes in Scheme 2a–c). In addition, doped chains show moderately perturbed regions at the periphery of the isolated polarons and between adjacent polarons in doped co-crystals (Scheme 2a,b). In Figure 7 we report the results obtained by means of spectral subtractions: for each Raman spectrum of solid P3HT at a given dopant percentage we have subtracted the two spectra of the pristine polymer and of the highly doped material. The weight of the spectra to be subtracted has been determined in a supervised manner, in such a way that band asymmetries or shoulders are well compensated. The result is very impressive: after the subtraction all the spectra collapse into only one component, peaking at 1430 cm^{-1}. We can make the hypothesis that this common component originates from ECC-like modes mainly involving the slightly perturbed rings at the periphery of the polarons (i.e., the orange segments of Scheme 2a,b). The result obtained convinced us to carry out a spectral deconvolution of RaAVs, which has been conducted in a supervised way according to the method illustrated in SI (Figures S4–S11 and discussion).

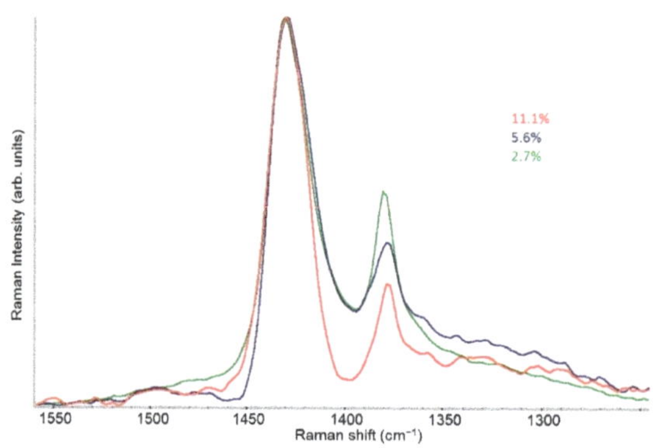

Figure 7. FT-Raman spectra of doped P3HT films after subtraction of the spectra of solid samples of pristine P3HT and of the heavily doped P3HT.

Very shortly, the fit implements the following:

- Two Lorentzian components for a good fit of pristine, neutral domains. They correspond to the hairy A and hairy B crystalline phases of P3HT illustrated in ref. [28], with ECC modes at slightly different frequencies (1453 and 1443 cm^{-1}, respectively) (modes of "yellow segments").

- A Lorentzian component at about 1430 cm^{-1} which describes the RaAV component identified by means of spectra subtraction illustrated in Figure 7 (modes of "orange" segments).
- Two Lorentzian components for the highly perturbed phase, peaking at 1420 and 1410 cm^{-1} respectively (modes of "red segments").

Figure 8 illustrates the result of the band deconvolution, showing the relative importance of the different components. The integrals (areas) of each component are normalized to the area of the CS stretching band at 720 cm^{-1}, which is taken as reference. Unlike the solutions, all the solid samples show a non-negligible contribution from neutral segments, which can be ascribed either to crystalline domains of neutral chains (Scheme 1d) or to the neutral segments of chains belonging to crystalline domains hosting diluted polarons (Scheme 1e). These contributions (yellow and green symbols in the plot) show a steep decrease with increasing dopant percentage; however, because of the very high Raman activity of the ECC modes of the polymer regions not affected by the doping, these components can be detected even at the higher dopant concentration. This observation is particularly relevant if one considers that the Raman experiments are carried out with λ_{exc} = 1064 nm, in the spectral range of the P2 polaron electronic transition: in principle, doped segments should be favored over the pristine polymer domain because of resonance enhancement. The intensity of the band at 1430 cm^{-1} (orange symbols) shows an increasing trend with dopant concentration: as doping increases, non-doped domains are replaced by doped crystals and co-crystals with a dense polarons population, which contribute to the rise of the 1430 cm^{-1} component (orange region of Scheme 2b). The low-frequency contribution at 1420 cm^{-1} (violet symbols) is quite constant, while at 11.1% the contribution of the 1410 cm^{-1} increases appreciably, thus suggesting that at this concentration some "orange" regions are lost and long sequences of "red" domains start to occur.

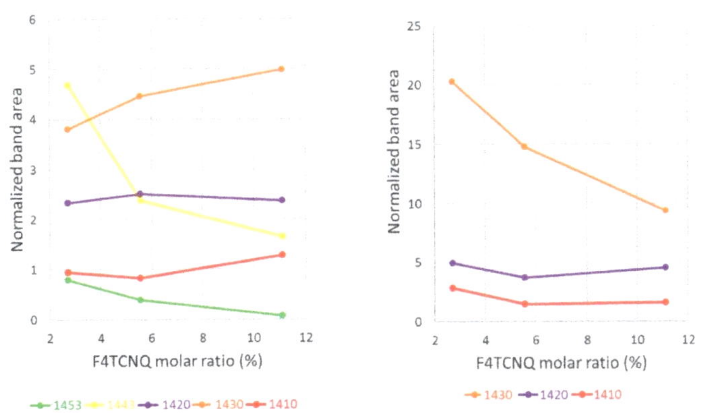

(**a**) (**b**)

Figure 8. (**a**) Plot of the intensities (bands areas normalized to the CS stretching band at 720 cm^{-1}) of the five main components from the band deconvolution of the Raman spectra of doped P3HT samples (solid state) in the region 1480–1400 cm^{-1}. Yellow and green lines: pristine components at 1443 and 1453 cm^{-1} respectively; orange line: 1430 cm^{-1} component; violet line: 1420 cm^{-1} component; red line: 1410 cm^{-1} component. (**b**) Plot of the intensities (bands areas normalized to the CS stretching band at 720 cm^{-1}) of the three main components from the band deconvolution of the Raman spectra of doped P3HT samples in solution in the region 1480–1400 cm^{-1}. Same color code for symbols as in panel (**a**).

The behavior of the RaAV of doped P3HT samples in solution differs from that of solid samples. The main difference concerns the absence of contributions from pristine domains, which was expected because of the conformational disorder that characterizes neutral P3HT chains—or neutral segments of doped chains—in solution. For the same reason, our internal standard—i.e., the intensity of the CS stretching band—does not contain contributions from pristine/neutral chains, as for the solids. Therefore, the trends in Figure 8a,b cannot be directly compared. Since Raman in solution can see only doped, planarized segments, going from lower to higher dopant concentrations, we obtain P3HT chains with an increasing density of polarons at the expense of the "orange" segments. This phenomenon can explain why the contribution of the band at 1430 cm^{-1} has a clear decreasing trend, while the other components are quite stable.

The idea that in solution only planar and doped domains contribute to the RaAV signal is fostered by the observation that the bandwidth of the main Raman band in the raw spectra (not deconvoluted) is systematically narrower in the solutions. The FWHM values for both solid samples and solutions are reported in Table 1.

Table 1. Peak wavenumber and full width at half maximum (FWHM) of the stronger Raman band (ECC mode) of pristine and doped P3HT in a $CHCl_3$ solution and solid samples.

Dopant Percentage	P3HT Samples in $CHCl_3$ Solution		Solid P3HT Samples	
	Peak Wavenumber (cm^{-1})	FWHM (cm^{-1})	Peak Wavenumber (cm^{-1})	FWHM (cm^{-1})
0% (pristine)	1474	42	1444	25
2.7%	1432	29	1439	42
5.6%	1428	29	1433	41
11.1%	1424	33	1430	39

2.4. *IR Spectra: Analysis of the IRAV Bands: Polarons on Isolated Chains and Polarons in 3D Domains*

Solid-state samples of doped regioregular P3HT show IR absorption features ascribed to vibrational transitions of the normal modes of the polarons, characterized by a remarkable ECC-like character and by a very strong absorption intensity [40,41,54–56]. Interestingly, in a wide range of doping percentages, the IRAVs pattern (peak frequencies and relative intensities) of doped P3HT is almost independent of the dopant, the doping method, and the sample morphology (e.g., films or fibers) [41,54]. These observations suggest that IRAVs are associated with vibrational modes localized on the doped polymer segments, which do not involve the dopant molecule, as indeed proven by the success of simple models (DFT calculations of isolated charged oligo-thiophenes) in predicting the IRAVs pattern and explaining the IRAVs activation mechanism [41]. Moreover, based on experimental observations, we can speculate that IRAVs are scarcely sensitive to the kind of 3D packing of the polymer chains, i.e., to the sample morphology, while it is the conformation of the polymer chain that plays the most relevant role. This is proven, for instance, by the fact that the IRAVs pattern of regio-random P3HT is significantly different from that of regio-regular P3HT [55,56].

The general features of the IR spectra of our doped P3HT samples (solutions and solids) are shown in Figure 1 and have already been commented. The following analysis will focus on the differences between the IRAVs trend of doped P3HT in solution at different dopant concentrations and the solid-state samples; this comparison allows highlighting some peculiar behavior not reported so far. A zoom in on the IRAVs region of the solutions (Figure 9a) shows an evolution of the spectral pattern from low to high dopant concentration. In particular, the strong IRAV band peaking at 1262 cm^{-1} in the sample at 2.7% dopant concentration shows an appreciable shift in its maximum at 11.1% concentration. The apparent band shift can be ascribed to the rise of a higher wavenumber band component; indeed, all the IRAV bands show structured band shapes resulting from more than one contribution from different ECC-like normal modes. Figure S12 illustrates in detail the evolution in intensity of the different band components in solutions. Very interestingly, at the highest dopant concentration (11.1%), the IRAVs pattern of the solution is almost superimposable to that of solid-state samples (see Figure S12), which instead is not affected at all by the degree of doping, as illustrated by Figure 9b, showing that the spectra of solid-state samples are indistinguishable if plotted in full scale.

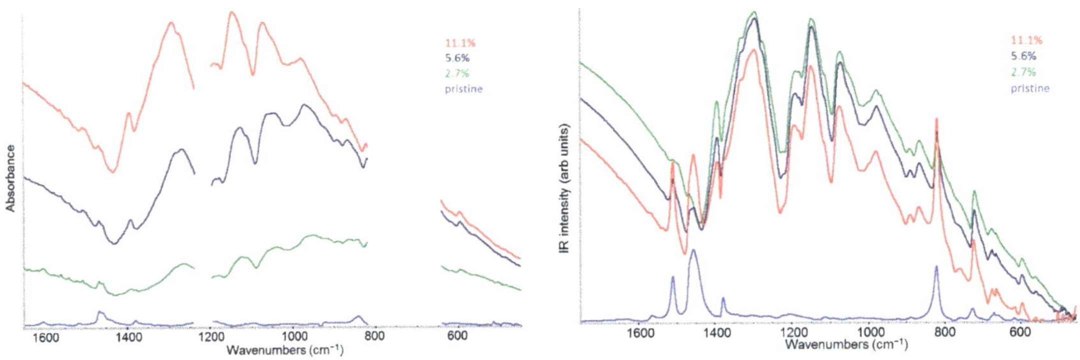

(a) (b)

Figure 9. Zoom in on the IRAVs region of the IR spectra of pristine P3HT (light blue line) and P3HT doped at 2.7% (green line), 5.6% (dark blue line), and 11.1% (red line) F4TCNQ/P3HT relative molar concentration. (**a**) Spectra of doped samples in solution. (**b**) Spectra of solid samples obtained from the same solutions of panel (**a**). Spectra of doped samples in panel (**b**) are displayed at full scale.

The modulation of the IRAVs pattern of the different solutions has to be ascribed to some peculiar structures that cannot be found in solid samples—or are totally shaded by the IR response of the more abundant phases present in the solid. In the lower doping regime, relatively few chains can be doped in solution, and polarons should be present mainly as diluted (or single) defects on isolated chains (Scheme 1b). Even if the hole injection leads to the planarization of a polymer segment, the lateral hexyl chains still take on the disordered conformation typical of solutions. At higher doping, long planar doped polymer segments carrying dense polarons sequences (Scheme 1c) are prone to forming small clusters consisting of a few P3HT chains that feel a local environment similar to that of doped chains in the solid state. The above picture suggests that the modulation of the IRAVs pattern in solution could be a probe of the onset of clustering phenomena. On the other hand, the IRAVs pattern is always the same in solid-state samples, and the simultaneous presence of different crystal structures (i.e., crystals hosting diluted polarons and co-crystals with different polaron density, possibly showing IL and/or Π-structures) has no appreciable effect on the IRAVs frequencies nor on their relative intensity. Unlike Raman, which is a powerful probe of the polaron environment, IRAVs are true polaron modes, mainly involving the highly perturbed charged polaron regions (red segment in Scheme 2). This results from the fact that only polarized CC bonds can develop strong IR transition dipoles and the strongly polarized regions share the same structural characteristic (geometry of charged thiophene rings) and the same vibrational dynamics.

In the light of the experimental findings for solids, the differences observed in the solutions can be rationalized considering that the formation of clusters of doped chains drastically reduces the conformational flexibility of the hexyl chains. ECC modes of the polarons are dynamically coupled to bending modes of the lateral chains, and a non-negligible effect of their disordered conformations on the IRAVs of isolated doped chains can be envisaged. It is interesting to notice that at 11.1%, doping also the behavior of the UV-vis absorption spectrum of the solution starts to show an evolution of the vibronic features, which indicates the occurrence of H-like interactions consistent with the presence of P3HT clusters.

A final observation derives from a joint analysis of IRAVs and RaAVs. In the doping regimes considered in this work, we have not obtained any evidence of the formation of charged defects other than polarons. The contributions to the Raman spectra of the doped domains are always given only by three Lorentzian components (at 1430, 1420, and 1410 cm^{-1}) both for solid samples and solutions and the IR spectra of the solid samples show no evolution from low to high dopant concentration. On the other hand, because of the different amount of charge transferred, the structure and the vibrational dynamics of a bipolaron should be significantly different from those of polarons, thus yielding distinct IR and Raman responses. We can thus conclude that it is reasonable to discard the hypothesis that bipolarons are formed, even in heavily doped samples.

3. Materials and Methods

Poly(3-hexylthiophene-2,5-diyl) (Mw = 74.000 g mol^{-1}, RR $\geq$ 97.3%) (P3HT) and 2,3,5,6-tetrafluoro-7,7,8,8-tetracyanoquinodimethane (sublimed grade) (F4TCNQ) were purchased from Ossila (Ossila Ltd., Solpro Business Park Windsor Street, Sheffield S4 7WB, UK). Chloroform ($CHCl_3$) was purchased from Merck (Merck KGaA, Frankfurter Straße 250, 64293 Darmstadt, Germany). All chemicals were used without any further purification.

Pristine P3HT and samples doped with F4TCNQ were prepared in chloroform solution with percent molar ratios (P3HT moles are referred to one 3-hexylthiophene unit) of 2.7%, 5.6%, and 11.1%, respectively. Each doped sample was obtained by adding to 1 mL of a 5 mg/mL P3HT/$CHCl_3$ solution the amount of a 0.25 mg/mL F4TCNQ/$CHCl_3$ needed to reach the desired molar ratio. The solutions are kept in glass vials preventively cleaned and heated at 350 °C to remove the adsorbed water. To obtain samples with the same P3HT concentration, chloroform was added to the vials, achieving the final volume of 2 mL.

FT-IR spectroscopy. The IR absorption spectra were recorded using a Nicolet 6700 FT-IR spectrometer (Thermo Fisher Scientific Inc., 168 Third Avenue, Waltham, MA, USA), 4 cm^{-1} resolution, 128 scans. The IR spectra of pristine and doped P3HT solutions were recorded using a sealed KBr windows IR cell for liquids (0.1 mm optical path); the spectra are plotted in the figures after removing saturated signals due to the strong absorptions of chloroform. Spectra of solid samples were recorded in transmission mode on thin films obtained by drop casting of the same solutions on a KBr window.

FT-Raman spectroscopy. Raman spectra of pristine and doped samples were recorded with a Nicolet NXR9600 FT-Raman spectrometer (Thermo Fisher Scientific Inc., 168 Third Avenue, Waltham, MA, USA), 1064 nm exciting laser line, 100 mW, 1024 scans at 4 cm^{-1} resolution in backscattering geometry. Solutions were analyzed in an NMR quartz tube (5 mm). Solid samples were recorded on the film formed on the inner walls of the NMR tube after $CHCl_3$ evaporation.

UV-vis-NIR absorption spectroscopy. Electronic absorption spectra were recorded with a Jasco V-570 (JASCO Corporation, 2967-5, Ishikawa-machi, Hachioji, Tokyo 192-8537, Japan) in the wavelength range 2500–200 nm (1 nm data pitch and 200 nm/min scan speed). Liquid samples were measured in a quartz cuvette (1 cm of optical path) by adding 3 µL of the prepared P3HT/F4TCNQ solutions to 2 mL of $CHCl_3$. Spectra of solid samples were recorded on drop-cast thin films on glass windows.

Spectra manipulation was performed with Omnic 8.2.0.387 software (Thermo Fisher Scientific Inc., 168 Third Avenue, Waltham, MA, USA), and spectral deconvolution was performed with the program fityk 1.3.1 [57,58].

4. Conclusions

A wide experimental campaign on doped P3HT samples in solution and in the solid state allowed highlighting the peculiarities of the different spectroscopic techniques in unfolding several different features of the intramolecular and intermolecular architectures that develop upon doping.

- The evidence of the presence of different structures in solutions is revealed by the evolution of the IR pattern depending on the level of doping. We observe (i) incipient IRAVs in the low doping case, i.e., dilute polarons on single P3HT chains; (ii) IRAVs of ordered doped domains (chains with a dense polaron population forming small aggregates in the solution) starting from 11.1% dopant concentration.
- In solid films obtained by casting from solutions, for all doping levels, the IR spectra converge to the same IRAVs pattern that is attributed to the normal modes of the polarons belonging to solid crystalline domains, irrespective of the polaron density along the chains. IRAVs do not distinguish co-crystals of kind П or IL nor domains with diluted polaron defects.
- The spectra of solutions suggest that there is an optimal dopant concentration (~10%) that facilitates the formation of pre-aggregates with ordered local structure already in solution.
- The existence of planar (doped) chains in solution is proven by the appearance of the vibronic structure in the UV-vis absorption spectra, which shows an evolution highlighting clustering phenomena.
- Unlike IRAV bands, which probe charged domains and are not sensitive to sample morphology, the Raman spectra are rich in information about the polaron environment. The analysis of the different components that give rise to the main RaAV band provides a description of the doped P3HT chains as made by domains of different kinds, characterized by thiophene rings featuring a different geometry and more or less polar bonds. While the evolution of the Raman spectra with increasing doping gives meaningful insights into the evolution of the sample morphology, a quantitative estimate of the relative abundance of the different domains is not affordable so far. Indeed, the intrinsic Raman activity of the ECC normal modes confined in different domains is different, and a reliable estimate of its value is not yet available. Experiments strongly corroborate the hypothesis that it drastically decreases with increasing CC bond polarity.
- The joint analysis of IRAVs and RaAV features at increasing doping concentration allows excluding the formation of bipolarons, at least in the doping regimes considered in this study.

Supplementary Materials: The following supporting information can be downloaded at https://www.mdpi.com/article/10.3390/molecules30071403/s1, Figure S1: Comparison of the FT-Raman spectra (λ_{exc} = 1064 nm) of P3HT in chloroform solution and as solid film; Figure S2: IR spectra of doped P3HT (11.1% F4TCNQ molar ratio): comparison of chloroform solution and solid sample; Figure S3: Comparison of the UV-vis-NIR absorption spectra of solution vs. solid samples of pristine and doped P3HT at different F4TCNQ molar ratios; Figures S4–S11: Fitting results of FT-Raman spectra of F4TCNQ-doped P3HT with different dopant/polymer molar ratios; Figure S12: Infrared spectra of pristine (green line) and doped P3HT samples in solution, with different dopant/polymer molar ratio.

Author Contributions: Conceptualization, C.C., L.B. and M.D.Z.; methodology, K.H., S.D. and L.B.; validation, K.H., C.C. and L.B.; formal analysis, C.C. and M.D.Z.; investigation, K.H., S.D. and L.B.; resources, C.C.; data curation, K.H. and L.B; writing—original draft preparation, C.C. and M.D.Z.; writing—review and editing, C.C., M.D.Z. and G.Z.; visualization, S.D. and L.B.; supervision, C.C. and G.Z.; funding acquisition, C.C. All authors have read and agreed to the published version of the manuscript.

Funding: This research received no external funding.

Institutional Review Board Statement: Not applicable.

Informed Consent Statement: Not applicable.

Data Availability Statement: The data presented in this study are (i) available in the article or the Supplementary Materials or (ii) available upon request from the corresponding author.

Acknowledgments: Hu Kaiyue acknowledges the financial support from the China Scholarship Council (CSC, No. 202006560017). C.C. acknowledges Carlo Saporiti for the useful discussion and for accurate reading, which contributed to the improvement of the manuscript.

Conflicts of Interest: The authors declare no conflicts of interest.

Abbreviations

The following abbreviations are used in this manuscript:

P3HT	poly(3-hexylthiophene-2,5-diyl)
F4TCNQ	2,3,5,6-tetrafluoro-7,7,8,8-tetracyanoquinodimethane
IRAV	IR Activated Vibrations
RaAV	Raman Activated Vibrations
ICT	Integer Charge Transfer Complex
CTC	Charge Transfer Complex
ECC	Effective Conjugation Coordinate

References

1. Ito, T.; Shirakawa, H.; Ikeda, S. Simultaneous polymerization and formation of polyacetylene film on the surface of concentrated soluble Ziegler-type catalyst solution. *J. Polym. Sci. Polym. Chem. Ed.* **1974**, *12*, 11. [CrossRef]
2. Bright, A.A.; Cohen, M.J.; Garito, A.F.; Heeger, A.J.; Mikulski, C.M.; MacDiarmid, A.G. Epitaxial crystalline films of the metallic polymer (SN)x. *Appl. Phys. Lett.* **1975**, *26*, 612. [CrossRef]
3. Shirakawa, H.; Louis, E.J.; MacDiarmid, A.G.; Chiang, C.K.; Heeger, A. Synthesis of electrically conducting organic polymers: Halogen derivatives of polyacetylene, (CH)x. *J. Chem. Commun.* **1977**, *16*, 578–580. [CrossRef]
4. Chiang, C.K.; Fincher, C.R., Jr.; Park, Y.W.; Heeger, A.J.; Shirakawa, H.; Louis, E. Electrical Conductivity in Doped Polyacetylene. *J. Phys. Rev. Lett.* **1977**, *39*, 1098. [CrossRef]
5. Skotheim, T.A. *Handbook of Conducting Polymers*; Dekker: New York, NY, USA, 1986.
6. Brédas, J.L.; Silbey, R. *Conjugated Polymers: The Novel Science and Technology of Highly Conducting and Nonlinear Optically Active Materials*; Springer: Dordrecht, The Netherlands, 1991.
7. Zhao, X.; Zhan, X. Electron transporting semiconducting polymers in organic electronics. *Chem. Soc. Rev.* **2011**, *40*, 3728–3743. [CrossRef]
8. Fratini, S.; Nikolka, M.; Salleo, A.; Schweicher, G.; Sirringhaus, H. Charge transport in high-mobility conjugated polymers and molecular semiconductors. *Nat. Mater.* **2020**, *19*, 491–502. [CrossRef]
9. Kim, T.; Yang, S.J.; Sung, S.J.; Kim, Y.S.; Chang, M.S.; Jung, H.; Park, C.R. Highly reproducible thermocontrolled electrospun fiber based organic photovoltaic devices. *ACS Appl. Mater. Interfaces* **2015**, *7*, 4481–4487. [CrossRef]
10. Zhao, W.; Ding, J.; Zou, Y.; Di, C.A.; Zhu, D. Chemical doping of organic semiconductors for thermoelectric applications. *Chem. Soc. Rev.* **2020**, *49*, 7210–7228. [CrossRef] [PubMed]
11. Lim, E.; Peterson, K.A.; Su, G.M.; Chabinyc, M.L. Thermoelectric Properties of Poly(3-hexylthiophene) (P3HT) Doped with 2,3,5,6-Tetrafluoro-7,7,8,8-tetracyanoquinodimethane (F4TCNQ) by Vapor-Phase Infiltration. *Chem. Mater.* **2018**, *30*, 998–1010. [CrossRef]

12. Lüssem, B.; Keum, C.M.; Kasemann, D.; Naab, B.; Bao, Z.; Leo, K. Doped Organic Transistors. *Chem. Rev.* **2016**, *116*, 13714–13751. [CrossRef]
13. Untilova, V.; Biskup, T.; Biniek, L.; Vijayakumar, V.; Brinkmann, M. Control of Chain Alignment and Crystallization Helps Enhance Charge Conductivities and Thermoelectric Power Factors in Sequentially Doped P3HT:F4TCNQ Films. *Macromolecules* **2020**, *53*, 2441–2453. [CrossRef]
14. Su, W.P.; Schrieffer, J.R.; Heeger, A. Solitons in Polyacetylene. *J. Phys. Rev. Lett.* **1979**, *42*, 1698.
15. Ehrenfreund, E.; Vardeny, Z.; Brafman, O.; Horovitz, B. Amplitude and phase modes in trans-polyacetylene: Resonant Raman scattering and induced infrared activity. *Phys. Rev. B* **1987**, *36*, 1535–1553.
16. Painelli, A.; Girlando, A.; Del Freo, L.; Soos, Z.G. Infrared intensity and local vibrations of charged solitons. *Phys. Rev. B* **1997**, *56*, 15100–15108.
17. Gussoni, M.; Castiglioni, C.; Zerbi, G. Vibrational spectroscopy of polyconjugated materials: Polyacetylene and polyenes. In *Spectroscopy of Advanced Materials*; Clark, R.J.H., Hester, R.E., Eds.; Wiley: New York, NY, USA, 1991; pp. 251–353.
18. Castiglioni, C.; Lopez Navarrete, J.T.; Zerbi, G.; Gussoni, M. A simple interpretation of the vibrational spectra of undoped, doped and photoexcited polyacetylene: Amplitude mode theory in the GF formalism. *Solid State Commun.* **1988**, *65*, 625–630.
19. Brinkmann, M. Structure and morphology control in thin films of regioregular poly(3-hexylthiophene). *J. Polym. Sci. B* **2011**, *49*, 1218–1233. [CrossRef]
20. Hamidi-Sakr, A.; Biniek, L.; Bantignies, J.-L.; Maurin, D.; Herrmann, L.; Leclerc, N.; Lévêque, P.; Vijayakumar, V.; Zimmermann, N.; Brinkmann, M. A Versatile Method to Fabricate Highly In-Plane Aligned Conducting Polymer Films with Anisotropic Charge Transport and Thermoelectric Properties: The Key Role of Alkyl Side Chain Layers on the Doping Mechanism. *Adv. Funct. Mater.* **2017**, *27*, 1700173. [CrossRef]
21. Lee, S.; Moon, G.D.; Jeong, U. Continuous production of uniform poly(3-hexylthiophene) (P3HT) nanofibers by electrospinning and their electrical properties. *J. Mater. Chem.* **2009**, *19*, 743–748.
22. Chen, J.Y.; Kuo, C.C.; Lai, C.S.; Chen, W.C.; Chen, H.L. Manipulation on the morphology and electrical properties of aligned electrospun nanofibers of poly(3-hexylthiophene) for field-effect transistor applications. *Macromolecules* **2011**, *44*, 2883–2892.
23. Watts, K.E.; Neelamraju, B.; Ratcliff, E.L.; Pemberton, J.E. Stability of Charge Transfer States in F4TCNQ-Doped P3HT. *Chem. Mater.* **2019**, *31*, 6986–6994.
24. Hase, H.; Berteau-Rainville, M.; Charoughchi, S.; Bodlos, W.; Orgiu, E.; Salzmann, I. Critical dopant concentrations govern integer and fractional charge-transfer phases in doped P3HT. *J. Phys. Mat.* **2022**, *6*, 014004.
25. Jacobs, I.E.; Cendra, C.; Harrelson, T.F.; Bedolla Valdez, Z.I.; Faller, R.; Salleo, A.; Moulé, A.J. Polymorphism controls the degree of charge transfer in a molecularly doped semiconducting polymer. *Mater. Horiz.* **2018**, *5*, 655–660.
26. Scholes, D.T.; Yee, P.Y.; Lindemuth, J.R.; Kang, H.; Onorato, J.; Ghosh, R.; Luscombe, C.K.; Spano, F.C.; Tolbert, S.H.; Schwartz, B.J. The Effects of Crystallinity on Charge Transport and the Structure of Sequentially Processed F4TCNQ-Doped Conjugated Polymer Films. *Adv. Funct. Mater.* **2017**, *27*, 1702654.
27. Nightingale, J.; Wade, J.; Moia, D.; Nelson, J.; Kim, J.S. Impact of Molecular Order on Polaron Formation in Conjugated Polymers. *J. Phys. Chem. C* **2018**, *122*, 29129–29140.
28. Brambilla, L.; Kim, J.-S.; Kim, B.J.; Hernandez, V.; Lopez Navarrete, J.T.; Zerbi, G. Poly(3-hexylthiophene-2.5-diyl): Evidence of different polymer chain conformations in the solid state from a combined study of regioregularity control and Raman spectroscopy. *J. Mol. Struct.* **2020**, *1221*, 128882.
29. Casalegno, M.; Famulari, A.; Meille, S.V. Modeling of Poly(3-hexylthiophene) and Its Oligomer's Structure and Thermal Behavior with Different Force Fields: Insights into the Phase Transitions of Semiconducting Polymers. *Macromolecules* **2022**, *55*, 2398–2412.
30. Stanfield, D.A.; Wu, Y.; Tolbert, S.H.; Schwartz, B.J. Controlling the Formation of Charge Transfer Complexes in Chemically Doped Semiconducting Polymers. *Chem. Mater.* **2021**, *33*, 2343–2356.
31. Pingel, P.; Zhu, L.; Park, K.S.; Vogel, J.O.; Eung-Gun Kim, S.J.; Rabe, J.P.; Brédas, J.-L.; Koch, N. Charge-Transfer Localization in Molecularly Doped Thiophene-Based Donor Polymers. *J. Phys. Chem. Lett.* **2010**, *1*, 2037–2041.
32. Zhu, L.; Kim, E.-G.; Yi, Y.; Brédas, J.-L. Charge Transfer in Molecular Complexes with 2,3,5,6-Tetrafluoro-7,7,8,8-tetracyanoquinodimethane (F4-TCNQ): A Density Functional Theory Study. *Chem. Mater.* **2011**, *23*, 5149–5159.
33. Duong, D.T.; Wang, C.; Antono, E.; Toney, M.F.; Salleo, A. The chemical and structural origin of efficient p-type doping in P3HT. *Org Electron.* **2013**, *14*, 1330–1336.
34. Müller, L.; Nanova, D.; Glaser, T.; Beck, S.; Pucci, A.; Kast, A.K.; Schröder, R.R.; Mankel, E.; Pingel, P.; Neher, D.; et al. Charge-Transfer-Solvent Interaction Predefines Doping Efficiency in p-Doped P3HT Films. *Chem. Mater.* **2016**, *28*, 4432–4439. [CrossRef]
35. Gao, J.; Roehling, J.D.; Li, Y.; Guo, H.; Moulé, A.J.; Grey, J.K. The effect of 2,3,5,6-tetrafluoro-7,7,8,8-tetracyanoquinodimethane charge transfer dopants on the conformation and aggregation of poly(3-hexylthiophene). *J. Mater. Chem. C* **2013**, *1*, 5638–5646. [CrossRef]

36. Gao, J.; Niles, E.T.; Grey, J.K. Aggregates promote efficient charge transfer doping of Poly(3- hexylthiophene). *J. Phys. Chem. Lett* **2013**, *4*, 2953–2957. [CrossRef] [PubMed]
37. Tashiro, K.; Kobayashi, M.; Kawai, T.; Yoshino, K. Crystal structural change in poly(3-alkyl thiophene)s induced by iodine doping as studied by an organized combination of X-ray diffraction, infrared/Raman spectroscopy and computer simulation techniques *Polymer* **1997**, *38*, 2867–2879. [CrossRef]
38. Castiglioni, C.; Tommasini, M. Infrared active vibrations in doped π-conjugated materials: The mechanism of activation of Raman modes. *Asian J. Phys.* **2022**, *31*, 211–225.
39. Castiglioni, C.; Tommasini, M.; Zerbi, G. Raman spectroscopy of polyconjugated molecules and materials: Confinement effect in one and two dimensions. *Philos. Trans. R. Soc. Lond. A* **2004**, *362*, 2425–2459. [CrossRef]
40. Lopez Navarrete, J.T.; Zerbi, G. Lattice dynamics and vibrational spectra of polythiophene. II. Effective coordinate theory, doping induced, and photoexcited spectra. *J. Chem. Phys.* **1998**, *94*, 965. [CrossRef]
41. Saporiti, C.; Brambilla, L.; Fazzi, D.; Castiglioni, C. Insights into the Structural and Vibrational Properties of Polaron in Doped Poly(3-alkyl-thiophene), P3HT. *J. Phys. Chem. C.* **2024**, *128*, 5189–5205. [CrossRef]
42. Jacobs, I.E.; Aasen, E.W.; Oliveira, J.L.; Fonseca, T.N.; Roehling, J.D.; Li, J.; Zhang, G.; Augustine, M.P.; Mascal, M.; Moulé, A.J Comparison of solution-mixed and sequentially processed P3HT:F4TCNQ films: Effect of doping-induced aggregation on film morphology. *J. Mater. Chem. C* **2016**, *4*, 3454–3466. [CrossRef]
43. Stanfield, D.A.; Mehmedović, Z.; Schwartz, B.J. Vibrational Stark Effect Mapping of Polaron Delocalization in Chemically Doped Conjugated Polymers. *Chem. Mater.* **2021**, *33*, 8489–8500. [CrossRef]
44. Ghosh, R.; Chew, A.R.; Onorato, J.; Pakhnyuk, V.; Luscombe, C.K.; Salleo, A.; Spano, F.C. Spectral Signatures and Spatial Coherence of Bound and Unbound Polarons in P3HT Films: Theory Versus Experiment. *J. Phys. Chem. C* **2018**, *122*, 18048–18060 [CrossRef]
45. Wu, E.C.; Salamat, C.Z.; Ruiz, O.L.; Qu, T.; Kim, A.; Tolbert, S.H.; Schwartz, B.J. Counterion Control and the Spectral Signatures of Polarons, Coupled Polarons, and Bipolarons in Doped P3HT Films. *Adv. Funct. Mater.* **2023**, *33*, 2213652. [CrossRef]
46. Enengl, C.; Enengl, S.; Pluczyk, S.; Havlicek, M.; Lapkowski, M.; Neugebauer, H.; Ehrenfreund, E. Doping-Induced Absorption Bands in P3HT: Polarons and Bipolarons. *ChemPhysChem* **2016**, *17*, 3836–3844. [CrossRef]
47. Nayyar, I.H.; Batista, E.R.; Tretiak, S.; Saxena, A.; Smith, D.L.; Martin, R.L. Role of Geometric Distortion and Polarization in Localizing Electronic Excitations in Conjugated Polymers. *J. Chem. Theory Comput.* **2013**, *9*, 1144–1154. [CrossRef] [PubMed]
48. Mosca, S.; Milani, A.; Castiglioni, C.; Hernández Jolín, V.; Meseguer, C.; López Navarrete, J.T.; Zhao, C.; Sugiyasu, K.; Ruiz Delgado, M.C. Raman Fingerprints of Pi-Electron Delocalization in Polythiophene-Based Insulated Molecular Wires. *Macromolecules* **2022**, *55*, 3458–3468. [CrossRef]
49. Saporiti, C.; Brambilla, L.; Tommasini, M.; Del Zoppo, M.; Castiglioni, C.; Zerbi, G. Probing the Charge State and the Intermolecular Environment by Vibrational Spectroscopy: The Peculiar Modulation of Frequencies and Band Intensities of F4TCNQ and Its Anion. *Spectrosc. J.* **2024**, *2*, 264–286. [CrossRef]
50. Fano, U. Effects of configuration interaction on intensities and phase shifts. *Phys. Rev.* **1961**, *124*, 1866–1878. [CrossRef]
51. Tek, G.; Hamm, P. A Correction Scheme for Fano Line Shapes in Two-Dimensional Infrared Spectroscopy. *J. Phys. Chem. Lett.* **2020**, *11*, 6185–6190. [CrossRef]
52. Spano, F.C.; Silva, C. H- and J-Aggregate Behavior in Polymeric Semiconductors. *Annu. Rev. Phys. Chem.* **2014**, *65*, 477–500. [CrossRef]
53. Mansour, A.E.; Valencia, A.M.; Lungwitz, D.; Wegner, B.; Tanaka, N.; Shoji, Y.; Fukushima, T.; Opitz, A.; Cocchi, C.; Koch, N. Understanding the evolution of the Raman spectra of molecularly p-doped poly(3-hexylthiophene-2,5-diyl): Signatures of polarons and bipolarons. *Phys. Chem. Chem. Phys.* **2022**, *24*, 3109–3118. [CrossRef]
54. Arrigoni, A.; Brambilla, L.; Bertarelli, C.; Saporiti, C.; Castiglioni, C. Conducting Electrospun Poly(3-hexylthiophene-2,5-diyl) Nanofibers: New Strategies for Effective Chemical Doping and its Assessment Using Infrared Spectroscopy. *Appl. Spectrosc.* **2024**, *78*, 1279–1294. [CrossRef] [PubMed]
55. Yin, J.; Wang, Z.; Fazzi, D.; Shen, Z.; Soci, C. First-Principles Study of the Nuclear Dynamics of Doped Conjugated Polymers. *J. Phys. Chem. C* **2016**, *120*, 1994–2001. [CrossRef]
56. Wang, S.; Fazzi, D.; Puttisong, Y.; Jafari, M.J.; Chen, Z.; Ederth, T.; Andreasen, J.W.; Chen, W.M.; Facchetti, A.; Fabiano, S. Effect of Backbone Regiochemistry on Conductivity, Charge Density, and Polaron Structure of n-Doped Donor-Acceptor Polymers. *Chem. Mater.* **2019**, *31*, 3395–3406. [CrossRef] [PubMed]
57. Fityk—Curve Fitting and Data Analysis. Available online: https://fityk.nieto.pl/ (accessed on 10 February 2025).
58. Wojdyr, M. Fityk: A general-purpose peak fitting program. *J. Appl. Cryst.* **2010**, *43*, 1126–1128. [CrossRef]

molecules

MDPI

Article

Reversal in Solvatochromism, enol-imine/keto-amine Tautomerism and (*E*)-(*Z*) Photoisomerizations in a Benzylidene Aniline Schiff Base Derivative in Different Solvents

İsa Sıdır [1,2,*], Yadigar Gülseven Sıdır [1,2], Halil Berber [3], Maria L. Ramos [2], Licínia L. G. Justino [2] and Rui Fausto [2,4,*]

1 Department of Physics, Faculty of Sciences and Letters, Bitlis Eren University, Bitlis 13000, Türkiye; ygsidir@beu.edu.tr
2 CQC–IMS, Department of Chemistry, University of Coimbra, 3004-535 Coimbra, Portugal; mlramos@ci.uc.pt (M.L.R.); liciniaj@ci.uc.pt (L.L.G.J.)
3 Department of Chemistry, Faculty of Sciences, Eskişehir Technical University, Eskişehir 26470, Türkiye; hlberber@eskisehir.edu.tr
4 ERA-Chair Spectroscopy@IKU, Faculty of Sciences and Letters, Department of Physics, Istanbul Kultur University, Ataköy Campus, Bakirköy, Istanbul 34156, Türkiye
* Correspondence: isidir@beu.edu.tr (İ.S.); rfausto@ci.uc.pt (R.F.)

Abstract: A novel Schiff base, (*E*)-4-acetyl-*N*-(4-hydroxy-3-methoxybenzylidene)aniline (abbreviated as ANHMA), was synthesized and characterized using infrared and ^{1}H- and ^{13}C-NMR spectroscopies. Optical properties in different solvents were evaluated using UV-vis absorption spectroscopy. The compound is shown to exhibit both positive and negative solvatochromism with reversal occurring for solvents with $E_T(30)$~45 (e.g., DMSO). The solvatochromic behavior of the compound was found to be strongly dependent on the hydrogen bond abilities and polarizability of the solvent, the observed reversal in solvatochromism being explained by the change in the dominant solvent effects in non-polar and polar–aprotic solvents (H-bond acceptor ability of the solvent and polarizability) compared to polar–protic solvents (H-bond donor ability), according to the developed Catalán multiparametric solvatochromic model. In all freshly prepared solutions studied, the (*E*)-enol-imine tautomer of the compound was found to strongly predominate over the keto-amine tautomeric forms, the latter increasing their populations over time in the presence of H-bond donor/acceptor species. Irradiation of ANHMA with UV light ($\lambda \geq 311$ nm) was also investigated in several solvents and shown to follow a general pattern, with the conversion of the (*E*)-enol-imine tautomer into the keto-amine forms in a solvent-mediated enol-imine/keto-amine tautomerism, and (*Z*)→(*E*) C=C isomerization between the keto-imine forms. The experimental results received support from an extensive series of calculations on the structure and UV-vis spectra of the relevant tautomeric/isomeric forms of the compound performed at the DFT(B3LYP)/6-311++G(d,p) level of approximation (including time-dependent DFT calculations and solvent consideration).

Keywords: Solvatochromism; isomerism; tautomerism; hydrogen bonding; benzylidene aniline Schiff base-type imine dye

1. Introduction

Solvent effects play a major role in chemistry, affecting kinetics, equilibria, and the reaction mechanisms and determining solubility and acid–base and redox strengths for example; they also affect physical properties of the solutes as, for instance, spectroscopic responses [1–3]. Solute–solvent interactions can be of different types and determined both by the physical and chemical properties of the solvents, such as their hydrogen bond donor/acceptor abilities, polarizability, polarity, and dipolarity. The combination of these types of interactions expresses the overall solvent effect on a given property of the solute as solubility or optical properties [1–4]. In the latter case, for example, it is well known that the position and/or intensity of the bands observed in the absorption and fluorescence electronic spectra of a dye can be remarkably influenced by solvent properties. In the simplest cases, such solvent-induced changes in the characteristics of the spectra of the dye reflect the perturbation imposed by the solvent to the electronic energy levels of the dye, and result in the variation of the color of the solutions of the same solute in different solvents, or solvatochromism [1–3] (The term solvatochromism does not strictly apply to solvent effects on the spectral properties of the solute in the visible region, so that the word "color" is here used lato sensu). To understand the solvatochromic behavior of a molecule, solvatochromic models have been developed, which describe the properties of the solvents in terms of a series of empirically determined parameters [1–3,5–9]. In this realm, solvent polarity scales are of fundamental importance, and many solvent polarity scales have been proposed, among which the $E_T(30)$ scale, based on Reichardt's pyridinium *N*-phenolate betaine dye [1,10–14], is the most used.

The Reichardt's $E_T(30)$ solvent polarity scale expresses the polarity of a given solvent in terms of the energy of the maximum of the band corresponding to the HOMO → LUMO charge-transfer electronic absorption of the *N*-phenolate betaine dye in that solvent (Equation (1), where λ_{max} is the wavelength of maximum of the absorption band of the dye) [1]. Other dyes can be used to assess the polarity of the solvents using a similar approach and an equation similar to Equation (1), but where $E_T(30)$ is replaced by E_T(dye) [5,12–15].

$$E_T(30)\ (\text{kcal mol}^{-1}) = 28{,}591.5/(\lambda_{max},\ \text{nm}) \tag{1}$$

Solvatochromic organic dyes are important functional materials for a large number of practical applications [16,17]. They present the advantage over other systems in that their solvatochromic properties can be tuned for a given application by tailoring appropriate π-conjugated electron-acceptor/electron-donor systems through the design of molecules, which bear different substituents in a basic molecular scaffold. By changing judiciously the molecular architecture, a family of dyes may comprehend members displaying negative as well as positive solvatochromism, i.e., exhibiting bathochromic (red) or hypsochromic (blue) shifts in the absorption (or fluorescence) maxima with increasing solvent polarity, respectively. Some species can show a reversal in solvatochromism, where red and blue shifts are observed for the same molecule in different regions of solvent polarity. Reversals in solvatochromism have been described over the past four decades, in particular for imine dyes [13,18,19], and have been explained on the basis of different factors: (i) protonation/deprotonation and solvent-assisted prototropism [11,19,20], (ii) differential intermolecular hydrogen bonding with solvent [21], (iii) changes in molecular conformation [22,23], (iv) self-aggregation [24], and (v) variable solvent polarity effects on electronic ground and excited states of the solute [12,13,18,25].

Schiff bases derived from salicylic aldehydes and benzylidene aniline-type imine dyes have been given special attention in relation to solvatochromism [9,26–31]. This is mostly because these types of compounds possess specific structural characteristics that can be easily manipulated to fine-tune their properties. In the case of benzylidene aniline-type imine dyes, both intramolecular proton transfer leading to enol-imine/keto-amine tautomerism and (*E*)-(*Z*) isomerization around the imino group (C=N) have been found to be relevant in determining solvatochromism [32–39], and these processes are well known to be strongly dependent on the specific substituents present in the aromatic rings. On these topics, several comprehensive reviews have been published, including those by Minkin and coworkers [38], Flores-Leonar et al. [40], Rahimova [41], and Brewer and coworkers [42]. Because the relevance of the (*E*)-(*Z*) isomerism, which in these compounds can in general be efficiently trigged by light, benzylidene aniline-type imine dyes have also been proposed as molecular switches [43,44] (mimicking the naturally occurring (*E*)-(*Z*) photoisomerization of retinal [45,46]). Due to the sensitivity of the optical properties of these dyes to the solvents' characteristics and their influence on the relative importance of the different tautomeric forms of the compound, solvents play a critical role in the photochemistry of benzylidene aniline-type imine dyes and, then, in their possible applications as molecular switches and as photochromic sensors, for example.

In this study, the solvatochromic properties exhibited by a novel benzylidene aniline-type imine dye, the Schiff base derived from 4-acetylaniline and 4-hydroxy-3-methoxybenzaldehyde, (*E*)-4-acetyl-*N*-(4-hydroxy-3-methoxybenzylidene)aniline (abbreviated as ANHMA), was investigated in 19 solvents. The compound was synthesized and characterized using infrared and ^{1}H- and ^{13}C-NMR spectroscopies, and its solvatochromism was investigated in the light of the Catalán multiparametric empirical model [1,6]. As shown below, ANHMA was found to show a reversal in solvatochromism for a solvent $E_T(30)$ ~45 kcal mol^{-1}, which is a consequence of the different dominant solvent effects in non-polar and polar–aprotic solvents (H-bond acceptor ability of the solvent and polarizability) compared to polar–protic solvents (H-bond donor ability), as shown by the developed Catalán multiparametric solvatochromic model.

Furthermore, (*E*)-(*Z*) photoisomerizations about the C=C bond and enol-imine/keto-amine phototautomerization of ANHMA upon UV light ($\lambda \geq 311$ nm) irradiation were also investigated in several solvents, using UV-vis spectroscopy complemented by density functional theory (DFT) calculations (including time-dependent DFT (TD-DFT) calculations and solvent consideration), demonstrating the sensitivity of the photochemical reactivity of these dyes to the solvents' characteristics.

2. Results and Discussion

2.1. Reversal in Solvatochromism of ANHMA

The studied compound is a white dust material that gives rise to colorless solutions. Figure 1 shows the room temperature UV-vis absorption spectra of ANHMA in solutions of three types of solvents: non-polar, polar–aprotic, and polar–protic. Above 250 nm, the spectrum of the compound exhibits two clearly discernible strong bands, with maxima in the ca. 310–345 nm range (λ_{max}^{II}; most intense band II) and around 285–300 nm (λ_{max}^{I}), in all studied solvents except in the cases of chlorobenzene and tetrahydrofuran (THF) where the shorter wavelength band is overlapped by an intense solvent band (in benzene and in toluene solvent bands also overlap this band extensively, but it is still noticeable).

As discussed in detail in Section 2.3, these bands are ascribed to the (*E*)-enol-imine tautomer of the studied compound, which was found to strongly predominate in all investigated solvents over the keto-amine forms. Table 1 lists the maximum absorption wavelengths of the two bands in the different solvents and the corresponding E_T(ANHMA) values.

In the studied non-polar solvents (solvents 1 to 6; see Table 1) λ_{max}^{I} and λ_{max}^{II} are observed between 285 and 291 nm and between 326 and 335 nm, respectively, with a trend to exhibit a red shift with the increase of Reichardt $E_T(30)$ solvent parameter that is more noticeable for λ_{max}^{II}, which globally shows a stronger dependence on the solvent.

For the majority of the polar–aprotic solvents investigated (7–13), λ_{max}^{II} is observed nearly in the same range as in the case of non-polar solvents, 333–339 nm, exhibiting larger red shifts (to 342 and 344 nm) for the two polar–aprotic solvents with the highest $E_T(30)$ values, dimethylformamide (DMF) and dimethylsulfoxide (DMSO), respectively. Taking as reference the shortest wavelength observed in *n*-hexane solution (326 nm), a bathochromic shift of 18 nm in the $\lambda_{max}{}^{II}$ maximum of the most intense ANHMA absorption band is observed for DMSO. On the other hand, in all polar–aprotic solvents $\lambda_{max}{}^{I}$ is observed within the range 293–299 nm, which shows a general red shift compared to the region where $\lambda_{max}{}^{I}$ is observed in non-polar solvents (285–291 nm), the longer wavelength being observed in DMSO, following the pattern also seen for $\lambda_{max}{}^{II}$.

Table 1. Reichardt's solvent parameter ($E_T(30)$, kcal mol^{-1}) for the used solvents, maximum UV absorption wavelengths (λ_{max}, nm) of ANHMA in the different solutions, and E_T solvent parameter obtained from the spectroscopic data of ANHMA (E_T(ANHMA), kcal mol^{-1}).

N	Solvent	$E_T(30)$	$\lambda_{max}{}^{I}$	$E_T(ANHMA)^{I}$	$\lambda_{max}{}^{II}$	$E_T(ANHMA)^{II}$
1	cyclohexane	30.9	288	99.3	329	86.9
2	*n*-hexane	31.0	285	100.3	326	87.7
3	toluene	33.9	288	99.3	335	85.3
4	benzene	34.3	288	99.3	335	85.3
5	diethyl ether	34.5	291	98.3	334	85.6
6	1,4-dioxane	36.0	288	99.3	335	85.3
7	chlorobenzene	36.8			336	85.1
8	THF	37.4			339	84.3
9	ethyl acetate	38.1	295	96.9	334	85.6
10	*n*-butyl acetate	38.5	297	96.3	335	85.3
11	chloroform	39.1	293	97.6	335	85.3
12	DMF	43.2	295	96.9	342	83.6
13	DMSO	45.1	299	95.6	344	83.1
14	1-octanol	48.1	292	97.9	342	83.6
15	2-propanol	48.4	290	98.6	337	84.8
16	1-butanol	49.7	291	98.3	337	84.8
17	ethanol	51.9	288	99.3	317	90.2
18	methanol	55.4	288	99.3	317	90.2
19	EG	56.3	287	99.6	314	91.1

E_T(ANHMA) (kcal mol^{-1}) = $hcN_A/(\lambda_{max}$, nm) = 28,591.5/(λ_{max}, nm).

In the case of polar–protic solvents (14–19), $\lambda_{max}{}^{II}$ is within the 337–342 nm range for 1-octanol, 2-propanol, and 1-butanol. The solvents in this group have smaller $E_T(30)$, which do not differ much from that observed in non-polar and polar–aprotic solvents, and shift to shorter wavelengths (314–317 nm) in the more polar solvents, ethanol, methanol, and ethylene glycol (EG). A maximum blue shift of 28 nm is observed when comparing the position of $\lambda_{max}{}^{II}$ in EG with that observed in 1-octanol (see Table 1). Though the differences in the $\lambda_{max}{}^{I}$ values are comparatively smaller than those found for $\lambda_{max}{}^{II}$, a similar general trend is observed for the values of $\lambda_{max}{}^{I}$ in polar–protic solvents: it is observed in the 290–292 nm range for the solvents with smaller $E_T(30)$ in this group and within the 287–288 nm interval for those with a higher $E_T(30)$, the maximum blue shift of 5 nm being observed between the position of $\lambda_{max}{}^{I}$ in EG and in 1-octanol.

The E_T(ANHMA) values obtained from the UV absorption spectra in the different solvents are plotted as a function of the solvents' $E_T(30)$ parameter in Figure 2. This plot highlights the reversal in solvatochromism exhibited by the dye. In going from the non-polar solvent cyclohexane ($E_T(30)$ = 30.9 kcal mol^{-1}) to the polar–aprotic solvent DMSO ($E_T(30)$ = 45.1 kcal mol^{-1}), the $E_T(ANHMA)^{II}$ values decrease from 86.9 to 83.1 kcal mol^{-1} and the $E_T(ANHMA)^{I}$ values from ca. 100 to 95.6 kcal mol^{-1} with the increase of the polarity of the solvent (positive solvatochromism). On the other hand, in going from DMSO to the polar–protic solvent with higher $E_T(30)$, EG ($E_T(30)$ = 56.3 kcal mol^{-1}), the $E_T(ANHMA)^{II}$ and $E_T(ANHMA)^{I}$ values increase from 83.1 to 91.1 kcal mol^{-1} and from 95.6 to 99.6 kcal mol^{-1}, respectively (negative solvatochromism). Thus, A*N*HMA exhibits a general positive solvatochromism for solvents with $E_T(30)$ values in the ~34–45 kcal mol^{-1} range and negative solvatochromism for solvents with $E_T(30)$ values higher than ~45 kcal mol^{-1}, the reversal occurring at an $E_T(30)$ value of *ca.* 45 kcal mol^{-1}, which is characteristic of DMSO.

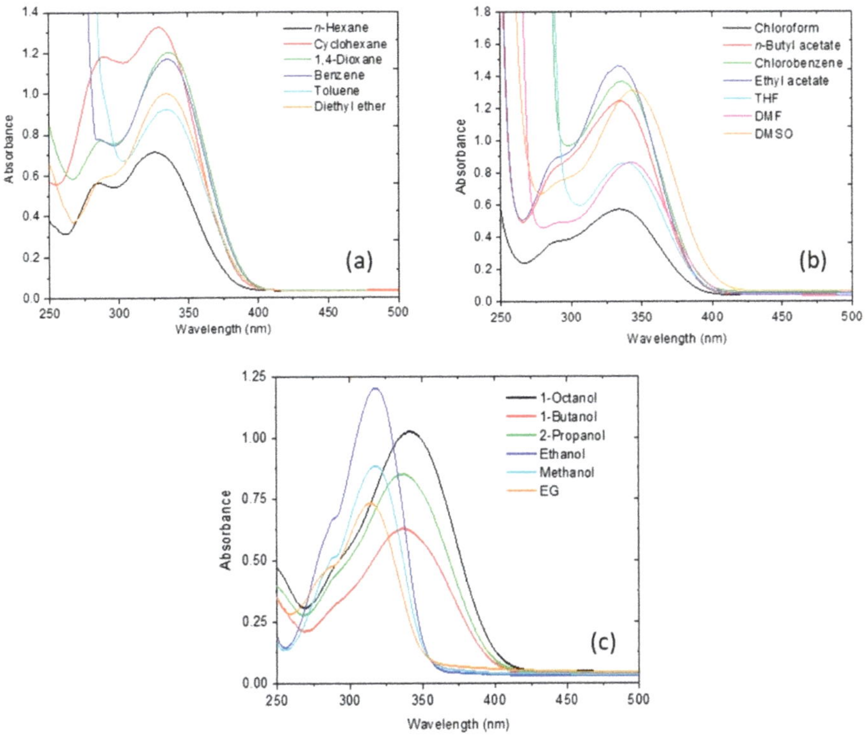

Figure 1. UV-vis absorption spectra of A*N*HMA in non-polar (**a**), polar–aprotic (**b**), and polar–protic (**c**) solvents. THF: Tetrahydrofuran, DMF: *N*,*N*-Dimethylformamide, DMSO: Dimethyl sulfoxide, EG: Ethylene glycol.

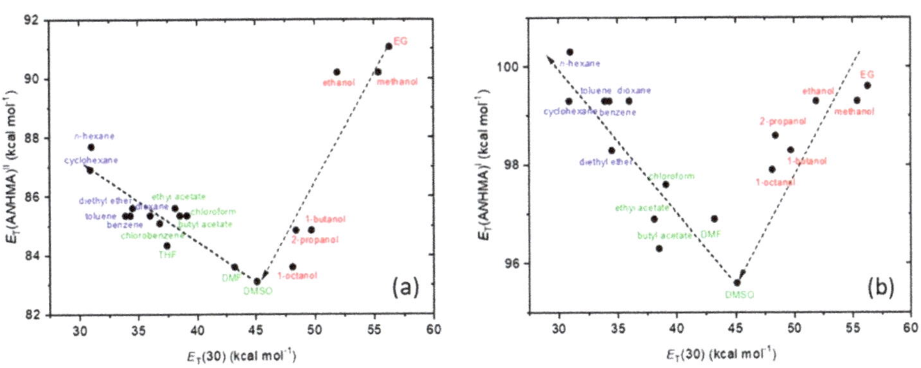

Figure 2. Plot of E_T(A*N*HMA)^{II} (**a**) and E_T(A*N*HMA)^{I} (**b**) values obtained from $\lambda_{max}{}^{II}$ and $\lambda_{max}{}^{I}$ vs. Reichardt's $E_T(30)$ solvents' parameter. Color code: blue, non-polar solvents; green, polar–aprotic solvents; red, polar–protic solvents. The arrows shall be taken as guidelines.

2.2. Solvatochromism in ANHMA at the Light of the Catalán Multiparametric Model

The Catalán approach to solvatochromism is based on a multiparametric empirical relation, as given by Equation (2) [1,6]:

$$E_T(\text{dye}) = E_T(\text{dye})_0 + a(\text{SA}) + b(\text{SB}) + d(\text{SP}) + e(\text{SdP}) \quad (2)$$

where $E_T(\text{dye})_0$ is a parameter representing the $E_T(\text{dye})$ value for a solvent-unaffected dye, and a, b, d, and e are the coefficients that reflect the contribution of the different solvent parameters to the $E_T(\text{dye})$ values determined from the observed band shifts in the different solvents. The parameters SA, SB, SP, and SdP relate to the hydrogen bond donor and hydrogen bond acceptor abilities, polarizability, and dipolarity of the solvent, respectively, and are provided in Table 2 for the relevant solvents [1,6].

Table 2. Reichardt's $E_T(30)$ (kcal mol^{-1}) and Catalán solvent parameters for the used solvents [1,6].

		E_T(30)	Catalán Parameters			
N	**Solvent**		**SA**	**SB**	**SP**	**SdP**
1	cyclohexane	30.9	0.000	0.073	0.683	0.000
2	*n*-hexane	31.0	0.000	0.056	0.616	0.000
3	toluene	33.9	0.000	0.128	0.782	0.284
4	benzene	34.3	0.000	0.124	0.793	0.270
5	diethyl ether	34.5	0.000	0.562	0.617	0.385
6	1,4-dioxane	36.0	0.000	0.444	0.737	0.312
7	chlorobenzene	36.8	0.000	0.182	0.833	0.537
8	THF	37.4	0.000	0.591	0.714	0.634
9	ethyl acetate	38.1	0.000	0.542	0.656	0.603
10	*n*-butyl acetate	38.5	0.000	0.525	0.674	0.535
11	chloroform	39.1	0.047	0.071	0.783	0.614
12	DMF	43.2	0.031	0.613	0.759	0.977
13	DMSO	45.1	0.072	0.647	0.830	1.000
14	1-octanol	48.1	0.299	0.923	0.713	0.454
15	2-propanol	48.4	0.283	0.830	0.633	0.808
16	1-butanol	49.7	0.341	0.809	0.674	0.655
17	ethanol	51.9	0.400	0.658	0.633	0.783
18	methanol	55.4	0.605	0.545	0.608	0.904
19	EG	56.3	0.717	0.534	0.777	0.910

Multiple linear regression using Equation (2), applied to the $E_T(\text{ANHMA})^{\text{II}}$ and $E_T(\text{ANHMA})^{\text{I}}$ experimental data, yields the following solutions:

$$E_T(\text{ANHMA})^{\text{II}} = 96.2(\pm 2.8) + 8.7(\pm 1.3)\text{SA} - 6.2(\pm 1.3)\text{SB} - 12.9(\pm 3.9)\text{SP} + 0.9(\pm 1.3)\text{SdP} \quad (3)$$

$$E_T(\text{ANHMA})^{\text{I}} = 101.5(\pm 2.1) + 4.7(\pm 1.0)\text{SA} - 1.7(\pm 0.9)\text{SB} - 2.1(\pm 2.9)\text{SP} - 2.9(\pm 0.9)\text{SdP} \quad (4)$$

where the correlation coefficient and the statistical parameter (R^2, F) equal to (0.848, 19.5) and (0.756, 9.3), respectively.

In both Equations (3) and (4), coefficient a is positive, while b and d are negative; coefficient e is positive in the model derived from $E_T(\text{ANHMA})^{\text{II}}$ (Equation (3), which refers to band II) and negative in the model derived from $E_T(\text{ANHMA})^{\text{I}}$ (Equation (4)). For band II, the absolute values of the coefficients follow the order $d > a > b >> e$, while for band I these values relate like $a > e \approx d \approx b$. It can then be concluded that the changes in the $E_T(\text{ANHMA})$ values in going from the non-polar to the polar–aprotic solvents, corresponding to bathochromic shifts in the maximum of the observed bands (associated with a reduction of the $E_T(\text{ANHMA})$ values), resulting from a higher polarizability and

H-bond acceptor ability of the solvent, while those observed in going from the polar–aprotic to the polar–protic solvents (corresponding to hypsochromic shifts, associated with an increase of the E_T(ANHMA) values) are dominated by the H-bond donor ability of the solvent. Overall, these changes in the dominant solvent effects in non-polar and polar–aprotic solvents on one side and in a polar–protic solvent, on the other side, explain the reversal in solvatochromism observed for the studied compound. The dipolarity term has a residual relevance in the case of band II, being relatively more important in the case of band I. The dipolarity term is more relevant for the polar solvents (see Table 2) and since, in this case, the coefficient *e* is negative, this term contributes the most to reducing the $E_T(\text{ANHMA})^{\text{I}}$ values of polar (both aprotic and protic) solvents and, thus, to distinguish the effect of these solvents on the position of the maximum of band I from that of the non-polar solvents.

It is also interesting to mention that wavelengths corresponding to $E_T(\text{ANHMA})_0{}^{\text{II}}$ and $E_T(\text{ANHMA})_0{}^{\text{I}}$ values in Equations (3) and (4) are 297.2 (±10.2) and 281.7 (±2.1) nm, respectively. These values are hypsochromically shifted in comparison with the observed position of the bands in all solvents and are then in consonance with the predicted relevance of the effect due to the polarizability of the solvent in all types of solvents. Indeed, the coefficient *d* affecting the term of polarizability of the solvent in the Catalán equation, has the largest absolute value among all coefficients in the model developed based on the band II data (Equation (3)), being also important, according to Equation (4), as a contributor to shift hypsochromically band I even for the most non-polar solvents, cyclohexane and *n*-hexane, for which the Catalán dipolarity and H-bond donor parameters, SdP and SA, are zero, and the H-bond acceptor parameter SB is also very small.

Figure 3 shows plots of the E_T(ANHMA) values calculated from the developed Catalán models vs. experimental values, which highlight the fairly good general predictive ability of the models but also show that specific solute–solvent interactions are also relevant for several of the solvents, which limit in some extent the goodness of the fits, as revealed by the R^2 values of the fittings, in particular for band I whose maximum of absorption varies the least.

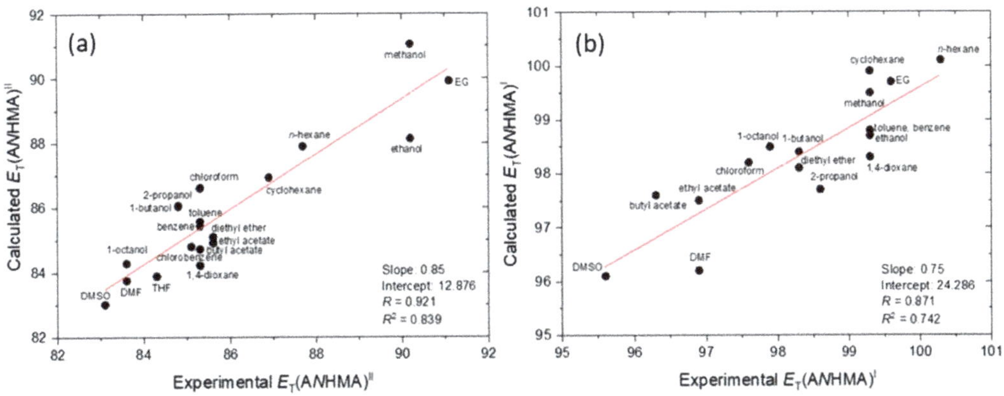

Figure 3. Plots of calculated $E_T(\text{ANHMA})^{\text{II}}$ (**a**) and $E_T(\text{ANHMA})^{\text{I}}$ (**b**) using the developed Catalán models as a function of the corresponding E_T(ANHMA) values.

2.3. DFT and TD-DFT Calculations: Structures and Spectra of Isomeric Species of ANHMA

ANHMA has four major basic isomeric forms, which are depicted schematically in Scheme 1: two enol-imine forms with (*E*) and (*Z*) configuration about the imine C=N bond, and two keto-amine forms with (*E*) and (*Z*) arrangements around the exocyclic C=C bond. All these species have several conformers that result from the different possible orientations of the OH, OCH_3, $C(=O)CH_3$ substituents and of the whole hydroxy-methoxyphenyl and acetylphenyl fragments, in the case of the enol-imine forms, and from the different orientations of the OCH_3 and $C(=O)CH_3$ substituents and of the whole acetylphenyl and acetylanilino fragments, in the case of the keto-amine forms. The DFT(B3LYP)/6-311++G(d,p) structures of the conformers of the different forms are provided in the Supporting Information Figures S1–S4 and their relative energies (in the gas phase) are given in Table S1.

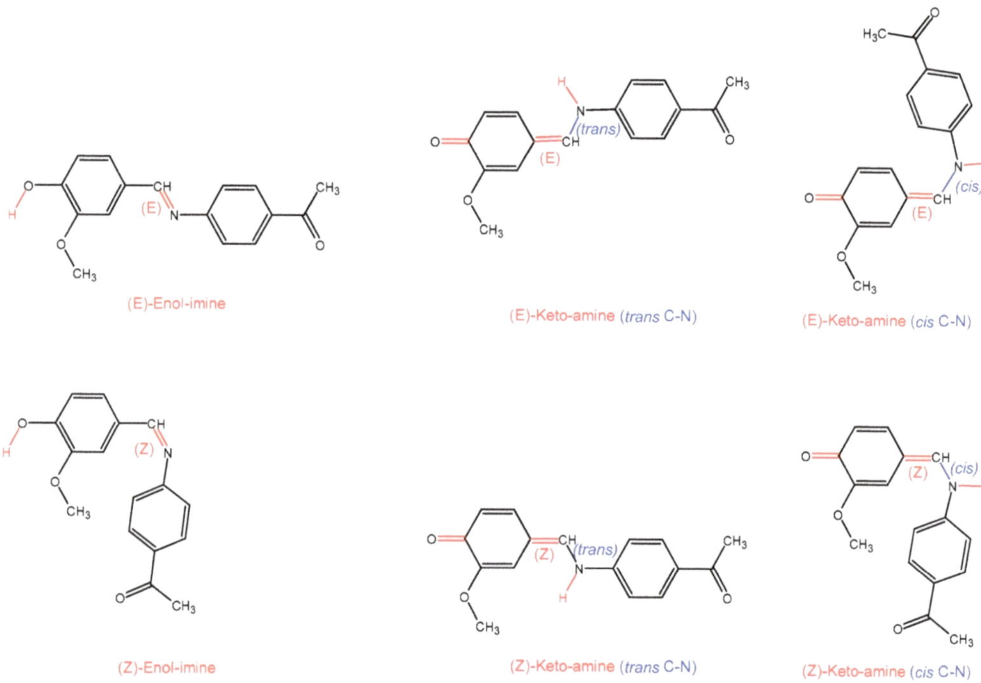

Scheme 1. *E*- and *Z*- forms of enol-imine and keto-amine tautomers of ANHMA.

The calculations indicate that in the gas phase, the (*E*)-enol-imine form is the most stable species, with the (*Z*)-enol-imine conformers being at least 22 kJ mol^{-1} higher in energy, and all keto-amine species having energies at least 42 kJ mol^{-1} above that of the most stable conformer of the (*E*)-enol-imine form. These results are in consonance with the (*E*)-enol-imine form being the species existing in the synthesized polycrystalline material, as indicated by its IR spectrum (Figure S5).

The (*E*)-enol-imine form of ANHMA is predicted by the calculations to possess 16 different conformers. The four lowest energy conformers can be grouped in two pairs of nearly degenerated conformers ((*E*)-EI1, (*E*)-EI2, and (*E*)-EI3, (*E*)-EI4), with the conformers of the second pair having energies ca. 5–6 kJ mol^{-1} above those belonging to the first pair (see Table S1). These four low-energy conformers are stabilized by an intramolecular $OH\cdots OCH_3$ hydrogen bond, the calculated $H\cdots O$ hydrogen bond distance in these conformers being in the range 2.088–2.099 Å. The lowest energy pair of conformers ((*E*)-EI1, (*E*)-EI2) have the methoxy substituent and the nitrogen atom of the imine bridge pointing to the same side, while the second pair of conformers ((*E*)-EI3, (*E*)-EI4) have these groups pointing to opposite sides; the conformers in each pair differ in the relative orientation of the acetyl substituent (see Figure S1). The relative energies of the remaining (*E*)-enol-imine conformers, which differ from the most stable ones by the orientation of the OH and OCH_3

groups, have calculated energies in the range of ~20–30 kJ mol^{-1} compared to that of the most stable conformer.

In the (*E*)-enol-imine conformers, the two rings make an angle of 43–45°, all forms having a symmetry-equivalent structure where the rings are rotated to the opposite direction around the N–C aniline bond. The acetyl and hydroxyl substituents are always aligned with the corresponding rings, while the methoxy substituent stays aligned with the phenolic ring in the first eight conformers in order of energy and is tilted ca. 67–70° from the plane of the ring in the last eight conformers.

The conformers of (Z)-enol-imine form have structures where the two rings are nearly perpendicular (to within 5°) to minimize steric repulsions (which are the major reason for the considerably higher energy of these forms compared to the (*E*)-enol-imine structures). Two (Z)-enol-imine conformers have an intramolecular OH···OCH_3 hydrogen bond and have energies ca. 22 and 27 kJ mol^{-1} above that of the most stable conformer of the (*E*)-enol-imine form ((*E*)-EI1), all the other conformers having much higher energies (above 42 kJ mol^{-1}, compared (*E*)-EI1). In the four lower energy (Z)-enol-imine conformers, the substituents attached to the rings are aligned with the corresponding ring plane, while the remaining four conformers have the methoxy group deviated ~67–69° out of ring plane, all conformers being doubly-degenerated by symmetry (Figure S2).

The conformers of the (Z)-keto-amine tautomer can be divided into two groups, depending on the arrangement about the central C(H)–N bridging bond. The C-N *trans* conformers have substantially lower energy than the C-N *cis* ones (~42–49 kJ mol^{-1}, vs. ~70–77 kJ mol^{-1} relatively to the most stable (*E*)-enol-imine form) due to the strong steric repulsions in the latter (see Figure S3). The common characteristic of these two groups of conformers is that within each group, the energies of the conformers are quite similar: in the C-N *trans* group of conformers, the energies vary within 4 kJ mol^{-1}, with the two most stable members with energies differing less than 1 kJ mol^{-1}; in the C-N *cis* group, the numbers are similar (see Table S1). Two of the four (Z)-keto-amine C-N *trans* conformers have a planar skeleton and correspond to single structures (C_s symmetry), while in the remaining two conformers, the methoxy substituent is tilted out of the ring plane by ca. 43° and have one symmetry-equivalent form each. In turn, all six (Z)-keto-amine C-N *cis* conformers are non-planar due to the steric repulsions between the two rings (see Figure S6) and correspond to pairs of symmetry-equivalent structures. In these forms, the C_{ph}=C–N–C_{ph} central bridge dihedral angle is within the 27–35° range, and the C–N–$C_{ph}C_{ph}$ dihedral stays in the 20–27° range, while in the four of these conformers where the methoxy group is out of the ring plane, the $C_{ph}C_{ph}$–O–C(H_3) dihedral angle is within the 43–46° range.

The conformational space of the (*E*)-keto-amine tautomer is similar to that of the (Z)-keto-amine one, with C-N *trans* and C-N *cis* conformers having energies ~48–54 kJ mol^{-1} and ~76–80 kJ mol^{-1} relatively to the most stable (*E*)-enol-imine form. Also, like for (Z)-keto-amine, the conformers of each of the two groups of conformers of the (*E*)-keto-amine tautomer have very similar energies (equal to within ca. 4 kJ mol^{-1}, with the two most stable members of each group differing in energy by less than ca. 1 kJ mol^{-1}). The geometries of the C-N *trans* and C-N *cis* conformers of the (*E*)-keto-amine tautomer (Figure S4) also follow the same trends as the equivalent conformers of (Z)-keto-amine. The molecular skeleton of the four C-N *trans* conformers is very close to planarity (both the C_{ph}=C–N–C_{ph} central bridge and C–N–$C_{ph}C_{ph}$ dihedral angles are tilted by less than ~3° from planarity), with the methoxy group in two of these conformers being ~53° tilted out of the ring plane. These latter conformers are doubly-degenerated by symmetry, while the remaining two correspond to single structures. The six (*E*)-keto-amine C-N *cis* conformers are non-planar doubly-degenerated structures, where the C_{ph}=C–N–C_{ph} central bridge dihedral angle stays within the 28–34° range, the C–N–$C_{ph}C_{ph}$ dihedral angle in the 22–29° range, and the $C_{ph}C_{ph}$–O–C(H_3) dihedral angle is within the 52–54° interval in the four conformers where the methoxy group is out of the ring plane.

The dipole moments of the different forms were also calculated (gas-phase values; Table S1). The range of values of the dipole moments for (*E*)-enol-imine, (*Z*)-enol-imine, and both (*Z*) and (*E*)-keto-amine C-N *trans* conformers are within 4.0 and 6.0 Debye, with very few exceptions, while those calculated for the high-energy keto-amine C-N *cis* conformers are larger (5–10 Debye).

Calculations for the most stable forms of the (*E*)-enol-imine and (*Z*)-keto-amine tautomers in DMSO, chloroform, and methanol solution using the integral equation formalism variant of the polarized continuum model (IEFPCM), within the self-consistent reaction field (SCRF) theoretical framework [47,48], were undertaken to evaluate the influence of the non-specific effects of these solvents on the relative energies of these forms. Compared to the gas phase, the relative energy of the keto-amine forms reduces to about a half (from ~40 kJ mol^{-1} to ca. 20 kJ mol^{-1}) in the case of methanol and DMSO and to ca. 30 kJ mol^{-1} in chloroform. We also tried to estimate the effect of the specific interactions (H-bonds or H-bond-like interactions) between the molecules of these solvents and the two tautomeric forms of the studied compound. The procedure used allows only a very rough estimation of these effects and is described in Figure S6. It essentially used average values of the different possible types of H-bond (or H-bond-like) interactions and calculated for each molecule the change in energy resulting from establishing intermolecular interactions at the expense of intramolecular ones. Then, the difference between these values for the different solvents is calculated. This very approximate procedure revealed that the keto-amine tautomer is stabilized relatively to the (*E*)-enol-imine form in all the solvents examined so that the energies of the two tautomers become very similar. This allows us to predict that, if a solvent-assisted mechanism exists that allows for proton transfer from the OH phenol moiety of the enol tautomer to the imine group (and vice versa), the keto-imine forms may be formed and exist in some amount in solution. For that, protic solvents and non-protic polar solvents that can act as proton acceptors immediately appear as suitable candidates to favor the possibility of experimental observation of the keto-amine forms. As shown in Section 2.4, this is indeed the case.

It shall also be mentioned that other tautomeric species might be conceivable for the basic ANHMA structure, for example those associated with the keto-enol tautomerism within the acetyl substituent (CH_3C=O vs. CH_2=COH) or with the (keto-imine)-(enol-amine) tautomerism involving the acetyl substituent and the imine bridge, but these forms are not expected to be important in the present context, since enolization of the acetyl moiety requires strong acid catalysis to be formed [49]. B3LYP/6-311++G(d,p) calculations performed on the enols of the acetyl tautomeric form derived from the most stable conformers of ANHMA (*E*)-enol-imine ((*E*)-EI1) and (*Z*)-keto-amine C-N *trans* forms revealed that the energies of these two species are over 150 and 60 kJ mol^{-1} above that of (*E*)-EI1.

Figure 4 shows the TD-DFT(B3LYP)/6-311++G(d,p) calculated UV-vis spectra for selected low-energy conformers of the different ANHMA isomeric species.

According to the calculations (Tables S2–S5), in the gas phase, the longest wavelength absorption maxima in (*E*)-enol-imine and (*Z*)-enol-imine forms showing significant oscillator strength are predicted at 343–354 nm (HOMO → LUMO) and 292–303 nm (HOMO → LUMO+1 or HOMO–1 → LUMO, depending on the specific conformer), respectively, while those of the (*Z*)-keto-amine and (*E*)-keto-amine forms are predicted at 397–407 nm (HOMO → LUMO) and 410–434 nm (HOMO → LUMO), respectively. Considering only the C-N *trans* conformers of the keto-amine tautomeric forms, the wavelength ranges reduce to 397–399 nm ((*Z*)-keto-amine forms) and 410–420 nm ((*E*)-keto-amine forms). As shown in Figure 4, shorter wavelength bands are predicted for the different species above 200 nm. These bands include contributions from several transitions (in Tables S2–S5, only those involving the S_2 and S_3 states are indicated, but states up to S_{20} have predicted energies that stay in this range). For the (*E*)-enol-imine form, one band resulting from all these possible contributions is observed above 250 nm, which, according to the calculations, stays nearly at the same position as one of the two bands originated in other than the transition to S_1 predicted for the keto-amine tautomers (both (*Z*) and (*E*) forms). TD-DFT calculations performed using solvent simulation (IEFPCM method) for the specific cases of chloroform, methanol, and DMSO indicate, in agreement with the solvatochromic results presented in Sections 2.1 and 2.2, that the lowest energy transition of the (*E*)-enol-imine form undergoes a batochromic shift in going from the gas phase to solution, from ca. 345 nm to ca. 365 nm (TD-DFT calculated values; the values predicted from the solvatochromic model based on the experimental data are from about 300 nm to ca. 330 nm) with the largest shift predicted (also in agreement with the observations) for DMSO (ca. 20 nm as calculated by the TD-DFT method; predicted by the solvatochromic model: ca. 40 nm). It shall, however, be noticed that, while the TD-DFT calculations with solvent simulation follow the general trend regarding the relative values of the band maxima in gas phase vs. solution, the predictions are far from being quantitative, as it could be anticipated because the theoretical model ignores the specific interactions between the solute and the solvent molecules (e.g., H-bonds), which have been shown to be very important for the studied system by the performed analysis based on the application of the solvatochromic Catalán model described above.

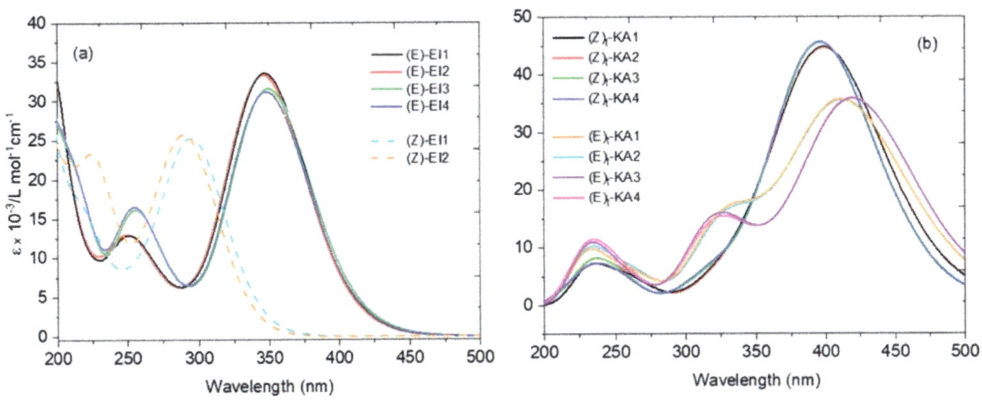

Figure 4. TD-DFT(B3LYP)/6-311++G(d,p) calculated UV-vis absorbance spectra of relevant ANHMA isomeric forms (gas phase): (*E*)- and (*Z*)-enol–imine lowest energy conformers (**a**), and (*Z*)- and (*E*)-keto–amine C–N *trans* conformers (**b**).

As a whole, the results of the calculations (relative energies, electronic spectra) provide strong evidence for the assignment of the two major bands observed experimentally, in the 285–300 nm (band I) and 310–345 nm (band II) ranges, to the (*E*)-enol-imine tautomer, with band II corresponding to the HOMO → LUMO ($S_1 \leftarrow S_0$) transition. The UV-vis spectral data thus indicate that the (*E*)-enol-imine tautomer strongly predominates (or is the unique form) in the different studied solvents.

2.4. NMR Investigation of the ANHMA Tautomerism in Solution

DMSO, methanol, and chloroform were chosen for a detailed investigation of the tautomerism of ANHMA in solution as representative solvents. DMSO was chosen because of its unique characteristics within the context of the present study (it is the solvent for which the observed reversal in solvatochromism takes place; see Sections 2.1 and 2.2), and methanol as a typical polar–protic solvent. The reason for choosing also chloroform will become clear later on, but it was essentially motivated by the fact that this solvent was the one where the studied compound was found to be photochemically less stable (Section 2.5).

2.4.1. ANHMA in DMSO

Figures 5 and 6 show the ^{1}H and ^{13}C NMR spectra of ANHMA in DMSO-d_6 solution. The correspondent bidimensional spectra are shown in Figures S7–S10, and the ^{1}H and ^{13}C NMR chemical shifts are presented in Tables 3 and 4, respectively, where they are compared with theoretically predicted values calculated at the GIAO/DFT(B3LYP)/6-311++G** level (where GIAO refers to the Gauge-Independent Atomic Orbital method [50,51]).

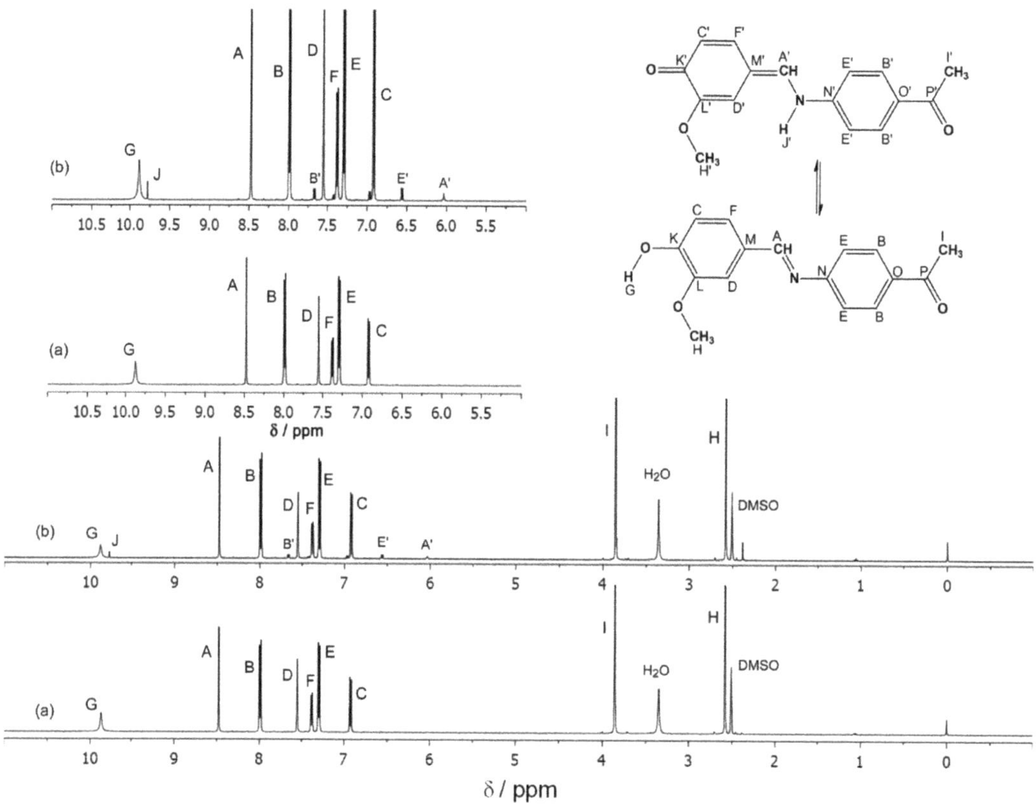

Figure 5. ^{1}H NMR spectra of ANHMA in DMSO-d_6 obtained (**a**) immediately and (**b**) one week after the dissolution; one inset containing the expansion from 5.00 to 11.00 pm for better visualization of the proton signals for the predominant form of ANHMA (the enol-imine tautomer) (92%) and the minor signals assigned to an additional species (the keto-amine tautomer) (8%), in a slow exchange equilibrium, in the NMR time scale. The signals at 3.36 and 2.51 ppm are assigned to traces of H_2O and DMSO, respectively.

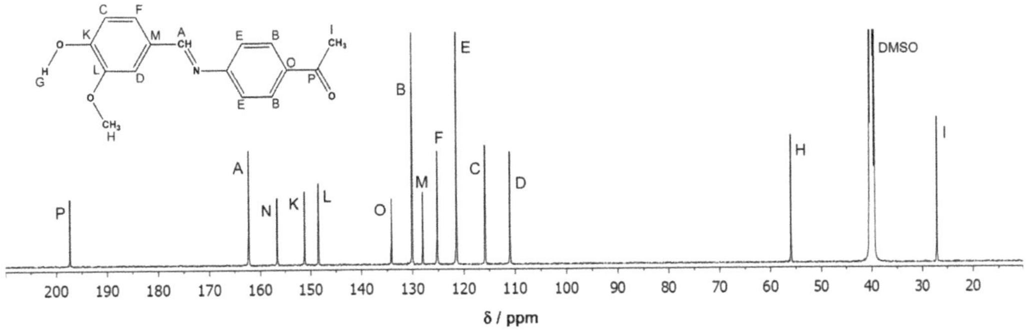

Figure 6. ^{13}C NMR spectrum of ANHMA in DMSO-d_6. The signals detected are assigned to the predominant form (the enol-imine tautomer). The spectrum obtained one week after the preparation of the solution is not shown because the signals of the minor form (the keto-amine tautomer) are not detected in ^{13}C NMR due to its low concentration. The signal detected at 40 ppm is assigned to DMSO.

Table 3. ^{1}H NMR chemical shifts [a] for ANHMA in DMSO, CD_3OD, and $CDCl_3$ (298 K).

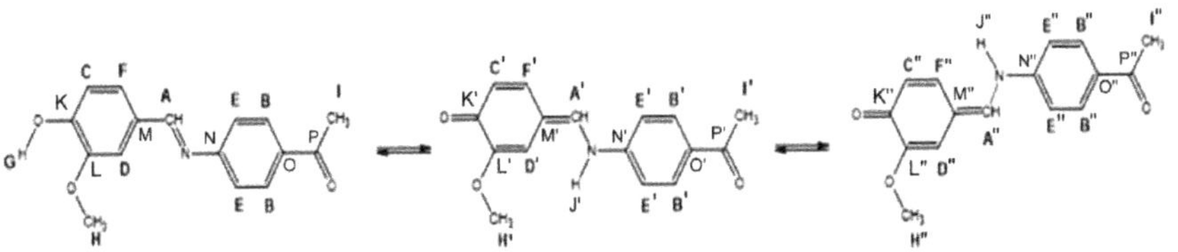

DMSO										
	H(A)	H(B)	H(C)	H(D)	H(E)	H(F)	H(G)	H(H)	H(I)	
δ (*E*)-enol-imine	8.48 (8.56)	7.99 (8.48/8.31) [8.40]	6.92 (7.28)	7.55 (8.01)	7.30 (7.62/7.35) [7.48]	7.38 (7.50)	9.87 (6.27)	3.85 (4.14)	2.58 (2.69)	
	H(A′)	H(B′)	H(C′)	H(D′)	H(E′)	H(F′)		H(H′)	H(I′)	H(J′)
δ (*Z*)-keto-amine	6.02 (7.90)	7.66 (8.53/8.27) [8.40]	-(b) (6.58)	-(b) (7.57)	6.57 (7.70/7.33) [7.52]	-(b) (8.51)		3.86 (4.03)	2.38 (2.66)	9.78 (7.73)
CD_3OD										
	H(A)	H(B)	H(C)	H(D)	H(E)	H(F)	H(G)	H(H)	H(I)	
δ (*E*)-enol-imine	8.44 (8.56)	8.05 (8.48/8.31) [8.40]	6.94 (7.28)	7.64 (8.01)	7.29 (7.62/7.35) [7.48]	7.37 (7.50)	4.60 (6.27)	3.98 (4.14)	2.63 (2.69)	
	H(A′)	H(B′)	H(C′)	H(D′)	H(E′)	H(F′)		H(H′)	H(I′)	H(J′)
δ (*Z*)-keto-amine	5.54 (7.73)	7.76 (8.54/8.26) [8.40]	6.93 (6.58)	7.45 (7.56)	6.85 (7.68/7.33) [7.50]	7.34 (7.49)		3.86 (4.04)	2.50 (2.65)	9.76 (7.70)
	H(A″)	H(B″)	H(C″)	H(D″)	H(E″)	H(F″)		H(H″)	H(I″)	H(J″)
δ (*E*)-keto-amine	5.27 (7.85)	7.74 (8.53/8.20) [8.37]	6.78 (8.62)	6.98 (7.15)	6.85 (7.62/7.24) [7.43]	7.44 (7.80)		3.67 (4.91)	2.48 (2.62)	9.77 (7.65)
$CDCl_3$										
	H(A)	H(B)	H(C)	H(D)	H(E)	H(F)	H(G)	H(H)	H(I)	
δ (*E*)-enol-imine	8.36 (8.48)	8.03 (8.50/8.19) [8.35]	7.03 (7.24)	7.67 (8.00)	7.19 (7.56/7.25) [7.41]	7.37 (7.41)	6.05 (6.13)	4.05 (4.10)	2.65 (2.63)	

Table 3. *Cont.*

	H(A′)	H(B′)	H(C′)	H(D′)	H(E′)	H(F′)	H(H′)	H(I′)	H(J′)
δ (*Z*)-keto-amine	6.20 (7.70)	7.95 (8.65/8.14) [8.35]	7.06 (6.54)	-(b) (7.32)	6.67 (7.54/7.16) [7.35]	7.30 (7.32)	4.12 (4.14)	2.54 (2.59)	9.83 (7.33)
	H(A″)	H(B″)	H(C″)	H(D″)	H(E″)	H(F″)	H(H″)	H(I″)	H(J″)
δ (*E*)-keto-amine	4.13 (7.67)	7.83 (8.57/8.10) [8.33]	7.06 (6.61)	7.25 (7.04)	6.69 (7.52/7.09) [7.30]	7.45 (7.61)	4.03 (4.22)	2.54 (2.57)	9.86 (7.35)

[a] δ values, in ppm, relative to Me4Si, using the signals of solvents, DMSO, CD_3OD and $CDCl_3$ (δ_H = 2.51, 3.35 and 7.29 ppm, respectively) as internal references. [b] not assigned.

Table 4. ^{13}C NMR chemical shifts [a] for A*N*HMA in DMSO, CD_3OD and $CDCl_3$ (298 K).

DMSO										
	C(A)	C(B)	C(C)	C(D)	C(E)	C(F)	C(H)	C(I)	C(K)	C(L)
δ (*E*)-enol-imine	162.26 (170.23)	130.09 (139.23/136.99) [138.11]	115.85 (119.90)	111.02 125.20 (112.31)	121.47 (131.47/123.97) [127.72]	125.20 (136.19)	56.03 (58.78)	27.09 (29.92)	151.22 (159.85)	148.51 (155.36)
	C(M)	C(N)	C(O)	C(P)						
	128.00 (136.70)	156.62 (167.60)	134.39 (139.67)	197.36 (208.74)						
CD_3OD										
	C(A)	C(B)	C(C)	C(D)	C(E)	C(F)	C(H)	C(I)	C(K)	C(L)
δ (*E*)-enol-imine	163.06 (170.16)	130.03 (139.17/133.99) [138.08]	115.30 (119.89)	113.26 103.41 (112.30)	127.08 (131.44/123.97) [127.69]	125.44 (136.15)	51.34 (58.75)	25.47 (29.88)	153.49 (159.83)	147.82 (155.34)
	C(M)	C(N)	C(O)	C(P)						
	127.08 (136.69)	156.78 (167.58)	134.39 (139.67)	198.83 (208.61)						
	C(A″)	C(B″)	C(C″)	C(D″)	C(E″)	C(F″)	C(H″)	C(I″)	C(K″)	C(L″)
δ (*E*)-keto-amine	103.80 [b] (146.18)	131.17 (139.88/137.40) [138.64]	129.65 (136.45)	119.73 (133.32)	113.09 (123.96/117.14) [120.33]	114.32 (133.33)	55.52 (61.66)	24.89 (29.81)	198.19 (188.06)	154.37 (160.07)
	C(M″)	C(N″)	C(O″)	C(P″)						
	110.10 (117.34)	147.82 (150.82)	146.50 (138.99)	198.30 (205.75)						

[a] δ values, in ppm, relative to Me_4Si, using the signals of solvents, DMSO and CD_3OD (δ_C = 40 and 48 ppm, respectively) as internal references. [b] The experimental and the calculated chemical shifts are not in accordance. According to HSQC, COSY, and HMBC two-dimensional spectra, the proton observed at 5.25 ppm (A″), which is directly bonded to the carbon observed at 103.80 ppm, has long-range couplings with the E″ and F″ protons and has the carbons labeled M″, D″ and H″ (CH_3) as vicinal carbon atoms. Based on these data, we have assigned the signal observed at 103.80 ppm to the carbon atom A″. This assignment is further supported by the fact that the NH proton (J″) is also vicinal to the A″ carbon atom.

For a freshly prepared solution, the spectra exhibited only signals due to the (*E*)-enol-imine form, which confirmed the UV-vis data and was in agreement with the results of the calculations. We noticed, however, that with time, additional peaks emerged in the ^{1}H spectrum. After one week of storage of the solution in the dark, the ^{1}H spectrum shows a set of new peaks that can be ascribed as the keto-amine tautomer (8% vs. 92% of the enol tautomer). The OH proton of the (*E*)-enol-imine form is observed at 9.87 ppm, in the expected position for a phenolic OH proton in DMSO [52] (the calculated chemical shift in the gas phase is lower), as a relatively broad feature, while the emerging NH proton is observed at 9.78 ppm. The assignment of these peaks to the OH and NH protons is doubtless, taking into account also the results of the bidimensional spectra shown in the Supporting Information, while the assignment of the other peaks is straightforward. The observation of the keto-amino form in the DMSO solution after one week of storage demonstrates that the equilibrium between the two forms is very slow, which also indicates that the lability of the OH and NH protons is, in this solvent, very reduced. This also justifies the clear observation of the peaks due to these protons, which would not be possible if they were exchanging fast with solvent deuterium atoms. In fact, it seems very much possible that the formation of the keto-amine tautomer in DMSO is essentially due to the presence of traces of water (that could not be eliminated in spite of our efforts), suggesting that in the absence of it the (*E*)-enol tautomer would be the sole form present in solution.

2.4.2. ANHMA in Methanol

Figures 7 and 8 show the ^{1}H and ^{13}C NMR spectra of ANHMA in a methanol-d$_4$ solution. The correspondent bidimensional spectra are shown in Figures S11–S14. The ^{1}H and ^{13}C NMR chemical shifts are given in Tables 3 and 4.

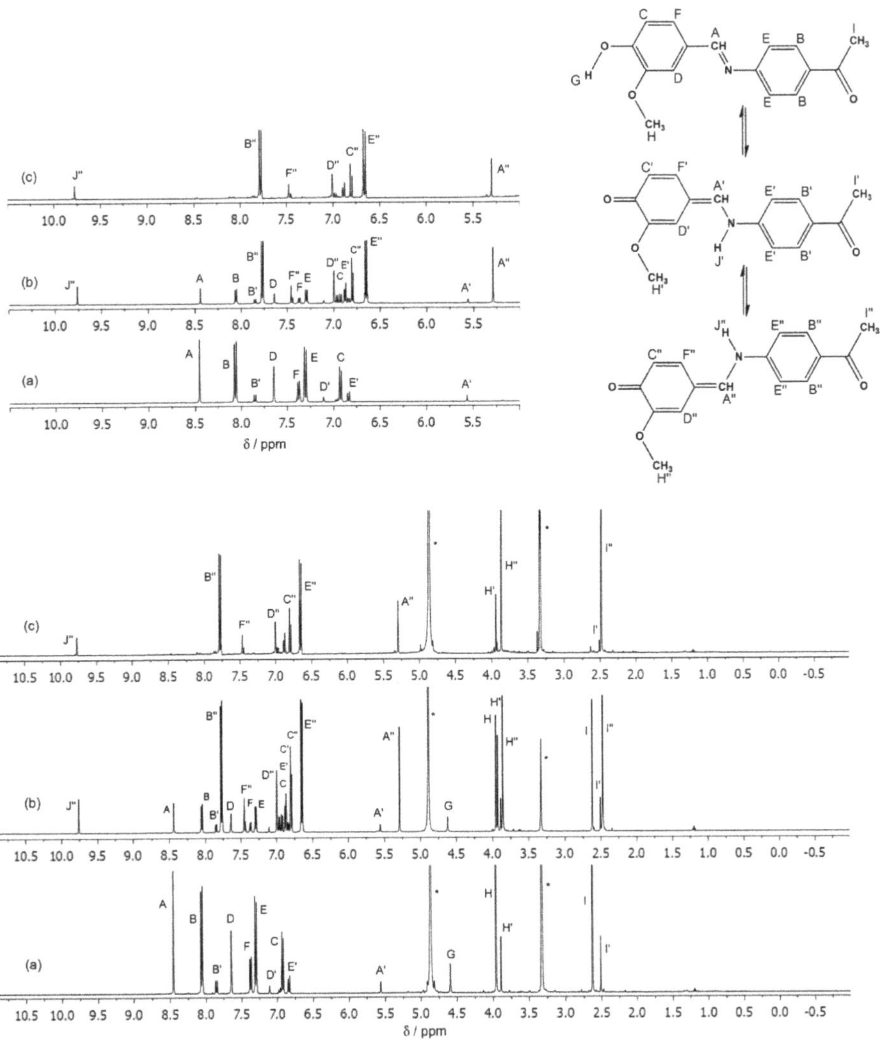

Figure 7. ^{1}H NMR spectra of ANHMA in CD_3OD obtained (**a**) immediately, (**b**) 1 h after the preparation, and (**c**) after the $\lambda \geq 311$ nm irradiation of the initial solution (>30 min); one inset containing one expansion from 5.00 to 11.00 pm for better visualization of the signals. The signals at 3.35 and 4.87 ppm (*) are assigned to the residual protons of methanol.

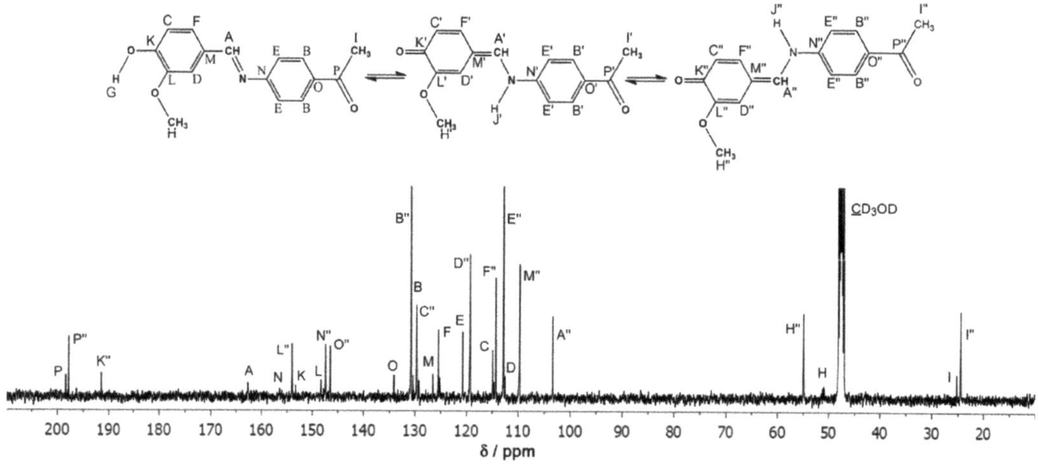

Figure 8. ^{13}C NMR spectrum of ANHMA in CD_3OD obtained 1 h after the preparation. The signals have been assigned to a mixture of the keto-amine tautomeric forms and the most abundant enol-imine tautomer. The signal detected at 48 ppm is assigned to the solvent.

The ^{1}H NMR spectrum obtained immediately after the dissolution (lower spectrum in Figure 7) shows as the predominant species the (*E*)-enol-imine tautomer, but low-intensity peaks due to the (*Z*)-keto-amine form are also visible, demonstrating that this species is also present in the studied solution. After 1 h of storage of the solution in the dark, the ^{1}H NMR spectrum changed considerably, with two sets of bands ascribable to the (*Z*)- and (*E*)-keto-amine forms being clearly visible, while those originated in the (*E*)-enol-imine tautomer reduced of intensity. In this (protic) solvent, the enol-imine/keto-amine equilibrium is attained more easily than in DMSO since the solvent can mediate the proton transfer. Nevertheless, the fact that both OH and NH protons can be observed in their expected positions (note that in this case, the OH proton appears at a much lower chemical shift—at 4.60 ppm—than in DMSO, as usually occurs for phenols [48]), reveals that the two protons are still not exchanging very much efficiently. In the case of the OH proton, this may indicate that the intramolecular hydrogen bond in the (*E*)-enol-imine tautomer subsists in significant amounts in methanol solution.

Another interesting observation is that both (*Z*)- and (*E*)-keto-amine forms participate in the equilibrium. There is nothing that can be expected to preclude the formation of both keto-amine isomers from the (*E*)-enol-imine species (the two keto forms are just the geometrically most favored products for different starting conformations of the (*E*)-enol-imine form; on the other hand, they cannot interconvert because this would require surpassing a high energy barrier associated with the rotation around the C=C double bond), but since the gas-phase energy of the (*E*)-keto-amine form is somewhat higher in gas phase than that of the (*Z*) form, the results suggest the (*E*) isomer is somehow more stabilized in methanol (probably because of a slightly larger average dipole moment, when the lowest energy C-N *trans* conformers of the two keto-amine isomers are compared; see Table S1).

2.4.3. ANHMA in Chloroform

Figure 9 shows the ^{1}H spectra of ANHMA in the $CDCl_3$ solution. The bidimensional spectra are shown in Figures S15–S18, and the chemical shifts are presented in Table 3.

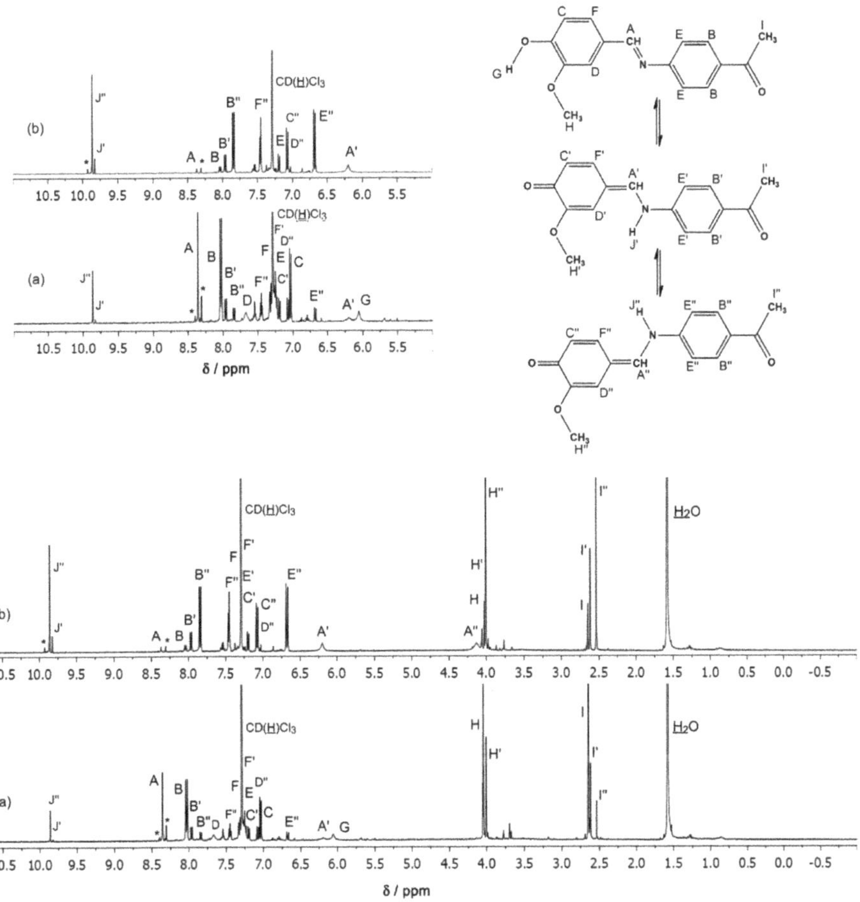

Figure 9. ^{1}H NMR spectra of ANHMA in $CDCl_3$ obtained (**a**) immediately and (**b**) 1 h after the preparation; one inset containing the expansion from 5.00 to 11.00 ppm for better visualization of the signals for the forms of the compound present in solution, the (*E*)-enol-imine tautomer and both (*Z*)- and (*E*)-isomer of the keto-amine tautomer, in a slow exchange equilibrium in the NMR time scale. Minor signals marked with * have uncertain origins. The signals at 1.58 and 7.29 ppm are assigned to traces of H_2O and residual protons of $CHCl_3$.

Like in methanol solution, in a freshly prepared chloroform solution, the compound exists predominantly in the (*E*)-enol-imine tautomeric form, with the keto-tautomer (in both (*Z*) and (*E*) isomeric forms) also present in a significant amount. In addition, like for the other solutions investigated, the amount of keto-amine tautomer increases with the time of storage of the solution (in the dark) so that after ca. 1 h, this tautomer appears as the major constituent. This result is interesting to note because this can be taken as an indication that in this solution (as well as in methanol, where also after 1 h the keto-amine tautomer seems to be dominant), the keto-amine tautomer might be slightly more stable than the (*E*)-enol-imine form. The equilibrium is reached slowly in all cases, starting from the (*E*)-enol-imine form that is present in the solid before dissolution. In the DMSO solution, the enol form seems to be the most stable one. These results are also in agreement with the predicted most significant stabilization in methanol and chloroform of the keto-imine tautomer than in DMSO, in particular, due to specific interactions between the molecules of the first two solvents and the dye (see Figure S6).

2.5. Photoisomerization ($\lambda \geq 311$ nm) of ANHMA in Different Solvents

UV ($\lambda \geq 311$ nm) irradiation of ANHMA (for a maximum of 30 min) in all studied solvents was undertaken. Significant changes in the compound's UV-vis absorbance spectrum were only observed for solutions of the compound in chlorobenzene and chloroform (both chlorinated polar–aprotic solvents) and to a lesser extent in *n*-butyl acetate (polar–aprotic), and 2-propanol, 1-butanol, ethanol, and methanol (polar–protic). The results obtained in chlorobenzene and chloroform solutions were distinct because these were the only cases where the solution considerably changed its visual appearance. Being initially colorless and transparent, the solution quickly turned to light yellow (in the first case) or became translucent (chloroform). All other solutions are colorless and transparent after exposure to irradiation.

The changes in the UV-vis spectrum of ANHMA in chloroform solution along irradiation are shown in Figure 10. After 1 min of irradiation, a reduction of the intensity of the band II ($\lambda_{max}{}^{II}$ = 335 nm), ascribed to the HOMO → LUMO transition of the (*E*)-enol-imine, was observed. The change of the intensity of band I ($\lambda_{max}{}^{I}$ = 293 nm; also due to the (*E*)-enol-imine form) is difficult to judge because new bands emerge at nearby the same position. Globally, the intensity within the wavelength range 270–320 nm increases, obscuring the expected decrease of intensity of band I. Concomitantly, a broad band appears in the 370–440 nm region (with an absolute maximum at ca. 390 nm; band III). The bands undergoing intensity fit well the predictions for the bands of the keto-amine isomeric forms (see Figure 4), and the broadness of band III can be easily correlated with the simultaneous contribution to this band of the (*Z*)- and (*E*)-isomeric forms of the keto-amine tautomer. Hence, it can be concluded that irradiation starts by promoting the photochemical conversion of the (*E*)-enol-imine tautomer into the keto-amine forms. After 5 min of irradiation, these changes were observed to attain their maximum, and from that time on, all bands in the spectra ascribable to ANHMA reduced intensity, indicating photodegradation of the compound upon prolonged UV irradiation.

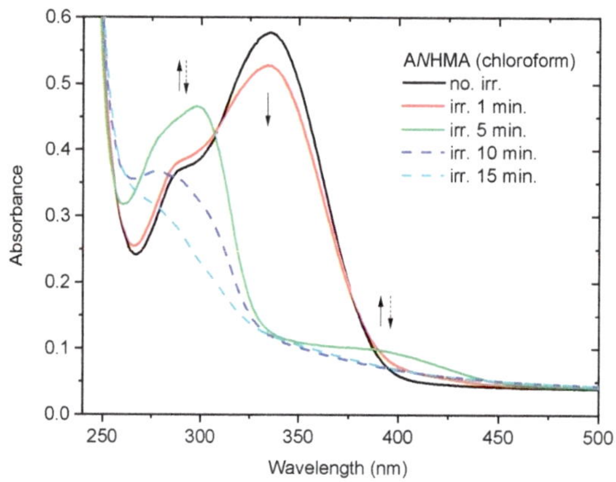

Figure 10. Absorbance spectrum of ANHMA in chloroform solution as a function of time of UV irradiation ($\lambda \geq 311$ nm).

The UV irradiation experiments carried out in chloroform solution were also attempted to be followed by NMR. However, the fast degradation of the compound upon irradiation and also the most difficult characterization by this technique of the compound in this solvent (including the fact that solvent bands appear in spectroscopically relevant spectral regions) posed practical difficulties that could not be overcome, precluding this investigation.

The results obtained in chlorobenzene are shown in Figure 11. In this solvent, band I, due to (*E*)-enol-imine tautomer, is not seen due to the overlap with a solvent band, but the changes in band II are clear: upon irradiation, this band reduces in intensity, while band III, ascribed to the keto-amine forms grows during the first 5 min of irradiation, indicating occurrence of phototransformation of the (*E*)-enol-imine tautomer into the keto-amine forms, as also observed in chloroform solution. This reaction, and also the (*Z*) to (*E*) C=C photoisomerization between the keto-amine isomers, takes place rather extensively and justifies the observed change of the color of the solution to yellow (see insert in Figure 11). Prolonged irradiation was found to lead to photodegradation of the compound, with all bands due to ANHMA reducing in intensity.

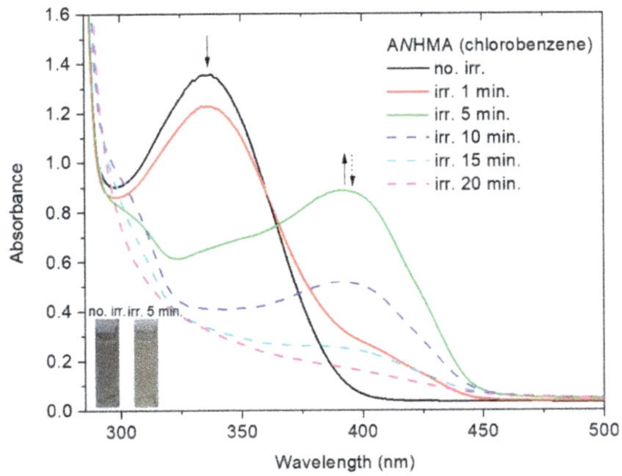

Figure 11. Absorbance spectrum of ANHMA in chlorobenzene solution as a function of time of UV irradiation ($\lambda \geq 311$ nm).

The photolysis results obtained in *n*-butyl acetate, 2-propanol, and 1-butanol are similar and follow the general trend observed for the two chlorinated solvents described above, i.e., irradiation promotes the conversion of (*E*)-enol-imine tautomer into both structural isomers of the keto-amine tautomer (Figure 12), but the efficiency of the processes, is lower.

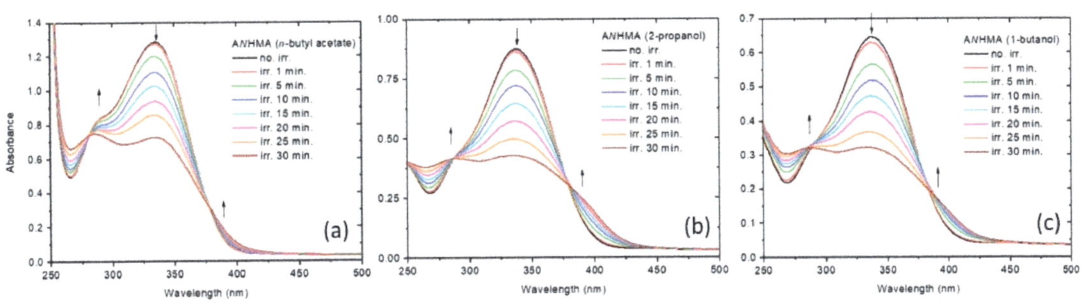

Figure 12. Absorbance spectrum of ANHMA in *n*-butyl acetate (**a**), 2-propanol (**b**), and 1-butanol (**c**) solutions as a function of time of UV irradiation ($\lambda \geq 311$ nm).

In the most polar and acidic solvents, ethanol and methanol, the investigation of the photochemistry of ANHMA revealed essentially the same panorama as for the other studied solvents. In this case, upon UV irradiation, the (*E*)-enol-imine bands (both bands

I and II) started to decrease visibly in intensity from the very beginning of irradiation (Figure 13), while the bands ascribed to keto-amine forms in the shorter wavelength range and band III increase of intensity. The efficiency of the processes is similar to those observed for the other alcohols. The experiment in methanol was followed by ^{1}H NMR. For that, photolysis was performed on the compound in a UV-vis transparent tube. The data confirm the general conclusions extracted from the UV-vis spectra but demonstrate that in this case, after prolonged irradiation (>30 min), the (*E*)-keto-amine isomer is the strongly dominant product (see Figure 7c). In this regard, the photochemically-induced process resembles the thermal process followed by NMR spectroscopy, and described above, where the solution was monitored along the storage time in the dark, which also tends to result in the (*E*)-keto-amine form as the most abundant species in methanol. However, the photochemical processes are much faster and more efficient.

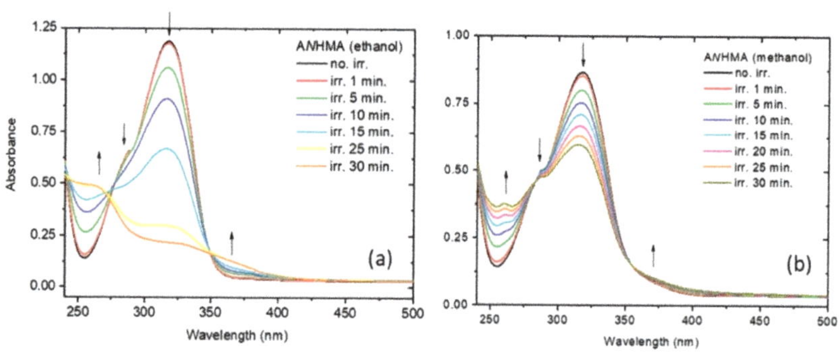

Figure 13. Absorbance spectrum of ANHMA in ethanol (**a**) and methanol (**b**) as a function of time of UV irradiation ($\lambda \geq 311$ nm).

3. Materials and Methods

3.1. Synthesis and Characterization (IR and NMR)

(*E*)-4-acetyl-*N*-(4-hydroxy-3-methoxybenzylidene)aniline (ANHMA) was synthesized from 4-hydroxy-3-methoxybenzaldehyde and 4-acetylaniline (Scheme 2), following a general procedure previously reported [28,53–57]. 4-Hydroxy-3-methoxybenzaldehyde (1.521 g and 0.01 mol) and 4-acetylaniline (1.352 g and 0.01 mol) were first dissolved in 25 mL ethanol in separate beakers at 40–45 °C, and the solutions were then mixed slowly. After a precipitate was formed, the mixture was stirred for another 1 h at the same temperature. The precipitated product was left cooling for 2–3 h at room temperature, and then it was filtered, purified by recrystallization from ethanol, and dried in a vacuum desiccator at room temperature. Its purity was checked by thin layer chromatography (TLC) in ethyl acetate: *n*-hexane (2:1). The infrared (IR) spectrum of the purified polycrystalline material (in a KBr pellet) and the ^{1}H- and ^{13}C-NMR spectra in deuterated dimethylsulfoxide (DMSO-d_6) solution were found to be compatible with the desired product (Supporting Information Figures S5, S7 and S8), the IR data demonstrating that in the solid phase the compound exists in its enol-imine tautomeric form:

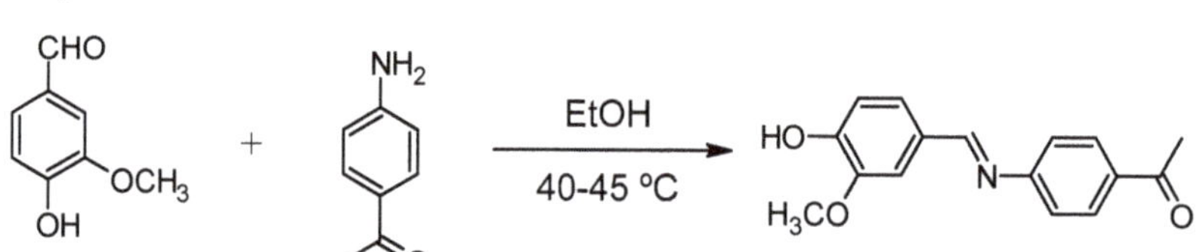

Scheme 2. Synthesis route for ANHMA.

ANHMA. IR (KBr pellet, ν/cm^{-1}): 3315 (O-H, H-bonded), 3083, 3029 (C-H, aromatic), 2972, 2940, 2904, 2875 (C-H, aliphatic), 1668 (C=O), 1583 (C=N), 1512–1463 (C=C, aromatic), 1390 (COH bend), 1274/1028 (C-O-C, alkyl aryl ether), 1250 (C-O phenol), 683 (OH torsion, phenol). ^{1}H NMR (400 MHz, DMSO-d$_6$, δ/ppm; (*E*)-enol-imine tautomer): 9.87 (s, 1H), 8.48 (s, 1H), 7.99 (d, *J* = 8.3 Hz, 2H), 7.55 (s, 1H), 7.38 (d, *J* = 8.2 Hz, 1H), 7.30 (d, *J* = 8.3 Hz, 2H), 6.92 (d, *J* = 8.1 Hz, 1H), 3.85 (s, 3H), 2.58 (s, 3H). ^{13}C NMR (101 MHz, DMSO-d$_6$, δ/ppm): 197.36 (s), 162.26 (s), 156.62 (s), 151.22 (s), 148.51 (s), 134.39 (s), 130.09 (s), 128.00 (s), 125.20 (s), 121.47 (s), 115.85 (s), 111.02 (s), 56.03 (s), 27.09 (s).

The IR spectrum was recorded in a Perkin–Elmer FTIR 100 spectrometer. The ^{1}H and ^{13}C NMR spectra were collected in deuterated DMSO, CD_3OD, and/or $CDCl_3$ solutions using a Bruker Avance III HD 500 MHz NMR spectrometer. The ^{13}C spectra were recorded using proton decoupling techniques, taking advantage of the nuclear Overhauser effect. The ^{1}H and ^{13}C NMR signals of the solvents were used as internal references for ^{1}H (δ 2.51, 3.35, 4.87, and 7.27 ppm) and ^{13}C (δ 40, 48, and 77.23 ppm) chemical shifts, respectively. The homo- and heteronuclear 2D NMR spectra, COSY, NOESY, HSQC, and HMBC were recorded on the same spectrometer. The assignments of the ^{1}H and ^{13}C NMR signals detected are based on bidimensional experiments, COSY, NOESY, HSQC, and HMBC, and compared with the calculated values. All experiments were performed at room temperature

3.2. UV-Vis Absorption Spectra

All solvents used in the experiments are of high purity and spectroscopic grade and were purchased from Sigma-Aldrich. The concentrations of the ANHMA solutions were kept as low as possible (<3 $\times 10^{-4}$ M) in order to avoid aggregation of the compound. The ultraviolet-visible (UV-vis) spectra were obtained in the 200–700 nm wavelength range using a Perkin–Elmer Lambda-35 UV-vis spectrometer. All measurements were performed using a quartz cell (standard cell, with 1 cm × 1 cm optical path) at room temperature, and then the solutions were kept in the dark for 5 days after preparation. Whenever required, the maximum wavelength values of the absorption bands were determined after band deconvolution, performed using OriginPro 2021 [58].

3.3. Quantum Chemical Calculations

All calculations were carried out at the DFT/B3LYP level of theory, with the 6-311++G(d,p) basis set [59–64] using GAUSSIAN 09 (revision C.01) [65]. Possible isomeric structures of ANHMA were subjected to geometry optimization in their electronic ground state, and their relative energies and vibrational frequencies (calculated using the same method and basis set) were obtained. The lack of imaginary frequencies indicates that all optimized geometries correspond to true minima. Excited state calculations were undertaken within the time-dependent DFT (TD-DFT) theoretical framework [66,67] for the isolated molecule *in vacuo* using the same functional and basis set as for the ground state calculations. The bulk solvent effects of chloroform, methanol, and DMSO were considered within the polarizable continuum model (PCM) framework using the integral equation formalism variant (IEFPCM) [47,48]. The NMR predicted data were obtained using the Gauge-Independent Atomic Orbital (GIAO) approach [50,51] at the B3LYP/6-311++G(d,p) level of theory in the different solvents. The ^{1}H and ^{13}C calculated GIAO absolute shieldings (σ; ppm) were converted into chemical shifts relative to tetramethylsilane (TMS) (δ values; ppm) by subtraction from the tetramethylsilane (TMS) GIAO calculated absolute shieldings.

The GaussView 5.0 [68] and ChemCraft (version 1.8) [69] programs were used both for input structure generation and visualization of the results.

4. Conclusions

In this study, a novel benzylidene aniline Schiff base derivative was synthesized and characterized using infrared and ^{1}H- and ^{13}C-NMR spectroscopies. The solvatochromic behavior of the compound was studied in 19 solvents with different characteristics. UV-vis absorption data analysis was done using the Catalán solvatochromic methodology and was supported by the thorough theoretical investigation of the structural characteristics of the different tautomeric/isomeric forms of the compound.

As for a previous bis-azo dye based on naphthalen-1-amine recently studied in our laboratories [5], the compound was shown to exhibit a reversal in solvatochromism for solvent Reichardt's $E_T(30)$~45 kcal mol^{-1} (DMSO), as a consequence of the change in the dominant solvent effects in non-polar and polar–aprotic solvents (H-bond acceptor ability of the solvent and polarizability) compared to polar–protic solvents (H-bond donor ability), as revealed by the Catalán analysis.

In all freshly prepared solutions of ANHMA, the (*E*)-enol-imine tautomeric form strongly dominates (or occurs exclusively), as shown by the UV-vis and NMR spectroscopic data. Upon irradiation of the ANHMA solutions with UV light ($\lambda \geq 311$ nm), the compound undergoes solvent-mediated (*E*)-enol-imino → keto-amino tautomerization, as well as (*Z*)-keto-amino → (*E*)-keto-amino C=C isomerization, the latter process being suggested by accumulation of the (*E*)-keto-amino form with time of irradiation. The photoreactions are particularly fast in chlorobenzene and chloroform, both chlorinated polar–aprotic solvents, and less effective in *n*-butyl acetate (polar–aprotic), and 2-propanol, 1-butanol, ethanol, and methanol (polar–protic), which may be considered an indication that triplet states might participate in the photoprocesses, since the ability of chlorine atoms to stabilize triplet states is well-known, which makes them energetically more accessible through spin-orbit coupling [70], but this conclusion requires confirmation (e.g., using time-resolved spectroscopy).

The studied molecule proved to be a versatile compound whose optical properties are very sensitive to solvent properties, which can be taken advantage of for its potential uses in different areas, such as sensing and detection, optoelectronic devices, and smart materials.

Supplementary Materials: The following supporting information can be downloaded at: https://www.mdpi.com/article/10.3390/molecules30030745/s1. Figures S1–S4: with the DFT(B3LYP)/6-311++G(d,p) optimized structures of the conformers of the (*E*)- and (*Z*)-isomers of the enol-imine and keto-amine tautomers of ANHMA; Figure S5: with the FTIR spectra of ANHMA in DMSO-d$_6$ and in KBr pellet; Figure S6: with estimation of the relative energies of specific interactions (H-bond or H-bond like) of (*E*)-enol-imine and (*Z*)-keto-amine forms in DMSO, chloroform and methanol, relative to gas phase; Figure S7: Bidimensional (H-H) NMR spectrum (COSY) of ANHMA in DMSO-d$_6$; Figures S8–S18, with the ^{1}H and ^{13}C NMR bidimensional data; Table S1: with the B3LYP/6-311++G(d,p) calculated dipole moments and energies for the different conformers of the tautomers of ANHMA; Tables S2–S5: with the results of TD-DFT(B3LYP)/6-311++G(d,p) for the different conformers of the tautomers of ANHMA.

Author Contributions: İ.S.: collection of spectroscopic experimental data, initial analysis of the results and interpretation, calculations, writing of the first draft of the manuscript, and funding acquisition. Y.G.S.: collection of spectroscopic experimental data and initial analysis of the results and interpretation. H.B.: synthesis and characterization of the studied compound and funding acquisition. M.L.R. and L.L.G.J.: collection and analysis of the NMR data. R.F.: data analysis and interpretation, writing of the final draft version of the manuscript, and funding acquisition. All authors have read and agreed to the published version of the manuscript.

Funding: Bitlis Eren University Research Foundation through projects BEBAP-2022.23, BEBAP-2024.14, BEBAP-2023.25. The Portuguese Science Foundation ("Fundação para a Ciência e a Tecnologia"—FCT, Projects CQC-IMS UIDB/00313/2025, UIDP/00313/2025 and and LA/P/0056/2020—National Funds). FEDER—European Regional Development Fund through the COM-

PETE Programme (Operational Programme for Competitiveness) and by National Funds through FCT—Fundação para a Ciência e a Tecnologia (Portuguese Foundation for Science and Technology), research grants REEQ/481/QUI/2006, RECI/QEQ-QFI/0168/2012, CENTRO-07-CT62-FEDER-002012, and Rede Nacional de Ressonância Magnética Nuclear (RNRMN). European Research Agency (HORIZON-WIDERA-2023-TALENTS-01-01 Program, ERA-Chair Spectroscopy@IKU—Manipulating and Characterizing Molecular Architectures: From Isolated Molecules to Molecular Crystals (Project number: 101184899).

Institutional Review Board Statement: Not applicable.

Informed Consent Statement: Not applicable.

Data Availability Statement: The original contributions presented in this study are included in the article/Supplementary Material. Further inquiries can be directed to the corresponding authors.

Acknowledgments: The authors greatly appreciate Bitlis Eren University Research Foundation for financial support through projects BEBAP-2022.23, BEBAP-2024.14, BEBAP-2023.25 and softwares, the Chemistry Department, Faculty of Sciences, Eskişehir Technical University, the Portuguese Science Foundation ("Fundação para a Ciência e a Tecnologia"—FCT, Projects CQC-IMS UIDB/00313/2025, UIDP/00313/2025 and and LA/P/0056/2020—National Funds). NMR data was collected at the UC-NMR facility which is supported in part by FEDER—European Regional Development Fund through the COMPETE Programme (Operational Programme for Competitiveness) and by National Funds through FCT—Fundação para a Ciência e a Tecnologia (Portuguese Foundation for Science and Technology) through grants REEQ/481/QUI/2006, RECI/QEQ-QFI/0168/2012, CENTRO-07-CT62-FEDER-002012, and Rede Nacional de Ressonância Magnética Nuclear (RNRMN). The ERA-Chair Spectroscopy@IKU—Manipulating and Characterizing Molecular Architectures: From Isolated Molecules to Molecular Crystals (Project number: 101184899) is funded by the European Research Agency under the HORIZON-WIDERA-2023-TALENTS-01-01 Program.

Conflicts of Interest: The authors declare no conflicts of interest.

References

1. Reichardt, C.; Welton, T. *Solvents and Solvent Effects in Organic Chemistry*; John Wiley & Sons: Hoboken, NJ, USA, 2011.
2. Suppan, P.; Ghoneim, N. *Solvatochromism*; Royal Society of Chemistry: Cambridge, UK, 1997; pp. 12–40.
3. Reichardt, C. Solvatochromic Dyes as Solvent Polarity Indicators. *Chem. Rev.* **1994**, *94*, 2319–2358. [CrossRef]
4. El Seoud, O.A. Understanding Solvation. *Pure Appl. Chem.* **2009**, *81*, 697–707. [CrossRef]
5. Sıdır, İ.; Kara, Y.E.; Sıdır, Y.G.; Berber, H.; Fausto, R. Reversal in Solvatochromism, Photochromism and Thermochromism in a New Bis-Azo Dye Based on Naphthalen-1-Amine. *J. Photochem. Photobiol. A Chem.* **2024**, *446*, 115138. [CrossRef]
6. Catalán, J. Toward a Generalized Treatment of the Solvent Effect Based on Four Empirical Scales: Dipolarity (SdP, a New Scale), Polarizability (SP), Acidity (SA), and Basicity (SB) of the Medium. *J. Phys. Chem. B* **2009**, *113*, 5951–5960. [CrossRef]
7. Filarowski, A.; Kluba, M.; Cieślik-Boczula, K.; Koll, A.; Kochel, A.; Pandey, L.; De Borggraeve, W.M.; Van der Auweraer, M.; Catalán, J.; Boens, N. Generalized Solvent Scales as a Tool for Investigating Solvent Dependence of Spectroscopic and Kinetic Parameters. Application to Fluorescent BODIPY Dyes. *Photochem. Photobiol. Sci.* **2010**, *9*, 996–1008. [CrossRef] [PubMed]
8. Catalán, J.; Catalán, J.P. On the Solvatochromism of the N↔Π* Electronic Transitions in Ketones. *Phys. Chem. Chem. Phys.* **2011**, *13*, 4072–4082. [CrossRef]
9. Jiménez-Sánchez, A.; Farfán, N.; Santillan, R. Multiresponsive Photo-, Solvato-, Acido-, and Ionochromic Schiff Base Probe. *J. Phys. Chem. C* **2015**, *119*, 13814–13826. [CrossRef]
10. Reichardt, C. Pyridinium N-Phenolate Betaine Dyes as Empirical Indicators of Solvent Polarity: Some New Findings. *Pure Appl. Chem.* **2004**, *76*, 1903–1919. [CrossRef]
11. de Melo, C.E.; Nicoleti, C.R.; Ferreira, M.; dos Santos, M.C.; Kreuz, A.; Schneider, F.S.; Oliboni, R.d.S.; Caramori, G.F.; Machado, V.G. Reverse Solvatochromism in a Family of Probes Having 2, 6–Di–Tert–Butylphenolate as Electron–Donor and 4–Nitrophenyl as Electron–Acceptor Groups. *Dye. Pigment.* **2022**, *203*, 110376. [CrossRef]
12. de Melo, C.E.A.; Nicoleti, C.R.; Rezende, M.C.; Bortoluzzi, A.J.; da S. Heying, R.; da S. Oliboni, R.; Caramori, G.F.; Machado, V.G. Reverse Solvatochromism of Imine Dyes Comprised of 5-Nitrofuran-2-yl or 5-Nitrothiophen-2-yl as Electron Acceptor and Phenolate as Electron Donor. *Chem. A Eur. J* **2018**, *24*, 9364–9376. [CrossRef] [PubMed]

13. De Melo, C.E.A.; Nandi, L.G.; Domínguez, M.; Rezende, M.C.; Machado, V.G. Solvatochromic Behavior of Dyes with Dimethylamino Electron-Donor and Nitro Electron-Acceptor Groups in Their Molecular Structure: SOLVATOCHROMIC BEHAVIOR OF N,N-DIMETHYLANILINE DYES. *J. Phys. Org. Chem.* **2015**, *28*, 250–260. [CrossRef]
14. Mukhopadhyay, A.; Mandal, K.J.; Moorthy, J.N. Anionic Merocyanine Dyes Based on Thiazol-2-Hydrazides: Reverse Solvatochromism, Preferential Solvation and Multiparametric Approaches to Spectral Shifts. *Phys. Chem. Chem. Phys.* **2018**, *20*, 4149–4159. [CrossRef]
15. Vidal, M.; Pastenes, C.; Rezende, M.C.; Aliaga, C.; Domínguez, M. The Inverted Solvatochromism of Protonated Ferrocenylethenyl Pyrimidines: The First Example of the Solvatochromic Reversal of a Hybrid Organic/Inorganic Dye. *Org. Chem. Front.* **2019**, *6*, 3896–3901. [CrossRef]
16. Machado, V.G.; Stock, R.I.; Reichardt, C. Pyridinium *N* -Phenolate Betaine Dyes. *Chem. Rev.* **2014**, *114*, 10429–10475. [CrossRef]
17. Klymchenko, A.S. Solvatochromic and Fluorogenic Dyes as Environment-Sensitive Probes: Design and Biological Applications. *Acc. Chem. Res.* **2017**, *50*, 366–375. [CrossRef] [PubMed]
18. de Melo, C.E.; Dominguez, M.; Rezende, M.C.; Machado, V.G. Solvatochromism of Dyes Inspired in Effenberger's Probe. *Dye Pigment.* **2021**, *184*, 108757. [CrossRef]
19. Nandi, L.G.; Facin, F.; Marini, V.G.; Zimmermann, L.M.; Giusti, L.A.; Silva, R.D.; Caramori, G.F.; Machado, V.G. Nitro-Substituted 4-[(Phenylmethylene)Imino]Phenolates: Solvatochromism and Their Use as Solvatochromic Switches and as Probes for the Investigation of Preferential Solvation in Solvent Mixtures. *J. Org. Chem.* **2012**, *77*, 10668–10679. [CrossRef]
20. Kalapos, P.P.; Kunfi, A.; Bogner, M.M.; Holczbauer, T.; Kochman, M.A.; Durbeej, B.; London, G. Salicylideneaniline/Dithienylethene Hybrid Molecular Switches: Design, Synthesis, and Photochromism. *J. Org. Chem.* **2024**, *89*, 16–26. [CrossRef] [PubMed]
21. Panigrahi, S.; Biswal, S.P.; Misra, P.K. Disclosure of the Solvatochromism and the Reversal Switch in Some Tailor-Made Electron Push-Push Anils. *J. Mol. Liq.* **2021**, *329*, 115536. [CrossRef]
22. Manzoni, V.; Coutinho, K.; Canuto, S. An Insightful Approach for Understanding Solvatochromic Reversal. *Chem. Phys. Lett.* **2016**, *655*, 30–34. [CrossRef]
23. Morley, J.O.; Morley, R.M.; Fitton, A.L. Spectroscopic Studies on Brooker's Merocyanine. *J. Am. Chem. Soc.* **1998**, *120*, 11479–11488. [CrossRef]
24. Martins, C.T.; Lima, M.S.; Bastos, E.L.; El Seoud, O.A. Thermo-Solvatochromism of Merocyanine Polarity Probes—What Are the Consequences of Increasing Probe Lipophilicity through Annelation? *Eur. J. Org. Chem.* **2008**, *2008*, 1165–1180. [CrossRef]
25. Aburto, I.; Muñoz, M.; Vidal, M.; Aliaga, C.; Domínguez, M. Extending the π-System of Inverted Solvatochromic Phenolate-Based Dyes Shifts Their Inversion Point Due to an Increment in Their Sensitivity to Solvent Polarizability. *J. Mol. Liq.* **2022**, *368*, 120784. [CrossRef]
26. Sıdır, Y.G.; Sıdır, İ.; Berber, H.; Türkoğlu, G. Solvatochromism and Electronic Structure of Some Symmetric Schiff Base Derivatives. *J. Mol. Liq.* **2015**, *204*, 33–38. [CrossRef]
27. Sıdır, I.; Sıdır, Y.G.; Berber, H.; Türkoğlu, G. Specific and Non-Specific Interaction Effect on the Solvatochromism of Some Symmetric (2-Hydroxybenzilydeamino) Phenoxy Schiff Base Derivatives. *J. Mol. Liq.* **2016**, *215*, 691–703. [CrossRef]
28. Sıdır, Y.G.; Sıdır, İ.; Berber, H. Optoelectronic Properties by Solution Technique and Comprehensive Solvatochromism of Novel Fluorescent Schiff Base Derivatives. *J. Mol. Liq.* **2022**, *357*, 119110. [CrossRef]
29. Tigineh, G.T.; Liu, L.-K. Solvatochromic Photoluminescence Investigation of Functional Schiff-Bases: A Systematic Study of Substituent Effects. *J. Photochem. Photobiol. A Chem.* **2017**, *338*, 161–170. [CrossRef]
30. Filarowski, A.; Lopatkova, M.; Lipkowski, P.; Van der Auweraer, M.; Leen, V.; Dehaen, W. Solvatochromism of BODIPY-Schiff Dye. *J. Phys. Chem. B* **2015**, *119*, 2576–2584. [CrossRef] [PubMed]
31. Bolz, I.; May, C.; Spange, S. Solvatochromic Properties of Schiff Bases Derived from 5-Aminobarbituric Acid: Chromophores with Hydrogen Bonding Patterns as Components for Coupled Structures. *New J. Chem.* **2007**, *31*, 1568–1571. [CrossRef]
32. Traven', V.F.; Ivanov, I.V.; Panov, A.V.; Safronova, O.B.; Chibisova, T.A. Solvent-Induced E/Z(C=N)-Isomerization of Imines of Some Hydroxy-Substituted Formylcoumarins. *Russ. Chem. Bull.* **2008**, *57*, 1989–1995. [CrossRef]
33. Luo, Y.; Utecht, M.; Dokić, J.; Korchak, S.; Vieth, H.; Haag, R.; Saalfrank, P. *Cis–Trans* Isomerisation of Substituted Aromatic Imines: A Comparative Experimental and Theoretical Study. *ChemPhysChem* **2011**, *12*, 2311–2321. [CrossRef]
34. Ziółek, M.; Burdziński, G.; Filipczak, K.; Karolczak, J.; Maciejewski, A. Spectroscopic and Photophysical Studies of the Hydroquinone Family of Photochromic Schiff Bases Analyzed over a 17-Orders-of-Magnitude Time Scale. *Phys. Chem. Chem. Phys.* **2008**, *10*, 1304–1318. [CrossRef]
35. Georgiev, A.; Todorov, P.; Dimov, D. Excited State Proton Transfer and E/Z Photoswitching Performance of 2-Hydroxy-1-Naphthalene and 1-Naphthalene 5,5′-Dimethyl- and 5,5′-Diphenylhydantoin Schiff Bases. *J. Photochem. Photobiol. A Chem.* **2020**, *386*, 112143. [CrossRef]
36. Güngör, Ö.; Nuralin, L. A New Novel Series of Vanillin-Based Hexaimine-Schiff Bases: Synthesis, Characterization and Investigation of Chromic and Fluorescence Properties. *J. Mol. Struct.* **2024**, *1310*, 138371. [CrossRef]

37. Ohshima, A.; Momotake, A.; Arai, T. Photochromism, Thermochromism, and Solvatochromism of Naphthalene-Based Analogues of Salicylideneaniline in Solution. *J. Photochem. Photobiol. A Chem.* **2004**, *162*, 473–479. [CrossRef]
38. Minkin, V.I.; Tsukanov, A.V.; Dubonosov, A.D.; Bren, V.A. Tautomeric Schiff Bases: Iono-, Solvato-, Thermo- and Photochromism. *J. Mol. Struct.* **2011**, *998*, 179–191. [CrossRef]
39. Georgiev, A.; Stoilova, A.; Dimov, D.; Yordanov, D.; Zhivkov, I.; Weiter, M. Synthesis and Photochromic Properties of Some *N*-Phthalimide Azo-Azomethine Dyes. A DFT Quantum Mechanical Calculations on Imine-Enamine Tautomerism and Trans-Cis Photoisomerization. *Spectrochim. Acta Part A Mol. Biomol. Spectrosc.* **2019**, *210*, 230–244. [CrossRef]
40. Flores-Leonar, M.; Esturau-Escofet, N.; Méndez-Stivalet, J.M.; Marı, A.; Amador-Bedolla, C. Factors Determining Tautomeric Equilibria in Schiff Bases. *J. Mol. Struct.* **2011**, *1006*, 600–605. [CrossRef]
41. Rahimova, A.R. A Short Review on Schiff Bases. *Inorg. Chem. Indian J.* **2023**, *18*, 1–16.
42. Brewer, G.; Brewer, C.; Butcher, R.J.; Zavalij, P. Formation of Ketimines from Aldimines in Schiff Base Condensation of Amino Acids and Imidazole-2-Carboxaldehydes: Tautomerization of Schiff Bases of Amino Acids Resulting in the Loss of Stereogenic Center. *Inorganics* **2023**, *11*, 381. [CrossRef]
43. Feringa, B.L.; Browne, W.R. *Molecular Switches*; John Wiley & Sons: Hoboken, NJ, USA, 2011.
44. Dattler, D.; Fuks, G.; Heiser, J.; Moulin, E.; Perrot, A.; Yao, X.; Giuseppone, N. Design of Collective Motions from Synthetic Molecular Switches, Rotors, and Motors. *Chem. Rev.* **2020**, *120*, 310–433. [CrossRef]
45. Feringa, B.L. The Art of Building Small: From Molecular Switches to Motors (Nobel Lecture). *Angew. Chem.-Int. Ed.* **2017**, *56*, 11059–11078. [CrossRef]
46. Tochitsky, I.; Helft, Z.; Meseguer, V.; Fletcher, R.B.; Vessey, K.A.; Telias, M.; Denlinger, B.; Malis, J.; Fletcher, E.L.; Kramer, R.H. How Azobenzene Photoswitches Restore Visual Responses to the Blind Retina. *Neuron* **2016**, *92*, 100–113. [CrossRef] [PubMed]
47. Miertus, S.; Scrocco, E.; Tomasi, J. Electrostatic Interaction of a Solute With a Continuum. A Direct Utilizaion of Ab Initio Molecular Potentials for the Prevision of Solvent Effects. *J. Chem. Phys.* **1981**, *55*, 117–129. [CrossRef]
48. Tomasi, J.; Mennucci, B.; Cammi, R. Quantum Mechanical Continuum Solvation Models. *Chem. Rev.* **2005**, *105*, 2999–3094. [CrossRef] [PubMed]
49. Chiang, Y.; Kresge, A.J.; O'ferrall, R.A.M.; Murray, B.A.; Schepp, N.P.; Wlrz, J. The Ketonization of Acetophenone Enol in Concentrated Aqueous Sulphuric and Perchloric Acid Solutions. Implication on *X* Acidity Function Correlations of the Enolization Reaction and Determination of the Keto–Enol Equilibrium Constant as a Function of Acidity. *Can. J. Chem.* **1990**, *68*, 1653–1656. [CrossRef]
50. Ditchfield, R. Self-Consistent Perturbation Theory of Diamagnetism: I. *A Gauge-Invariant LCAO Method for NMR Chemical Shifts, Mol. Phys.* **1974**, *27*, 789–807. [CrossRef]
51. Wolinski, K.; Hilton, J.F.; Pulay, P. Efficient Implementation of the Gauge-Independent Atomic Orbital Method for NMR Chemical Shift Calculations. *J. Am. Chem. Soc.* **1990**, *112*, 8251–8260. [CrossRef]
52. Abraham, R.J.; Mobli, M. An NMR, IR and Theoretical Investigation of ^{1} H Chemical Shifts and Hydrogen Bonding in Phenols. *Magn. Reson Chem.* **2007**, *45*, 865–877. [CrossRef]
53. Sıdır, İ.; Sıdır, Y.G.; Góbi, S.; Berber, H.; Ildiz, G.O.; Fausto, R. UV-Induced –OCH3 Rotamerization in a Matrix-Isolated Methoxy-Substituted Ortho-Hydroxyaryl Schiff Base. *Photochem. Photobiol. Sci.* **2022**, *21*, 835–847. [CrossRef] [PubMed]
54. Sıdır, İ.; Sarı, T.; Sıdır, Y.G.; Berber, H. Synthesis, Solvatochromism and Dipole Moment in the Ground and Excited States of Substitute Phenol Derivative Fluorescent Schiff Base Compounds. *J. Mol. Liq.* **2022**, *346*, 117075. [CrossRef]
55. Sıdır, İ.; Gülseven Sıdır, Y.; Gobi, S.; Berber, H.; Fausto, R. Structural Relevance of Intramolecular H-Bonding in Ortho-Hydroxyaryl Schiff Bases: The Case of 3-(5-Bromo-2-Hydroxybenzylideneamino) Phenol. *Molecules* **2021**, *26*, 2814. [CrossRef] [PubMed]
56. Sıdır, Y.G.; Aslan, C.; Berber, H.; Sıdır, İ. The Electronic Structure, Solvatochromism, and Electric Dipole Moments of New Schiff Base Derivatives Using Absorbance and Fluorescence Spectra. *Struct. Chem.* **2019**, *30*, 835–851. [CrossRef]
57. Gülseven Sıdır, Y.; Berber, H.; Sıdır, İ. The Dipole Moments and Solvatochromism of ((4-(Benzyloxy)Benzylidene)Amino)Phenol Compounds as Solvatochromic Materials. *J. Solut. Chem.* **2019**, *48*, 775–806. [CrossRef]
58. OriginPro 2021 (64 bits), v. *9.8.0.200*; OriginLab Corporation: Northamptom, MA, USA, 2021.
59. Becke, A.D. Density-Functional Exchange-Energy Approximation with Correct Asymptotic Behavior. *Phys. Rev. A* **1988**, *38*, 3098–3100. [CrossRef]
60. Lee, C.; Yang, W.; Parr, R.G. Development of the Colle-Salvetti Correlation-Energy Formula into a Functional of the Electron Density. *Phys. Rev. B* **1988**, *37*, 785–789. [CrossRef]
61. Vosko, S.H.; Wilk, L.; Nusair, M. Accurate Spin-Dependent Electron Liquid Correlation Energies for Local Spin Density Calculations: A Critical Analysis. *Can. J. Phys.* **1980**, *58*, 1200–1211. [CrossRef]
62. McLean, A.D.; Chandler, G.S. Contracted Gaussian Basis Sets for Molecular Calculations. I. Second Row Atoms, Z= 11–18. *J. Chem. Phys.* **1980**, *72*, 5639–5648. [CrossRef]
63. Clark, T.; Chandrasekhar, J.; Spitznagel, G.W.; Schleyer, P.V.R. Efficient Diffuse Function-augmented Basis Sets for Anion Calculations. III. The 3-21+G Basis Set for First-row Elements, Li–F. *J. Comput. Chem.* **1983**, *4*, 294–301. [CrossRef]

64. Frisch, M.J.; Pople, J.A.; Binkley, J.S. Self-Consistent Molecular Orbital Methods 25. Supplementary Functions for Gaussian Basis Sets. *J. Chem. Phys.* **1984**, *80*, 3265–3269. [CrossRef]
65. Frisch, M.J.; Trucks, G.W.; Schlegel, H.B.; Scuseria, G.E.; Robb, M.A.; Cheeseman, J.R.; Scalmani, G.; Barone, V.; Mennucci, B.; Petersson, G.A. *Gaussian 09, Revision A.02*; Gaussian, Inc.: Wallingford, CT, USA, 2016.
66. Bauernschmitt, R.; Ahlrichs, R. Treatment of Electronic Excitations within the Adiabatic Approximation of Time *Dependent* Density Functional Theory. *Chem. Phys. Lett.* **1996**, *256*, 454–464. [CrossRef]
67. Casida, M.E.; Jamorski, C.; Casida, K.C.; Salahub, D.R. Molecular Excitation Energies to High-Lying Bound States from Time-Dependent Density-Functional Response Theory: Characterization and Correction of the Time-Dependent Local Density Approximation Ionization Threshold. *J. Chem. Phys.* **1998**, *108*, 4439–4449. [CrossRef]
68. Frisch, A.; Hratchian, H.P.; Dennington, R.D., II; Keith, T.A.; Millam, J.; Nielsen, A.B.; Holder, A.J.; Hiscocks, J. *GaussView 5.0*; Gaussian, Inc.: Wallingford, CT, USA, 2009.
69. Chemcraft (Version 1.8), Graphical Software for Visualization of Quantum Chemistry Computations. Available online: https://www.chemcraftprog.com/ (accessed on 24 May 2024).
70. Ahmed, R.; Manna, A.K. Origins of Molecular-Twist-Triggered Intersystem Crossing in Functional Perylenediimides: Singlet–Triplet Gap versus Spin–Orbit Coupling. *J. Phys. Chem. A* **2022**, *126*, 6594–6603. [CrossRef]

molecules

MDPI

Review

Coherent Vibrational Anti-Stokes Raman Spectroscopy Assisted by Pulse Shaping

Kai Wang [1,*,†], James T. Florence [1], Xia Hua [2], Zehua Han [1], Yujie Shen [3], Jizhou Wang [1], Xi Wang [1] and Alexei V. Sokolov [1,*]

1 Department of Physics and Astronomy, Institute for Quantum Science and Engineering, Texas A&M University, College Station, TX 77843, USA; jtflorence6@physics.tamu.edu (J.T.F.); zehuahan@tamu.edu (Z.H.); wjz1993@physics.tamu.edu (J.W.); xwangphy@gmail.com (X.W.)

2 Innovation Academy for Precision Measurement Science and Technology, Chinese Academy of Sciences, Wuhan 430071, China; huaxia@apm.ac.cn

3 Laboratory of Neurotechnology and Biophysics, Rockefeller University, New York, NY 10065, USA; yshen01@rockefeller.edu

* Correspondence: kaiwang@tamu.edu or wangk289@mail.sysu.edu.cn (K.W.); sokol@tamu.edu (A.V.S.)

† Current address: School of Physics and Astronomy, Sun Yat-sen University, Zhuhai 519082, China.

Abstract: Coherent anti-Stokes Raman scattering (CARS) is a powerful nonlinear spectroscopic technique widely used in biological imaging, chemical analysis, and combustion and flame diagnostics. The adoption of pulse shapers in CARS has emerged as a useful approach, offering precise control of optical waveforms. By tailoring the phase, amplitude, and polarization of laser pulses, the pulse shaping approach enables selective excitation, spectral resolution improvement, and non-resonant background suppression in CARS. This paper presents a comprehensive review of applying pulse shaping techniques in CARS spectroscopy for biophotonics. There are two different pulse shaping strategies: passive pulse shaping and active pulse shaping. Two passive pulse shaping techniques, hybrid CARS and spectral focusing CARS, are reviewed. Active pulse shaping using a programmable pulse shaper such as spatial light modulator (SLM) is discussed for CARS spectroscopy. Combining active pulse shaping and passive shaping, optimizing CARS with acousto-optic programmable dispersive filters (AOPDFs) is discussed and illustrated with experimental examples conducted in the authors' laboratory. These results underscore pulse shapers in advancing CARS technology, enabling improved sensitivity, specificity, and broader applications across diverse scientific fields.

Keywords: coherent anti-Stokes Raman scattering; pulse shaper; spectral focusing

1. Introduction

Coherent anti-Stokes Raman scattering (CARS), appropriately first reported from the Ford Motor Company by P.D. Maker and R.W. Terhune in 1965, is a third-order optical nonlinear process that has proven to be one of the most powerful nonlinear spectroscopy techniques for probing vibrational and rotational modes of molecules with high specificity and sensitivity [1]. For CARS of molecular vibration, the signals provide a distinct chemical "fingerprint" of the molecular composition. For CARS of molecular rotation, it provides a unique spectral tool to study molecular structure in gas-phase molecules. In the rest of this paper, we will focus on CARS based on molecular vibration, which has advantages in biophotonics and chemical identifications. A typical CARS process involves three beams (pump, probe, Stokes) interacting with a medium to generate a coherent signal at the anti-Stokes frequency $\omega_{AS} = \omega_{pump} + \omega_{probe} - \omega_{stokes}$, or $\omega_{AS} = 2\omega_{pump} - \omega_{stokes}$ when the pump is also used as the probe beam (Figure 1). The signal is resonantly enhanced when the frequency difference between the pump and Stokes beams matches a molecular Raman resonance, providing intrinsic vibrational contrast. Compared to spontaneous Raman scattering that suffers from inherently weak signals due to its reliance on inelastic scattering, CARS benefits from coherent interactions that enhance the signal-to-noise ratio (SNR) by 10^4–10^5 times, significantly reducing spectrum acquisition time [2,3]. CARS offers several advantages over other Raman techniques. Unlike spontaneous Raman and other coherent Raman (stimulated Raman scattering, and coherent Stokes Raman scattering, or CSRS), the blue-shifted CARS signal minimizes fluorescence interference. Phase matching conditions among the pump, the Stokes, and the probe in CARS can be more easily achieved and thereby make CARS more advantageous in practice. These features make CARS invaluable for a wide range of applications ranging from gas sensing [4–9] to combustion diagnostics [10–14], biological imaging [15–24], material characterization [25–29], environmental monitoring [30,31], and the food industry [32].

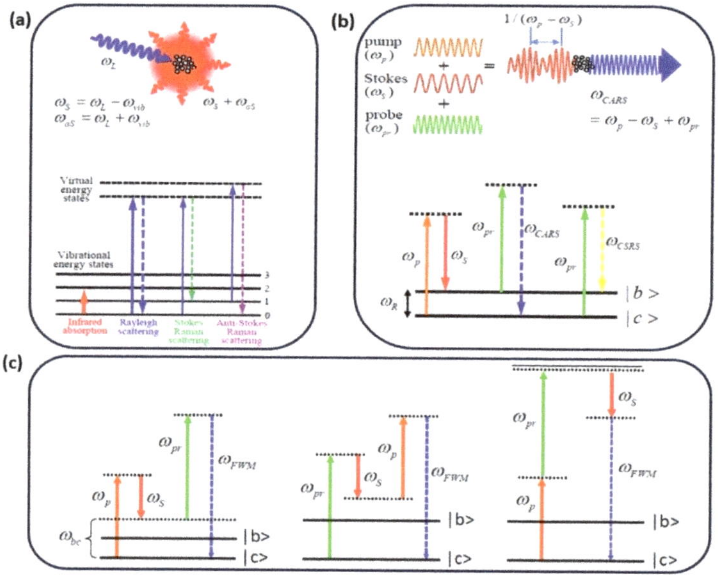

Figure 1. Principle of Raman. (**a**) Spontaneous Raman process. (**b**) Coherent Raman scattering including CARS and coherent Stokes Raman scattering (CSRS). (**c**) Energy diagram for NR FWM background including multiple off-resonant vibrational modes and the instantaneous electronic response. ω_p, ω_{pr}, ω_s, and ω_{as} are the frequencies of the pump light, the probe light, the Stokes light, and the anti-Stokes light.

Despite these strengths, detecting a CARS signal is often complicated by the presence of non-resonant (NR) four-wave mixing (FWM) background due to multiple off-resonant vibrational modes and the instantaneous electronic response from other molecules, or even from the same molecules. CARS itself is a specialized form of FWM that targets Raman vibrations. Non-resonant FWM frequencies are generated in the same manner as CARS, albeit not specific to Raman vibrations, thus resulting in a larger range of produced frequencies that do not correspond to specific molecular vibrations of the sample. This can cause the NR FWM background to overpower the CARS signal, and fluctuations of the NR background can also wash out the CARS signal, preventing detection. To suppress or avoid the unwanted NR background and improve signal-to-noise ratio (SNR) of CARS, various techniques have been proposed and employed. Some examples include the following: Epi-CARS detects signals in the backward direction where the NR FWM background is also reduced [33]. This technique requires the size of the sample to be smaller than the optical wavelength. Polarization-dependent CARS suppresses the NR background by leveraging Raman depolarization ratio of certain modes [34]. Interferometric schemes have been employed in detection to increase the SNR ratio and spectral resolution [35–37]. Heterodyne detection technique provides an alternative way to suppress the NR background [38]. Notably, combined with dual frequency technique, CARS signal can be obtained with high resolution (4 cm^{-1}) over a Raman spectrum spanning more than 1200 cm^{-1} recorded within less than 15 μs [39]. These aforementioned techniques rely on intrinsic properties of samples or a complicated laser system like dual frequency comb. They are less effective in the presence of strong multiple scattering in rough samples because scattering randomizes spectral phases and polarization, which hinders the practical application of CARS.

To circumvent these obstacles, techniques involving the manipulation of laser pulse properties, termed pulse shaping, have been adapted in CARS in the last two decades [40]. A pulse shaper is a versatile tool that can provide an ability to actively tailor the time or frequency structure of optical pulses to precisely meet the needs of the quantum system being manipulated [41,42]. This concept can also include methods developed specifically for optimizing CARS signals, including various approaches for broadband CARS generation, such as the high-speed platform achieving full fingerprint and CH-stretch imaging in biological tissues [22], along with other methods, where recent advances demonstrate label-free chemical imaging of cells and tissues across the full vibrational fingerprint and functional group regions [15]. Both simpler passive pulse shaping techniques (such as imposing linear chirp or spectral filtering) and more sophisticated active pulse shaping techniques (using programmable devices) have been exploited to manipulate the pulses in order to suppress the NR FWM background, optimize the spectral resolution, increase the Raman efficiency, and simplify the experimental setup. The remainder of this article is organized as follows: Section 2 gives a theoretical description of CARS using ultrafast laser pulses. Section 3 introduces passive pulse shaping techniques with two specific examples: hybrid CARS technique, which is established stemming from time-resolved CARS, and the spectral focusing CARS, which is performed by linearly chirping the broadband pulses in order to optimize the light-matter interaction, providing a rapid and selective CARS technique suitable for biological imaging. Section 4 covers active pulse shaping techniques. We firstly review the progress of CARS spectroscopy that relies on a spatial light modulator (SLM). Then we discuss adaptive pulse shaping by using another programmable pulse shaper with the assistance of acousto-optic programmable dispersive filters (AOPDFs), including some potential applications in encryption. Section 5 concludes the article and overlooks the prospect of pulse shaping techniques in CARS.

We are delighted to dedicate this review to our hero Professor Jaan Laane, a colleague and a friend who, through the years, has generously provided general wisdom and numerous insights into applications of vibrational spectroscopy as a tool used to study molecular structure and intramolecular dynamics.

2. Theory of Ultrafast CARS

As a third-order nonlinear optical effect, using ultrashort pulses has a significant benefit due to their high intensity and their capability for time-resolved measurements. It leads ultrafast time-resolved CARS spectroscopy where a pair of ultrashort (~50 fs) pulses is typically used to excite broadband Raman modes uniformly, and another ultrashort pulse is used to probe the coherence, resulting in the generation of CARS signal.

As a nonlinear phenomenon, the generation of CARS using ultrafast laser pulses can be described through third-order polarization, which can be written as the sum of the background and resonant contributions [43,44]:

$$P^{(3)}_{CARS}(\omega) = P^{(3)}_{B}(\omega) + P^{(3)}_{R}(\omega) = \int_{0}^{+\infty} d\Omega[\chi^{(3)}_{(B)} + N\chi^{(3)}_{(R)}(\Omega)]E_{pr}(\omega - \Omega)S(\Omega) \quad (1)$$

$$S(\Omega) = \int_{0}^{+\infty} d\omega' E_{pump}(\omega')E'_{Stokes}(\omega' - \Omega) \quad (2)$$

where $E_{pr}(\omega)$ is the probe field; $S(\Omega)$ is the convolution of the pump field $E_{pump}(\omega)$ and the Stokes field $E_{Stokes}(\omega)$; and N is the concentration of the Raman active molecule. In general, although not always, $\chi^{(3)}_{B}$ corresponds to the NR response and is purely real while the resonant susceptibility $\chi^{(3)}_{R}$ is complex and, for the case of Lorentzian lineshape, can be written as:

$$\chi^{(3)}_{R}(\Omega) = \sum_{j} \frac{A_j}{\Omega_j - \Omega - i\Gamma_j}, \quad (3)$$

where $\Omega = \omega_{pump} - \omega_{Stokes}$; A_j, j, and Γ_j are the amplitude, frequency, and spectral half width of the j th vibrational mode, respectively. The total CARS signal is given by:

$$I_{CARS}(\omega) \propto \left|P^{(3)}_{CARS}(\omega)\right|^2 = \left|P^{(3)}_{B}\right|^2 + \left|P^{(3)}_{R}(\omega)\right|^2 + 2Re\left[P^{(3)}_{B}P^{(3)*}_{R}(\omega)\right], \quad (4)$$

where the first term $\left|P^{(3)}_{B}\right|^2$ is the aforementioned NR FWM background that limits the sensitivity of the CARS measurements.

In theory, it is possible to eliminate the NR background by delaying the probe pulse, taking advantage of the instantaneous response of the NR FWM and the longer lifetime of the resonant coherent vibrations. This can be realized in experiments by using an ultrafast laser whose pulse duration is shorter than the lifetime of resonant vibrations. Delaying the probe pulse is effective for background suppression, even when using extremely short (<10fs) pump/Stokes pulses combined with a sub ps probe to cover ultrabroad bandwidths [45]. However, there is a trade-off in spectral resolution for only using ultrashort pulses. Equation (1) implies that the CARS spectrum is the convolution of the probe field and the third-order susceptibility. The linewidth of the CARS signal peak is determined by the probe bandwidth and the bandwidth of the vibrational mode, whichever is larger. In general, a narrowband probe can help to spectrally resolve multiple CARS lines. However, the bandwidth inherent in the ultrafast pulse is usually broader than the linewidth of Raman lines, so that multiple CARS lines close together could not be distinguished in the spectrum using ultrafast pulses, resulting in multimode interference, generally referred to as quantum beats, in time-resolved CARS spectra. This brings in the time-resolved CARS technique, which can not only identify CARS lines, but also explore dynamics of vibrational states [46–51]. Time-resolved CARS suppresses the NR background by delaying the probe pulse, but the technique remains vulnerable to fluctuations. Moreover, when multiple Raman modes are excited together, the multimode interference requires the ability to record high-quality spectral profiles over a relatively large probe-delay span that is challenging in the presence of both scattering and fluctuations.

There are many strategies to optimize CARS generation. A technique named FAST CARS (Femtosecond Adaptive Spectroscopic Techniques for CARS) was proposed for maximizing the coherent molecular oscillation with a sequence of femtosecond pulses and minimizing the NR background for rapid identification of bacterial spore [52,53]. To address a significant part of the Raman vibrational bands simultaneously, a broadband pump and Stokes and a narrowband probe are required [15,22,45]. The narrowband probe can be obtained by laser technique [45] or pulse shaping technique. The broadband spectrum can be achieved from supercontinuum (SC) generation [15,22], which can cover the entire Raman active region (400–300 cm^{-1}), or advanced laser technique [45]. When the duration of the broadband pulse is shorter than the period of molecular vibration and the bandwidth is larger than Raman shifts, it can lead to the concept of impulsive excitation [37,40,54] in which a single pulse is used as the pump/the Stokes. It significantly simplifies the experimental setup, but the broadband excitation of Raman bands usually results in strong NR background. Therefore, another selective excitation of Raman is proposed with manipulating both the phase and polarization [55,56].

Pulse shaping technique provides a method to tailor the pulse duration, the bandwidth, the spectral phase, and the temporal delay, aiming to achieve close-to-optimal resonant response with reasonable suppression of the NR background, higher spectral resolution, selective and optimized excitation, and so on. The actual optimal values depend on the specific Raman line width, the sensitivity of a setup employed, and the relative strengths of the resonant and NR susceptibilities. For example, the top-hat spectrum gives a sinc-squared temporal profile as $[\mathrm{Sin}(\omega t/2)/(\omega t/2)]^2$ for the probe pulse intensity. Putting the pump and Stokes pulses in the node of the sinc function, as shown in Figure 2, would result in effective suppression of the NR background while the resonant signal remains. Nevertheless, in practice, the background suppression is never perfect, and especially at low concentration of the target molecules the residual background often interferes with the CARS signal. According to Equation (3), we can have $I_{\mathrm{CARS}}(\omega) \propto \left|P_R^{(3)}(\omega)\right|^2 \propto N^2$, which means that in the ideal situation, the intensity of CARS would be proportional to the square of the number of molecules. However, due to the NR background, the interference term $2Re\left[P_B^{(3)} P_R^{(3)*}(\omega)\right]$ that is proportional to N tends to dominate, resulting in a linear relationship between the signal intensity and the number of molecules. By taking advantage of background interference, a proper tailored probe pulse can lead to a novel heterodyne detection strategy for CARS measurement [54,57].

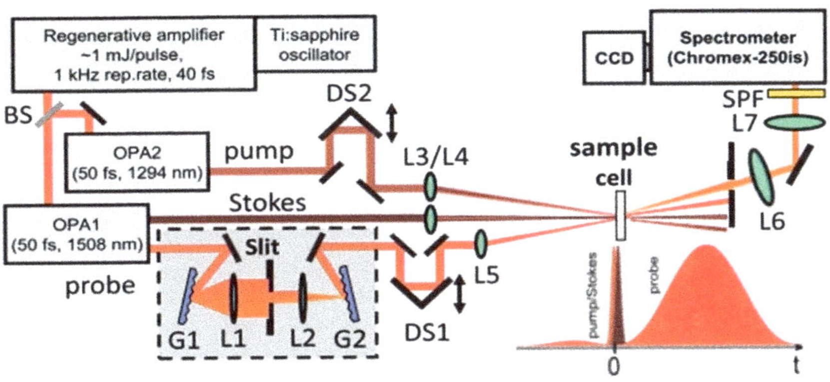

Figure 2. Schematics of the hybrid CARS setup from [58]. BS: beam splitter. OPA: optical parametric amplifier. DS: delay stage. G: grating. L: lens. SPF: short pass filter.

Pulse shaping techniques can be divided into two categories, passive pulse shaping and active pulse shaping, depending on the energy consuming. Both of these two techniques can be applied to manipulate spectral phase and spectral amplitude, which have various applications depending on laser intensity, laser wavelength, experimental condition, sample, and so on. We will discuss both passive pulse shaping and active pulse shaping in the next two sections.

3. Passive Pulse Shaping for CARS

Passive pulse shaping is usually based on apparatuses like optical slit based on knife-edge, glass rods, prism pair, and so on. It is simple to set up and can be used without energy consuming. It provides a straightforward way to manipulate laser pulses, which is implemented in CARS spectroscopy such as hybrid CARS and spectral focusing. We discuss these two techniques in detail.

3.1. Hybrid CARS

Hybrid CARS is a technique that combines the advantage of time-resolved CARS spectroscopy with the robustness of frequency-resolved CARS [44,53]. Specifically, besides femtosecond pump and Stokes pulses, a proper designed narrowband probe (it is usually a picosecond pulse) ensures a sufficient spectral resolution that can resolve different Raman modes. The probe can be obtained by a passive pulse shaper, which consists of a set of optics to spatially disperse the component of light into separate wavelengths, an optical slit based on knife edge to narrow the spectrum, and another set of optics to recombine the light. A multichannel detector (usually a CCD) is used to simultaneously record the anti-Stokes signal at all optical frequencies within the band of interest. These features allow us to discriminate between the resonant contribution and the NR background, extracting the CARS signal even at zero probe delay. It is worth mentioning that hybrid CARS can also be realized with advanced laser technology and active pulse shaping.

Figure 2 shows an experiment layout of hybrid CARS based on a homemade pulse shaper using knife-edge technique [58]. The experiment utilizes a Ti:Sapphire regenerative amplifier (Legend, Coherent Co., Saxonburg, PA, USA) with two evenly pumped optical parametric amplifiers (OPAs) (Coherent OPerA-VIS/UV and OPerA-SFG/UV) to output two tunable femtosecond pulses as the pump and the Stokes. A small fraction of the amplifier output is sent through a home-made pulse shaper with an adjustable slit to obtain a probe field with a sinc-squared temporal profile. The three beams are focused separately to make them overlapped at the focuses. The time delay of the pump and the probe are adjusted by two delay stages, respectively. The scattered light is collected and focused onto the entrance slit of the spectrometer with a liquid nitrogen-cooled charge-coupled device (CCD) (Spec 10, Princeton Instruments Co., Trenton, NJ, USA).

The bandwidth of probe that affects the spectral resolution is determined by the width of the optical slit in the pulse shaper [49] (Figure 3). The quantum beats effect occurs when the spectral resolution of the CARS setup cannot resolve both Raman peaks. By narrowing the bandwidth of the probe filed using pulse shaper, the beating pattern gradually transforms from quantum beats to two separate streak lines as shown in Figure 3d. Pyridine molecules that have two Raman modes at 992 and 1031 cm^{-1} are measured using CSRS, which is intrinsically equivalent to CARS, except wavelength (Quantum beats of CARS can be found in Ref. [59]). The two Raman modes are excited simultaneously by ultrashort pulses.

When the spectral resolution is high enough to resolve the Raman shift, the NR background is still there veiling the CARS signal. If modifying the probe field to be a sinc-squared temporal waveform, it can help to further suppress the NR background by adjusting the time delay of the probe. When the first node of the sinc-squared temporal profile overlaps the Stokes and pump pulses, multiple Raman lines stand out against a weak background (Figure 4b), in contrast to the solid (red) curve in Figure 4a, where the probe has the same bandwidth and background dominates. The background is still greater than all the signals because suppression is never perfect. In this work [54], a linear relationship between intensity of signal and concentration, which can be attributed to influence of background as discussed previously, is observed. Also, it highlights a sensitivity that can detect 5 mM glucose in liquid.

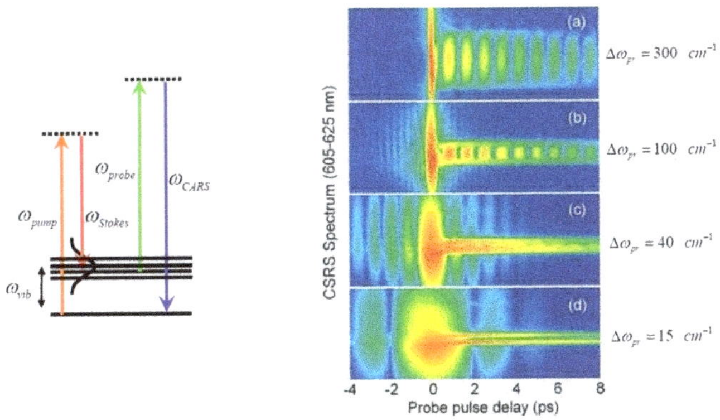

Figure 3. Transition from time-resolved to hybrid CSRS [44]. CSRS spectrograms for different spectral bandwidths of the probe pulse: (**a**) 300 cm^{-1}, (**b**) 100 cm^{-1}, (**c**) 40 cm^{-1}, (**d**) 15 cm^{-1}. Two Raman lines of pyridine, 992 and 1031 cm^{-1}, are excited via a pair of ultrashort laser pulses. Pump: λp = 737 nm, FWHM ≈ 260 cm^{-1}, 0.5 μJ/pulse. Stokes: λS = 801 nm, FWHM ≈ 480 cm^{-1}, 0.9 μJ/pulse. Probe: λpr = 577.9 nm, 0.15 μJ/pulse.

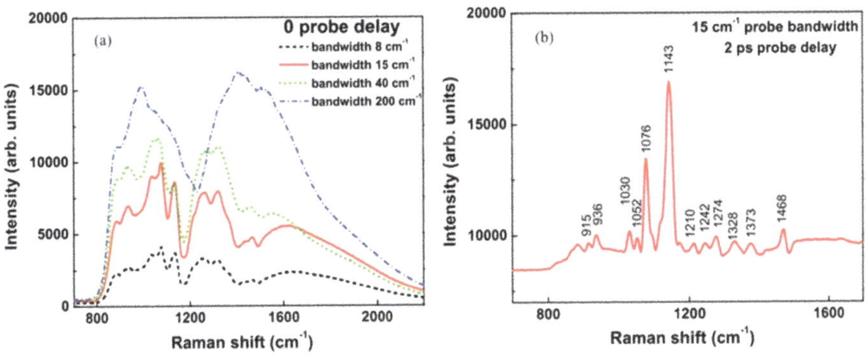

Figure 4. CARS spectra of D-glucose solution at 2680 mM. (**a**) Spectra at 0 probe delay with different probe bandwidths and the same power; (**b**) background suppression when probe is delayed to the node; note that the background level is still around 9000 units. Here the integration time is 0.2 s (From Ref. [58]). Considering the linewidth of Raman modes, the bandwidth of the probe pulse determines if the Raman peaks can be distinguished in CARS. In (**a**), the linewidth of glucose mode is around 20 cm^{-1}. Therefore, it can be spectrally resolved by the narrower bandwidth probe, but it cannot be resolved if the bandwidth of probe is broader than 20 cm^{-1}.

In general, hybrid CARS based on a passive pulse shaper is a versatile technique in which both collinear and non-collinear configuration for laser beams can be implemented to measure either transmission or reflection [44,60,61]. It expands the applicability of this technique for studying spectroscopy or imaging of both transparent samples and opaque samples. In a forward scheme to measure transmission, hybrid CARS can perform quantitative measurement and reveal concentrations of target molecules in various environments [45,58,62,63]. For example, organic fluid can be measured in a two-color-beam CARS [45], and glucose in aqueous condition down to 5 mM is demonstrated [58], which can be used for diabetes diagnosis. Molecules in gas phase can be well studied using hybrid CARS based on molecular vibration in the context of molecular concentration [62,63], temperature [64–66], and molecular dynamics [66]. Gaseous samples usually have weak

NR background, which can ensure a quadratic dependence between CARS signal and the number of molecules [62]. It has been demonstrated to measure the temperature up to 2400 K [64] with an accuracy better than the nano-second laser diagnostical technique. Alternatively, the temperature dependence of vibrational state populations in CO_2 has been probed using fs/ps CARS to perform thermometry from single spectra [66]. Impulsive CARS can also be adapted in hybrid CARS for thermometry [66–68] and bioimaging [69,70]. A narrowband probe light can be achieved using second harmonic generation of the fundamental light assisted by a passive pulse shaper [66], volume Bragg grating [67], laser etalon [68], the phase matching of nonlinear crystal [69,70], and so on. The bandwidth of the probe can be narrowed down to 0.37 cm^{-1} with the assistance of volume Bragg grating, which would be sufficient to study rotational Raman scattering [67]. Reflection measurements might not be able to perform quantitative measurements and are mainly applied in measuring opaque samples like bacteria [71] or microspectroscopy for bioimaging [72,73].

It is worth mentioning that, in addition to vibrational Raman spectroscopy, hybrid CARS for rotational Raman can also be implemented to explore temperature [68,74–77], concentration [68], and pressure [78] of gaseous samples, particularly for the study of flames and combustion. A technique called second harmonic band compression (SHBC) is developed to generate a narrowband probe field by mixing oppositely chirped pulses [75,76]. Hybrid CARS based on SHBC provides a gas-phase thermometry to diagnose flame and combustion primarily based on linewidth of rotational CARS spectra [74–77]. The gas pressure can be measured as demonstrated in Ref. [78], where an impulsive excitation is used for rotational CARS, and a single laser shot spectrum of N_2 is recorded to obtain the pressure.

3.2. Spectral Focusing CARS

Another passive pulse shaping technique that has been explored recently is called spectral focusing CARS, in which the quadratic spectral phases of the pump and the Stokes pulses are modified to have the same linear chirp. It can be realized with glass rods which would chirp the pulse passing through [79]. It provides a simple and efficient way to manipulate the spectral phase to achieve picosecond pulses. It reduces power loss compared with other technique to control quadratic phase like using grating-based stretchers [79], chirped mirrors [80], and pulse shapers [81].

By optimizing the length of glass rods, the pump and Stokes pulses are prepared with the same linear chirp, creating a constant instantaneous frequency difference between them that can be tuned to match specific molecular vibrations [79,82,83]. The stretched duration increases the interaction time between the pulses, extending probing of the molecular vibrations at the cost of lower peak power while maintaining the same total pulse energy. The chirped pulses have reduced peak power compared to transform-limited pulses, which consequently reduces the intensities of both NR FWM and CARS, but the contrast of the resonant Raman signal to the NR FWM background is increased according to the experimental results [82,83]. The spectral resolution of spectral focusing can be hundreds of times higher than the un-chirped pulse [84,85]. Spectral focusing CARS is particularly valuable when working with sensitive samples, such as living cells, where high peak powers could cause photodamage. Moreover, the instantaneous frequency difference can be quickly tuned by adjusting the time delay between the pump and the Stokes pulses. This allows for a rapid and selective excitation of narrow vibrational resonances, which is suitable for chemical imaging. If dual frequency comb technique is adapted with spectral focusing, it can further accelerate the imaging speed [86]. These features make spectral focusing CARS highly promising for rapid hyperspectral imaging [87–92], particularly in biology and medicine [93–95]. It can also be used to explore concentration and temperature for gaseous samples based on rotational Raman scattering by optimizing pulses with 4f pulse shapers in which Raman shifts of 100 cm^{-1} or more are attainable and allow for enhanced detection of high-energy (150–300 cm^{-1}) rotational Raman transitions at near-transform-limited optimum sensitivity [96].

We conduct an experiment to demonstrate spectral focusing CARS (Figure 5). A commercial Ti:Sapphire femtosecond (Astrella Co., New York, NY, USA) provides 70 fs pulse at 5KHz (central wavelength at 800 nm) with a pulse energy of 4 mJ. About 40% of the amplifier output is used to pump an OPA to generate Stokes pulses centered at 1040 nm. A portion of remaining 800 nm pulses is used as a pump and probe pulse for CARS. Several glass rods (N-SF 11, dispersion parameter D: −537.38 ps/(nm km) at 806 nm, −214.70 ps/(nm km) at 1040 nm) with a total length of 230 mm are inserted in the pump pulse path to chirp them to around 1.65 ps. Another 350 mm glass rod is inserted in the beam path of 1040 nm beam to stretch the pulse to 1.65 ps. The two pulses are then combined collinearly and focused on the sample to generate FWM, which include CARS and NR FWM. The total intensity of FWM varies as a function of relative time delay, and it would be enhanced when it is on resonant with a Raman mode. Hence, Raman shift can be deduced from the measurement of the light intensity of FWM by varying the time delay instead of measuring the spectra.

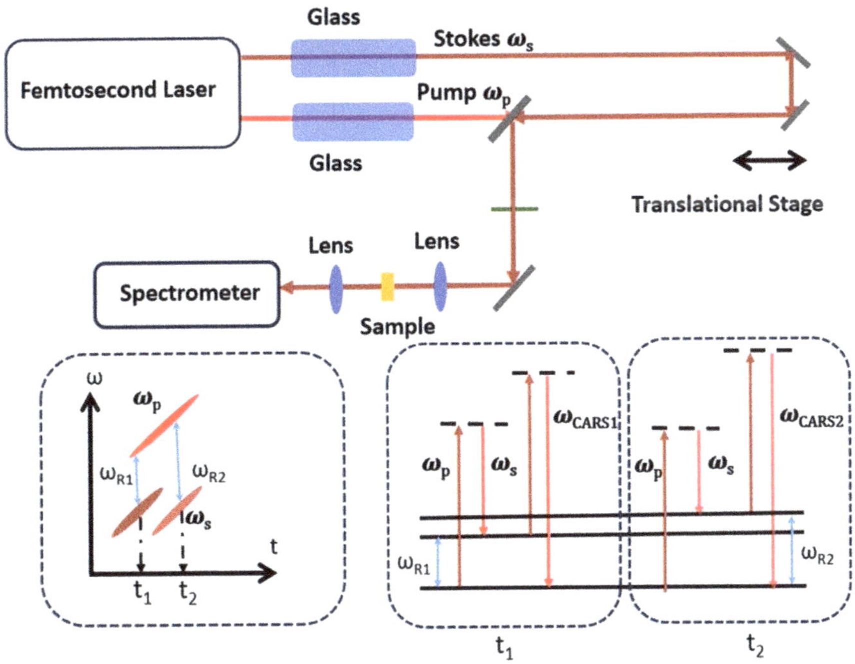

Figure 5. Experimental layout for spectral focusing. Insets are the principle of spectral focusing CARS. The pump field will also be the probe field. At different time delay t_1 and t_2, different Raman modes (ω_{R1} and ω_{R2}) are excited and different CARS are generated.

In the experiment, we record the spectra of FWM as a time delay between two pulses with a spectrometer (HR550, Horiba Co., Kyoto, Japan). In the data processing, the spectral intensity is summarized at each time delay to obtain the temporal intensity profile. We also record the spectra generated from a piece of glass slide, which is 150 µm, as a reference for the NR FWM background. Samples measured in our experiment are low density polyethylene (LDPE) film, plastic film (polyethylene terephthalate (PET)), and acetone in a cuvette. Figure 6a is the relationship between the measured Raman shift and the relative time between the pump pulse and the Stokes pulse. The data are processed considering the standard spontaneous Raman spectra. The linear fitting implies a linear chirp for both the pump and the Stokes pulses. Figure 6b is the measured temporal profile for LDPE (blue), PET (orange), and acetone (red). They are obtained by summarizing the spectral data recorded by spectrometer at different time delays. It shows that the Raman shift can be linearly correlated with time delay, which is attributed to a linear chirp generated by the glass rod. Even though the higher order chirp is unavoidable in practice, its impact on the result in our experiment is minimal. Figure 6c,d are the retrieved spectra of CARS for LDPE and PET according to the fitting result in Figure 6a. Although the spectral resolution of spectral focusing is insufficient to fully resolve the Raman shifts, the linear chirp enabled accurate mapping of vibrational modes. Discrepancies between the retrieved CARS spectrum and spontaneous Raman spectrum primarily result from the varying non-resonant contributions rather than higher order dispersion, as discussed in Ref. [80]. Our results have shown the potential of spectral focusing CARS as a simple method to achieve high speed imaging in applications such as bio-imaging.

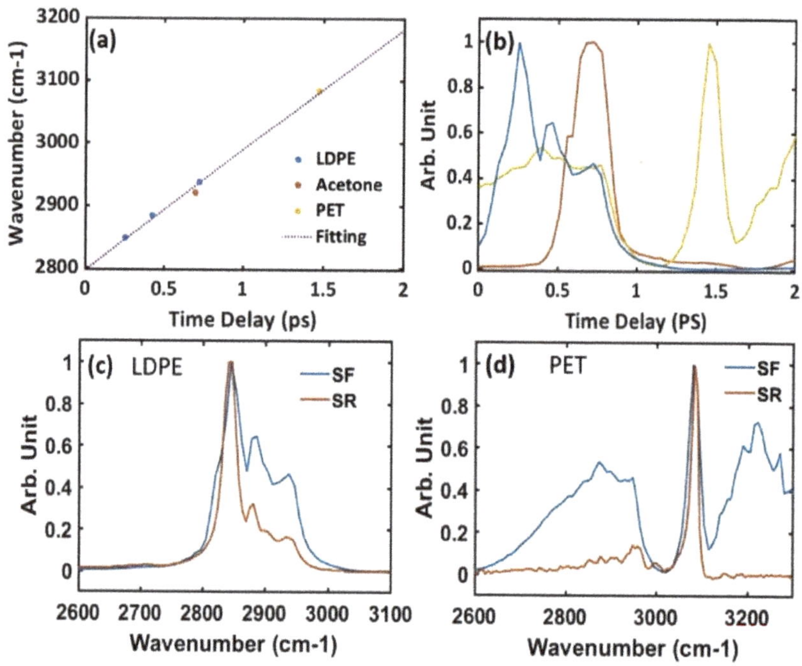

Figure 6. (**a**) The relation between relative time delay and Raman shifts of LDPE (blue), PET (orange) and acetone (red). The Raman shift for LDPE is at 2848.5 cm^{-1}, 2884.5 cm^{-1}, and 2937.8 cm^{-1}. The Raman shift for PET is at 3083.7 cm^{-1}. The Raman shift for acetone is at 2921 cm^{-1}. The time delay for x-axis is arbitrary and only for reference. It shows a linear dependence between the Raman shift and the temporal delay, which results from the linear chirped pulses. (**b**) Temporal intensity profile for different measurements: LDPE (blue), PET (orange), and acetone (red). It is constructed by recording the data at different time delays. (**c**) Spectrum of LDPE from spontaneous Raman (SR) (red line) and spectral focusing (SF) CARS (blue line). (**d**) Spectrum of PET from spontaneous Raman (SR) (red line) and spectral focusing (SF) CARS (blue line).

4. Active Pulse Shaping CARS

Active pulse shaping usually relies on programmable devices like spatial light modulators (SLMs), acousto-optic modulators, AOPDFs and so on. It can manipulate both spectral phase and spectral amplitude simultaneously, which are implemented for optimizing Raman excitation, selective Raman excitation, and reducing acquisition time. Active pulse shapers based on SLMs and AOPDFs are programmable devices controlled by a computer. There are two different control strategies: open loop control and feedback (adaptive) control [42]. In the open loop configuration, the desired output waveform is specified by the user with knowledge of the input pulse including necessary pre-compensation due to unknown distortion. Alternatively, the adaptive control, also known as a close loop control, relies on the ability to program a pulse shaper under computer control. In this strategy, it usually starts with a stochastically prepared pulse shaper, which would be updated iteratively according to a designed optimization algorithm based on the difference between desired and measured experimental output. The adaptive control scheme is less intuitive, but is suitable for optimization of strongly nonlinear processes, or for manipulation of quantum states with insufficient knowledge. We will cover pulse shaping with programmable pulse shapers such as SLMs and AOPDFs, which are adopted in hybrid CARS.

4.1. Pulse Shaping with SLM

An SLM is a widely used programmable device for pulse shaping in optics and photonics. The use of SLMs for programmable phase shaping in CARS was fundamentally demonstrated by Dudovich et al. [40], who showed that a computer-controlled SLM in a 4f pulse shaper enabled selective excitation of vibrational levels and effective suppression of the NR background in single-pulse CARS. An SLM is usually based on liquid crystal, whose pixel size is on the order of μm. The spectral resolution of this device can be close to several wavenumbers. Critically, they achieved high spectral resolution, significantly exceeding the pulse bandwidth limit, by precisely modulating the spectral phase using the SLM to exploit quantum interference pathways [40]. An SLM is commonly integrated into a 4f dispersion management system (pulse shaper, shown in Figure 7). In this configuration, the input laser pulse is spectrally dispersed onto the SLM, which applies a user-defined phase pattern onto the spectral components before they are recombined. Common implementations use programmable liquid crystal modulator arrays that allow independent, simultaneous gray-level control of both spectral amplitude and phase [42]. It provides a simple way to achieve the desired laser pulses, which is valuable to help achieve impulsive excitation CARS [40] and optimization of CARS for microscopy [97,98]. A two beam hybrid CARS combined with a spatial light modulator (SLM) to optimize the Raman excitation has been demonstrated to be able to precisely measure the temperature in a high pressure gas cell based on rotational Raman scattering [99]. Programmable hyperspectral CARS microscopy is achieved by using a 2D SLM to tailor the Stokes light to collect spectral information in a more rapid and efficient manner [100]. In a noise autocorrelation spectroscopy for coherent Raman scattering, an SLM can be used to add noise to the probe beam to achieve a spectral resolution without temporal scanning or spectral pulse shaping [101].

Liquid crystal SLMs can also offer phase-only modulation, which has advantages over other pulse shaping techniques because the power loss is reduced. It brings in unique programmability, which allows for dynamic optimization of complex light-matter interactions inherent in CARS. Phase controlling can lead to various techniques like phase cycling to suppressing the NR background [102], selective excitation of Raman mode [103–105], spectral phase optimization for high spectral resolution [106], optimizing CARS with frequency resolved optical gating technique [107], and so on. Temporal pulse shaping, such as generating delayed pulse sequences using an SLM, allows for time-gated detection methods that can suppress the instantaneous NR background, improving the sensitivity for detecting specific molecular in microscopy [103]. A dual-mask SLMs configuration which allows for combined phase and polarization control of the supercontinuum is originally proposed by Oron et. al. [55] and helps to achieve a broadband selective excitation of CARS for microscopy [54,55], manipulation of quantum states [108], thermometry [109,110], stand-off Raman spectroscopy up to a 12 m distance [111,112], and so on. Controlled relative phases using an SLM allows for "all-optical processing", coherently adding signals from multiple Raman lines of target molecules for enhanced detection or achieving destructive interference to cancel signals from specific (potentially background) molecules [113]. SLM-based shaping enables highly sensitive heterodyne detection in a single-beam configuration, where part of the shaped beam acts as a phase-controlled local oscillator, providing significant signal amplification and linear concentration dependence, ideal for detecting low-concentration species [114]. The achievable spectral resolution in spectral focusing is quantitatively linked to the applied chirp, i.e., the quadratic spectral phase. SLMs enable a precise manipulation of spectral phase. Using SLM-based tailored spectral focusing, Brückner et al. investigated this trade-off, showing experimentally and via simulations how the CARS linewidth could be actively narrowed (e.g., from >50 cm^{-1} down towards the probe limit of ~25 cm^{-1} for acetonitrile) by increasing the chirp magnitude. In principle, it can help to correct much higher order spectral phase [89]. The active control allows the resolution to be optimized for specific molecular features and allows for balancing signal strength against the resolution needed, for instance, to resolve CO_2 Fermi dyads (~6 cm^{-1}) [109] versus broader hot bands (~14 cm^{-1}) [66].

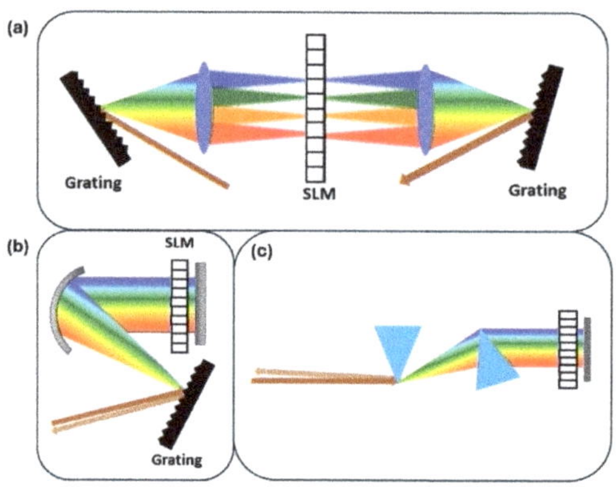

Figure 7. Schematics of pulse shaper with SLM. (**a**) A standard 4f dispersion management system. An input pulse is dispersed by the first grating, collimated by the first lens onto the SLM in the Fourier plane to independently control the spectral phase and amplitude of ultrashort pulses, and then recombined by the second lens and grating after spectral modulation. (**b**) A folding scheme for 4f system. A concave mirror is used to collect the light, and another flat mirror would reflect the beam back. (**c**) Pulse shaper with prism pairs.

4.2. AOPDF for CARS

AOPDF is another type of programmable spectral filter that can accomplish similar ends but with a different physical mechanism based on a co-propagating acousto-optic interaction [115,116]. It is capable of manipulating the spectral-temporal phase and amplitude shaping through the interaction between light waves and acoustic waves inside a birefringent crystal. It has been widely used with amplified femtosecond laser system for pulse optimization and stabilization [117–120], pulse characterization [121–123], higher order spectral phase manipulation [124], and selective detection in low frequency Raman imaging [125].

In CARS, we attempt to adopt pulse shaping assisted with AOPDF and a passive pulse shaper to selectively excite certain Raman modes and suppress other Raman lines. We demonstrate a selective excitation of Raman mode by making a molecule like cyclohexane perform "scissors" (Raman shift at 1470 cm^{-1}), but no "wagging" (1060 cm^{-1}) with its CH_2 groups of atoms. The experiment is based on the Hybrid-CARS technique with an AOPDF device inserting in the beam path to manipulate the pulse as shown in Figure 8. The device for AOPDF used in our experiment is called dazzler (Fastlite Co., Antibes, France), which can control temporal shapes of total duration anywhere from ~40 fs to ~8 ps and work with wavelength ranged from 1100 nm to 1500 nm. A cuvette filled with cyclohexane is put at the focal point. We use the adaptive control strategy to optimize the pump pulse according to the pre-set Stokes pulse to generate the best experiment results. Genetic algorithm (GA), which has been adopted to optimize CARS with SLMs [126,127], is implemented for the adaptive control of the pump pulse through a computer that remotely controls the AOPDF. The parameters of the spectral phase and the amplitude would be genetic representations of GA. Specifically, GA is implemented through LabVIEW (National Instruments Co., Austin, TX, USA) in the experiment. The fitness function of GA is based on the SNR of CARS at 1470 cm^{-1}, which is the Raman mode to be excited. To test our strategy, we pre-set the Stokes pulse with a positive chirp, a negative chirp, and a higher order spectral phase. The best result is shown in Figure 9a,b. Both the pump and Stokes pulses are characterized with cross-frequency resolved optical gating (X-FROG) technique with the assistance of probe light. After 87 generations of GA, the SNR of CARS at 1470 cm^{-1} is around 2000, which is 1000 times higher than that at 1060 cm^{-1}. All other Raman modes are suppressed greatly. We also attempt to achieve selective excitation by manually controlling the spectral phase without using a dazzler. The results are similar to those by using GA.

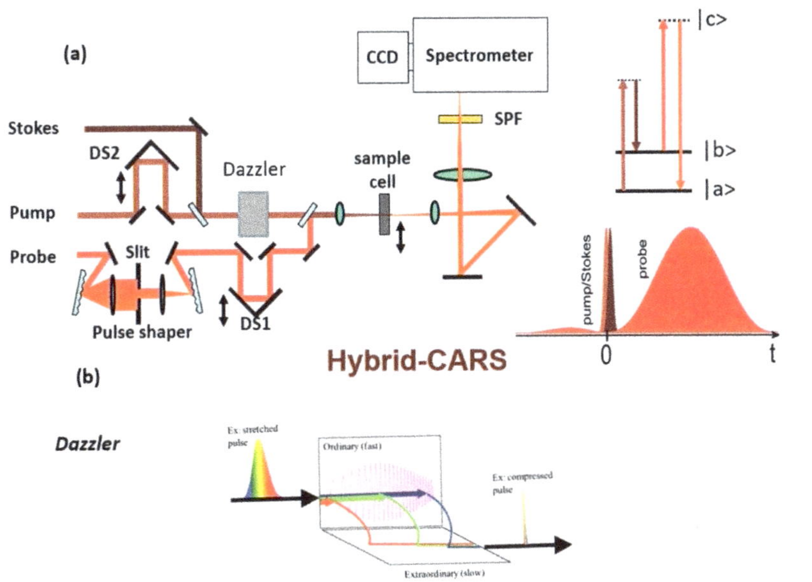

Figure 8. (**a**) Experimental setup for hybrid-CARS with AOPDF. A dazzler is inserted in the beam path to manipulate the Stokes pulse. (**b**) Principle of AOPDF.

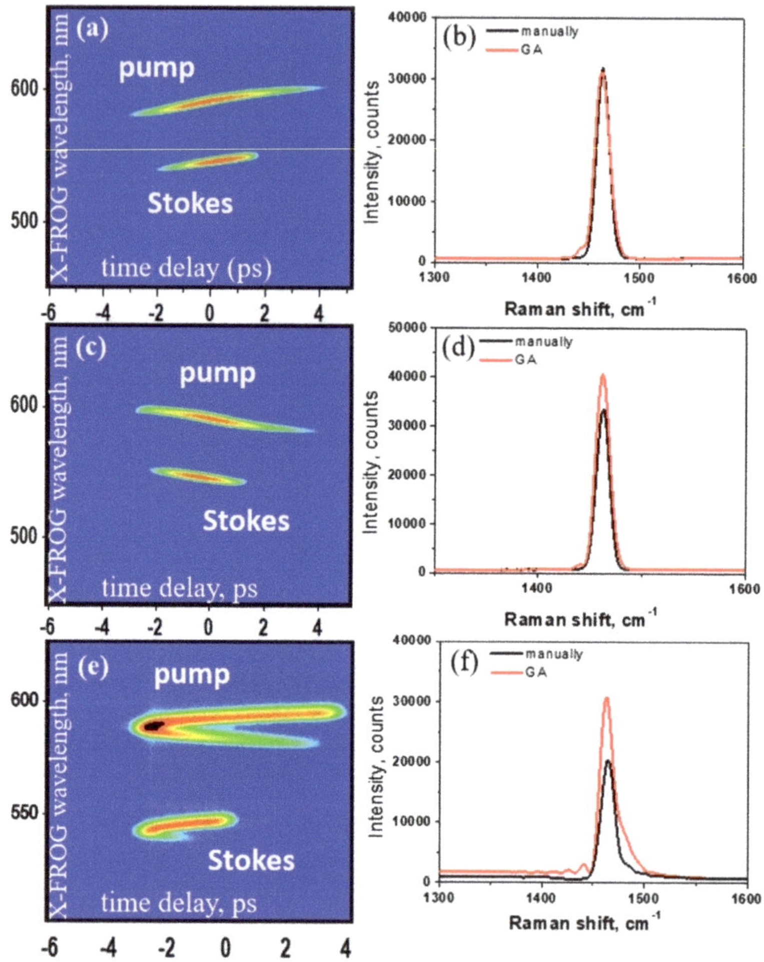

Figure 9. (**a**,**b**) X-FROG of the pump pulse and the Stokes pulse with a positive chirp and the spectrum of CARS. (**c**,**d**) X-FROG of the pump pulse and the Stokes pulse with a negative chirp and the spectrum of CARS. (**e**,**f**) X-FROG of the pump pulse and the Stokes pulse with a higher order spectral phase and the spectrum of CARS. In (**b**,**d**,**f**), black lines are the results using manually control and red lines are the results using GA.

In principle, people can also use pulse shapers to excite all the Raman modes with a flat, uniform pulse in the spectral domain. Selective molecular excitation hinders species-specific detection, while excitation uniformity comes at the expense of excitation strength. These tradeoffs can be clearly traced through theoretical analysis and need to be considered when designing a spectroscopic laser system for a particular goal. The selective excitations could be applied to effectively control the vibration of molecules, which might find applications in biology and chemistry.

In addition, since high order spectral phase cannot be measured by a linear method, we can develop a protocol for encryption, which can only be read through a nonlinear process. In this protocol, the information can be encrypted in the spectral phase of an ultrashort pulse. In order to decrypt the information, we need to generate a 2D spectrogram, which can correlate the spectral intensity to the time and frequency domains. The 2D spectrogram is generated between the encoded pulse and a designed read-out pulse through the four-wave mixing (FWM) process as shown in Figure 8a, which is based on the third-order nonlinearity. We experimentally demonstrate this protocol by using a dazzler to encode the information in the higher order spectral phase. We encode four letters "TAMU" in the spectral phase of a pump pulse whose central wavelength is 1257 nm and bandwidth is around 30 nm (Figure 10b). The read-out pulse is centered at 804 nm with a pulse duration around 200 fs. The experimental setup is shown in Figure 10c. The FWM is generated on a piece of glass by mixing two pulses. The FWM is recorded as a function of time delay between the encrypted pulse and the read-out pulse to generated 2D-spectrogram. The result is shown in Figure 10b where it clearly shows the "TAMU" word in the 2D-spectrogram, but if the pulse is only measured in either the frequency or time domain, we cannot obtain the four letters correctly.

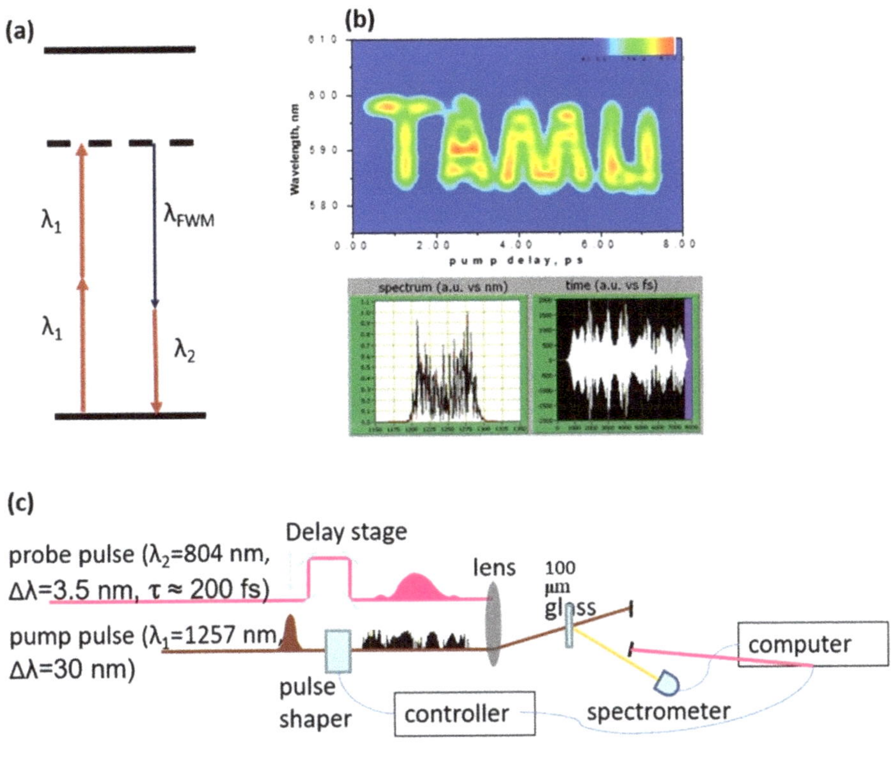

Figure 10. Protocol for encryption based on nonlinear spectroscopy using AOPDF. (**a**) The energy level scheme for FWM. (**b**) The decrypted letters and the corresponded pump pulse in both frequency domain and time domain designed by dazzler. (**c**) Experimental setup.

In this protocol, the pulse would be chirped if it propagates in the dispersive medium, but with pre-knowledge of the dispersion, people can design a scheme to compensate for the dispersion and still obtain the right information. This implies that the information can only be read out successfully by a designed read-out pulse, which can have a special design spectral phase as well. If an eavesdropping thief intercepted the encoded pulse, it would still be unreadable. For the data capacity, we can encrypt all the information in one pulse, but for security, we can encrypt the information in a series of pulses as shown in Figure 8. The complexity of the pulse-shape protocol is around 10^{200} in theory, which is sufficiently high for encryption. In Figure 11, the first column shows the decrypted four letters using four different pulses. Their encrypted spectra are shown in the middle column. The last column shows the pulse in the time domain predicted by dazzler.

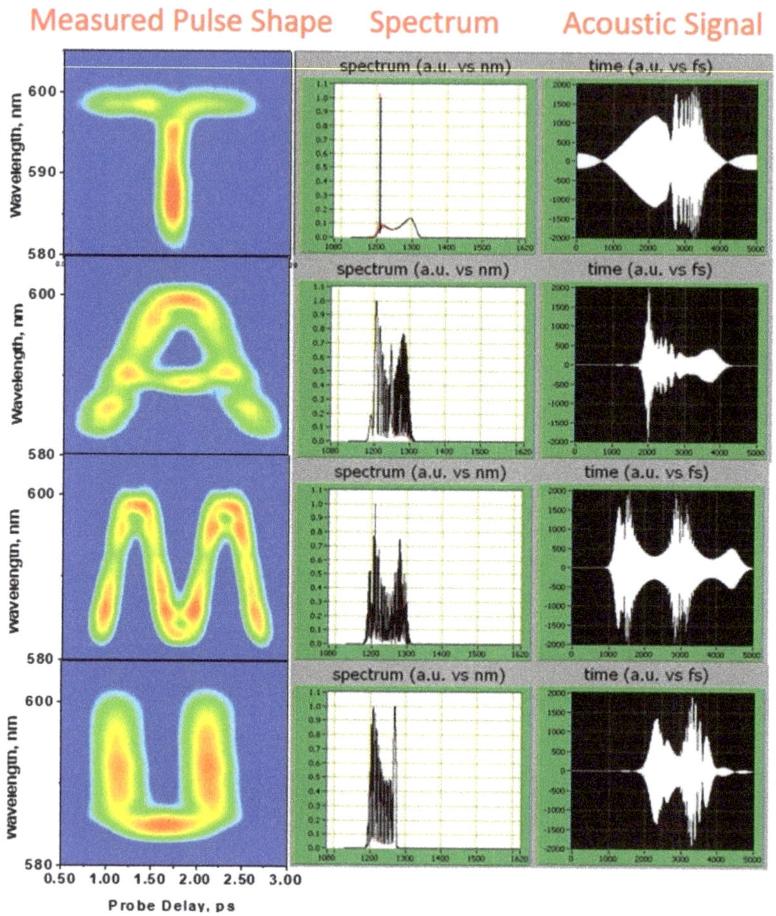

Figure 11. The first column shows the decoded four letters using four different pulses. Their encoded spectra are shown in the middle column. The last column shows the pulse in the time domain predicted by dazzler.

In addition, programmable pulse shapers such as AOPDF offer a way to manipulate and characterize the spectral phase for ultrafast pulses. It can precisely create higher order spectral phases, which would have many applications in various areas. In the next section, we will discuss a simpler and more practical way, called spectral focusing, to generate equal linear chirp for ultrashort pulses, which can help to achieve selective excitation in CARS spectroscopy.

5. Conclusions

In this paper, we review multiple strategies to implement pulse shaping techniques in CARS spectroscopy for molecular vibrations to increase the SNR and spectral resolution. Pulse shaping relies on pulse shapers, including passive pulse shapers and active pulse shapers. Particularly, two typical techniques relied on passive pulse shaping to modify the spectral linewidth (hybrid CARS) and spectral phase (spectral focusing CARS) are comprehensively examined. Furthermore, CARS based on active pulse shaping, which used the programmable devices such as SLMs and AOPDFs, is reviewed. Specifically, a scheme combining an AOPDF and another passive pulse shaper for CARS is experimentally demonstrated for selective excitation and a further application in encryption using ultrashort pulses is introduced as well. Pulse shaping techniques, which can manipulate spectral phase and amplitude of laser pulse, have shown its perspective to effectively optimize the CARS signal generation and detection in a complicated nonlinear optical process. It paves the way for implementing CARS spectroscopy in variety of applications.

Author Contributions: Conceptualization, K.W., Y.S., J.W., X.W. and A.V.S.; investigation, K.W., J.T.F., X.H., Z.H., Y.S., J.W. and X.W.; writing—review and editing, K.W., J.T.F., X.H., Z.H., Y.S., J.W. and A.V.S.; supervision, A.V.S.; funding acquisition, A.V.S. All authors have read and agreed to the published version of the manuscript.

Funding: This work is funded by the Robert A. Welch Foundation (Grant No. A-1547). J.T.F. is supported by the Herman F. Heep and Minnie Belle Heep Texas A&M University Endowed Fund held and administered by the Texas A&M Foundation. K.W. acknowledges support from the National Natural Science Foundation of China under grant 12174459. A.V.S. acknowledges support from his Stephen E. Harris endowed Professorship in Quantum Optics.

Data Availability Statement: The data that support the findings of this study are available from the corresponding authors upon reasonable request.

Conflicts of Interest: The authors declare no conflicts of interest.

Abbreviations

The following abbreviations are used in this manuscript:

CARS	Coherent anti-Stokes Raman scattering
AOPDF	Acousto-optic programmable dispersive filters
CSRS	Coherent Stokes Raman Scattering
NR	Non-resonant
FWM	Four wave mixing
SNR	Signal-to-noise ratio
FASTCARS	Femtosecond adaptive spectroscopic techniques for CARS
OPA	Optical parametric amplifier
CCD	Charge-coupled device
SLM	Spatial light modulator
GA	Genetic algorithm
X-FROG	Cross-frequency resolved optical gating
LDPE	Low density polyethylene
PET	Polyethylene terephthalate

References

1. Tolles, W.M.; Nibler, J.W.; McDonald, J.R.; Harvey, A.B. A Review of the Theory and Application of Coherent Anti-Stokes Raman Spectroscopy (CARS). *Appl. Spectrosc.* **1997**, *31*, 253–271. [CrossRef]
2. Petrov, G.I.; Arora, R.; Yakovlev, V.V.; Wang, X.; Sokolov, A.V.; Scully, M.O. Comparison of Coherent and Spontaneous Raman Microspectroscopies for Noninvasive Detection of Single Bacterial Endospores. *Proc. Natl. Acad. Sci. USA* **2007**, *104*, 7776–7779. [CrossRef] [PubMed]
3. Pestov, D.; Ariunbold, G.O.; Wang, X.; Murawski, R.K.; Sautenkov, V.A.; Sokolov, A.V.; Scully, M.O. Coherent versus Incoherent Raman Scattering: Molecular Coherence Excitation and Measurement. *Opt. Lett.* **2007**, *32*, 1725–1727. [CrossRef] [PubMed]
4. Régnier, P.R.; Taran, J.-P.E. Gas Concentration Measurement by Coherent Raman Anti-Stokes Scattering. In *Laser Raman Gas Diagnostics*; Springer: New York, NY, USA, 1974.
5. Bohlin, A.; Kliewer, C.J. Communication: Two-Dimensional Gas-Phase Coherent Anti-Stokes Raman Spectroscopy (2D-CARS): Simultaneous Planar Imaging and Multiplex Spectroscopy in a Single Laser Shot. *J. Chem. Phys.* **2013**, *138*, 221101. [CrossRef]

6. Bahari, A.; Sower, K.; Wang, K.; Han, Z.; Florence, J.T.; Wang, Y.Y.; Gao, S.; Lee, H.W.H.; Scully, M.O.; Zheltikov, A.; et al. Background-Penalty-Free Waveguide Enhancement of CARS Signal in Air-Filled Anti-Resonant Hollow-Core Fiber. *Opt. Lett.* **2022**, *47*, 4339–4342. [CrossRef]
7. Zhang, F.; Xie, H.; Yuan, L.; Zhang, Z.; Fu, B.; Yu, S.; Li, G.; Zhang, N.; Lu, X.; Yao, J.; et al. Background-free single-beam coherent Raman spectroscopy assisted by air lasing. *Opt. Lett.* **2022**, *47*, 481–484. [CrossRef]
8. Zhang, Z.; Zhang, F.; Xu, B.; Xie, H.; Fu, B.; Lu, X.; Zhang, N.; Yu, S.; Yao, J.; Cheng, Y.; et al. High-sensitivity gas detection with air-lasing-assisted coherent Raman spectroscopy. *Ultrafast Sci.* **2022**, *2022*, 9761458. [CrossRef]
9. Wu, Y.; Zhang, N.; Lu, X.; Li, Z.; Chen, Y.; Gong, R.; Huang, S.; Wang, Q.; Yao, J. Measurement of Carbon Dioxide Isotopes with Air-Lasing-Based Coherent Raman Spectroscopy. *J. Phys. Chem. Lett.* **2024**, *15*, 2944–2950. [CrossRef]
10. Marko, K.A.; Rimai, L. Space- and Time-Resolved Coherent Anti-Stokes Raman Spectroscopy for Combustion Diagnostics. *Opt Lett.* **1979**, *4*, 211–213. [CrossRef]
11. Hancock, R.D.; Bertagnolli, K.E.; Lucht, R.P. Nitrogen and Hydrogen CARS Temperature Measurements in a Hydrogen/Air Flame Using a Near-Adiabatic Flat-Flame Burner. *Combust. Flame* **1997**, *109*, 323–331. [CrossRef]
12. Vestin, F.; Sedarsky, D.; Collin, R.; Aldén, M.; Linne, M.; Bengtsson, P.-E. Rotational Coherent Anti-Stokes Raman Spectroscopy (CARS) Applied to Thermometry in High-Pressure Hydrocarbon Flames. *Combust. Flame* **2008**, *154*, 143–152. [CrossRef]
13. Roy, S.; Gord, J.R.; Patnaik, A.K. Recent Advances in Coherent Anti-Stokes Raman Scattering Spectroscopy: Fundamental Developments and Applications in Reacting Flows. *Prog. Energy Combust. Sci.* **2010**, *36*, 280–306. [CrossRef]
14. Song, K.; Xia, M.; Yun, S.; Zhang, Y.; Zhang, S.; Ge, H.; Deng, Y.; Liu, M.; Wang, W.; Zhao, L.; et al. Advances in Femtosecond Coherent Anti-Stokes Raman Scattering for Thermometry. *J. Phys. Chem.* **2024**, *11*, 622. [CrossRef]
15. Vernuccio, F.; Vanna, R.; Ceconello, C.; Bresci, A.; Manetti, F.; Sorrentino, S.; Ghislanzoni, S.; Lambertucci, F.; Motiño, O.; Martins, I.; et al. Full-Spectrum CARS Microscopy of Cells and Tissues with Ultrashort White-Light Continuum Pulses. *J. Phys. Chem. B* **2023**, *127*, 4733–4745. [CrossRef] [PubMed]
16. Nan, X.; Cheng, J.X.; Xie, X.S. Vibrational Imaging of Lipid Droplets in Live Fibroblast Cells with Coherent Anti-Stokes Raman Scattering Microscopy. *J. Lipid Res.* **2003**, *44*, 2202. [CrossRef]
17. Wang, H.; Fu, Y.; Zickmund, P.; Shi, R.; Cheng, J.X. Coherent Anti-Stokes Raman Scattering Imaging of Axonal Myelin in Live Spinal Tissues. *Biophys. J.* **2005**, *89*, 581–591. [CrossRef]
18. Huff, T.B.; Cheng, J.X. In Vivo Coherent Anti-Stokes Raman Scattering Imaging of Sciatic Nerve Tissue. *J. Microsc.* **2007**, *225*, 175–182. [CrossRef]
19. Evans, C.L.; Xie, X.S. Coherent Anti-Stokes Raman Scattering Microscopy: Chemical Imaging for Biology and Medicine. *Annu. Rev. Anal. Chem.* **2008**, *1*, 883–909. [CrossRef]
20. Okuno, M.; Kano, H.; Leproux, P.; Couderc, V.; Day, J.P.; Bonn, M.; Hamaguchi, H.O. Quantitative CARS Molecular Fingerprinting of Single Living Cells with the Use of the Maximum Entropy Method. *Angew. Chem. Int. Ed.* **2010**, *49*, 6773–6777. [CrossRef]
21. Shi, Y.; Kim, S.; Huff, T.B.; Borgens, R.B.; Park, K.; Shi, R.; Cheng, J.X. Effective Repair of Traumatically Injured Spinal Cord by Nanoscale Block Copolymer Micelles. *Nat. Nanotechnol.* **2010**, *5*, 80–87. [CrossRef]
22. Camp, C.H., Jr.; Lee, Y.J.; Heddleston, J.M.; Hartshorn, C.M.; Hight Walker, A.R.; Rich, J.N.; Lathia, J.D.; Cicerone, M.T. High-Speed Coherent Raman Fingerprint Imaging of Biological Tissues. *Nat. Photonics* **2014**, *8*, 627–634. [CrossRef] [PubMed]
23. Tu, H.; Boppart, S.A. Coherent Anti-Stokes Raman Scattering Microscopy: Overcoming Technical Barriers for Clinical Translation. *J. Biophotonics* **2014**, *7*, 9–22. [CrossRef] [PubMed]
24. Li, S.; Li, Y.; Yi, R.; Liu, L.; Qu, J. Coherent Anti-Stokes Raman Scattering Microscopy and Its Applications. *Front. Phys.* **2020**, *8*, 598420. [CrossRef]
25. Baryalay, S.; Byrne, S.; McNamara, T.; Muddiman, R.; Hennelly, B. Broadband Coherent Anti-Stokes Raman Spectroscopy for Small Scale Microplastic Detection. In *Nonlinear Optics and Its Applications 2024*; SPIE: St Bellingham, WA, USA, 2024; Volume 13004, pp. 165–175.
26. Choi, D.S.; Jeoung, S.C.; Chon, B.H. Thickness Dependent CARS Measurement of Polymeric Thin Films without Depth-Profiling. *Opt. Express* **2008**, *16*, 2604–2613. [CrossRef]
27. Yahng, J.S.; Jeoung, S.C. Thickness Determination with Chemical Identification of Double-Layered Polymeric Thin Film by Using Multiplex CARS. *Opt. Laser Eng.* **2011**, *49*, 66–70. [CrossRef]
28. Virga, A.; Ferrante, C.; Batignani, G.; De Fazio, D.; Nunn, A.D.G.; Ferrari, A.C.; Cerullo, G.; Scopigno, T. Coherent Anti-Stokes Raman Spectroscopy of Single and Multi-Layer Graphene. *Nat. Commun.* **2019**, *10*, 3658. [CrossRef]
29. Lin, X.-M.; Li, J.-F. Applications of In Situ Raman Spectroscopy on Rechargeable Batteries and Hydrogen Energy Systems. *ChemElectroChem* **2023**, *10*, e202201003. [CrossRef]
30. Takahashi, T.; Pawel, K.; Herdzik, K.; Bourdakos, N.; Arthur, J.; Mahajan, R.S. Selective Imaging of Microplastic and Organic Particles in Flow by Multimodal Coherent Anti-Stokes Raman Scattering and Two-Photon Excited Autofluorescence Analysis. *Anal. Chem.* **2021**, *93*, 5234–5240. [CrossRef] [PubMed]

31. Rhee, H.; Jeong, S.; Lee, H.; Cho, M.G.; Choi, D.S. Rapid Detection and Identification of Microplastics from Nonchemically Treated Soil with CARS Microspectroscopy. *Anal. Chim. Acta* **2024**, *342*, 123080. [CrossRef]
32. Zhu, H.; Xu, C.; Yakovlev, V.V.; Zhang, D. What Is Cooking in Your Kitchen: Seeing "Invisible" with Time-Resolved Coherent Anti-Stokes Raman Spectroscopy. *Anal. Bioanal. Chem.* **2023**, *415*, 6471–6480. [CrossRef]
33. Cheng, J.-X.; Volkmer, A.; Book, L.D.; Xie, X.S. An Epi-Detected Coherent Anti-Stokes Raman Scattering (E-CARS) Microscope with High Spectral Resolution and High Sensitivity. *J. Phys. Chem. B* **2001**, *105*, 1277–1280. [CrossRef]
34. De Vito, G.; Bifone, A.; Piazza, V. Rotating-Polarization CARS Microscopy: Combining Chemical and Molecular Orientation Sensitivity. *Opt. Express* **2012**, *20*, 29369–29377. [CrossRef]
35. Cheng, J.-X. Coherent Anti-Stokes Raman Scattering Microscopy. *Appl. Spectrosc.* **2007**, *61*, 197–208. [CrossRef]
36. Oron, D.; Dudovich, N.; Yelin, D.; Silberberg, Y. Narrow-Band Coherent Anti-Stokes Raman Signals from Broad-Band Pulses. *Phys. Rev. Lett.* **2002**, *88*, 063004. [CrossRef] [PubMed]
37. Lim, H.; Caster, A.G.; Leone, S.R. Single-Pulse Phase-Control Interferometric Coherent Anti-Stokes Raman Scattering Spectroscopy. *Phys. Rev. A* **2005**, *72*, 041803. [CrossRef]
38. Kee, T.W.; Zhao, H.; Cicerone, M.T. One-Laser Interferometric Broadband Coherent Anti-Stokes Raman Scattering. *Opt. Express* **2006**, *14*, 3631–3640. [CrossRef]
39. Potma, E.O.; Evans, C.L.; Xie, X.S. Heterodyne Coherent Anti-Stokes Raman Scattering (CARS) Imaging. *Opt. Lett.* **2006**, *31*, 241–243. [CrossRef]
40. Dudovich, N.; Oron, D.; Silberberg, Y. Single-pulse coherently controlled nonlinear Raman spectroscopy and microscopy. *Nature* **2002**, *418*, 51–514. [CrossRef] [PubMed]
41. Ideguchi, T.; Holzner, S.; Bernhardt, B.; Guelachvili, G.; Picqué, N.; Hänsch, T.W. Coherent Raman Spectro-Imaging with Laser Frequency Combs. *Nature* **2013**, *502*, 355–358. [CrossRef]
42. Weiner, A.M. Ultrafast Optical Pulse Shaping: A Tutorial Review. *Opt. Commun.* **2011**, *284*, 3669–3692. [CrossRef]
43. Pestov, D.S. *Detection of Bacterial Endospores by Means of Ultrafast Coherent Raman Spectroscopy*; Texas A&M University: College Station, TX, USA, 2008.
44. Wang, X. Coherent Anti-Stokes Raman Scattering (CARS) Optimized by Exploiting Optical Interference. Ph.D. Dissertation, Texas A&M University, College Station, TX, USA, 2011.
45. Koch, T.; Ackermann, R.; Stöcker, A.; Meyer-Zedler, T.; Gabler, T.; Lippoldt, T.; Missbach-Güntner, J.; Rußmann, C.; Popp, J.; Nolte, S. Ultrabroadband Two-Beam Coherent Anti-Stokes Raman Scattering and Spontaneous Raman Spectroscopy of Organic Fluids: A Comparative Study. *J. Biophotonics* **2024**, *17*, e202300505. [CrossRef]
46. Warren, W.S.; Rabitz, H.; Dahleh, M. Coherent Control of Quantum Dynamics: The Dream Is Alive. *Science* **1993**, *259*, 1581–1588. [CrossRef] [PubMed]
47. Bernhard, C. Response Function Theory of Time-Resolved CARS and CSRS of Rotating Molecules in Liquids under General Polarization Conditions. *Chem. Phys.* **1987**, *113*, 131–147.
48. Hogiu, S.; Werncke, W.; Pfeiffer, M.; Lau, A.; Steinke, T. Picosecond Time-Resolved CARS Spectroscopy of a Mixed Excited Singlet State of Diphenylhexatriene. *Chem. Phys. Lett.* **1998**, *287*, 8–16. [CrossRef]
49. Pestov, D.; Zhi, M.; Sariyanni, Z.-E.; Kalugin, N.G.; Kolomenskii, A.; Murawski, R.; Rostovtsev, Y.V.; Sautenkov, V.A.; Sokolov, A.V.; Scully, M.O. Femtosecond CARS of Methanol–Water Mixtures. *J. Raman Spectrosc.* **2006**, *37*, 392–396. [CrossRef]
50. Scully, M.O.; Kattawar, G.W.; Lucht, R.P.; Opatrny, T.; Pilloff, H.; Rebane, A.; Sokolov, A.V.; Zubairy, M.S. FAST CARS: Engineering a Laser Spectroscopic Technique for Rapid Identification of Bacterial Spores. *Proc. Natl. Acad. Sci. USA* **2002**, *99*, 10994–11001. [CrossRef] [PubMed]
51. Sidorov-Biryukov, D.A.; Serebryannikov, E.E.; Zheltikov, A.M. Time-Resolved Coherent Anti-Stokes Raman Scattering with a Femtosecond Soliton Output of a Photonic-Crystal Fiber. *Opt. Lett.* **2006**, *31*, 2323–2325. [CrossRef]
52. Seeger, T.; Kiefer, J.; Leipertz, A.; Patterson, B.D.; Kliewer, C.J.; Settersten, T.B. Picosecond Time-Resolved Pure-Rotational Coherent Anti-Stokes Raman Spectroscopy for N2 Thermometry. *Opt. Lett.* **2009**, *34*, 3755–3757. [CrossRef]
53. Pestov, D.; Murawski, R.K.; Ariunbold, G.O.; Wang, X.; Zhi, M.; Sokolov, A.V.; Sautenkov, V.A.; Rostovtsev, Y.V.; Dogariu, A.; Huang, Y.; et al. Optimizing the Laser-Pulse Configuration for Coherent Raman Spectroscopy. *Science* **2007**, *316*, 265–268. [CrossRef]
54. Lim, H.; Caster, A.G.; Leone, S.R. Fourier Transform Spectral Interferometric Coherent Anti-Stokes Raman Scattering (FTSI-CARS) Spectroscopy. *Opt. Lett.* **2007**, *32*, 1332–1334. [CrossRef]
55. Oron, D.; Dudovich, N.; Silberberg, Y. Femtosecond Phase-and-Polarization Control for Background-Free Coherent Anti-Stokes Raman Spectroscopy. *Phys. Rev. Lett.* **2003**, *90*, 213902. [CrossRef] [PubMed]
56. van Rhijn, A.C.W.; Offerhaus, H.L.; van der Walle, P.; Herek, J.L.; Jafarpour, A. Exploring, Tailoring, and Traversing the Solution Landscape of a Phase-Shaped CARS Process. *Opt. Express* **2010**, *18*, 2695–2709. [CrossRef] [PubMed]
57. Wang, X.; Wang, K.; Welch, G.R.; Sokolov, A.V. Heterodyne coherent anti-Stokes Raman scattering by the phase control of its intrinsic background. *Phys. Rev. A* **2011**, *84*, 021801(R). [CrossRef]

58. Wang, X.; Zhang, A.; Zhi, M.; Sokolov, A.V.; Welch, G.R. Glucose concentration measured by the hybrid coherent anti-Stokes Raman-scattering technique. *Phys. Rev. A* **2007**, *81*, 013813. [CrossRef]
59. Voronine, D.V.; Sinyukov, A.M.; Hua, X.; Wang, K.; Jha, P.K.; Munusamy, E.; Wheeler, S.E.; Welch, G.; Sokolov, A.V.; Scull, M.O. Time-resolved surface-enhanced coherent sensing of nanoscale molecular complexes. *Sci. Rep.* **2012**, *2*, 891. [CrossRef]
60. Prince, B.D.; Chakraborty, A.; Prince, B.M.; Stauffer, H.U. Development of Simultaneous Frequency- and Time-Resolved Coherent Anti-Stokes Raman Scattering for Ultrafast Detection of Molecular Raman Spectra. *J. Chem. Phys.* **2006**, *125*, 044502. [CrossRef] [PubMed]
61. Shen, Y.; Voronine, D.V.; Sokolov, A.V.; Scully, M.O. A versatile setup using femtosecond adaptive spectroscopic techniques for coherent anti-Stokes Raman scattering. *Rev. Sci. Instrum.* **2015**, *86*, 083107. [CrossRef] [PubMed]
62. Shutov, A.; Pestov, D.; Altangerel, N.; Yi, Z.; Wang, X.; Sokolov, A.V.; Scully, M.O. Collinear FAST CARS for chemical mapping of gases. *Appl. Sci.* **2017**, *7*, 705. [CrossRef]
63. Zhao, H.; Tian, Z.; Li, Y.; Wei, H. Hybrid fs/ps Vibrational Coherent Anti-Stokes Raman Scattering for Simultaneous Gas-Phase $N_2/O_2/CO_2$ Measurements. *Opt. Lett.* **2021**, *46*, 1688–1691. [CrossRef]
64. Miller, J.D.; Slipchenko, M.N.; Meyer, T.R. Probe-Pulse Optimization for Non-Resonant Suppression in Hybrid fs/ps Coherent Anti-Stokes Raman Scattering at High Temperature. *Opt. Express* **2011**, *19*, 13326–13333. [CrossRef]
65. Miller, J.D.; Slipchenko, M.N.; Meyer, T.R.; Stauffer, H.U.; Gord, J.R. Hybrid Femtosecond/Picosecond Coherent Anti-Stokes Raman Scattering for High-Speed Gas-Phase Thermometry. *Opt. Lett.* **2010**, *35*, 2430–2432. [CrossRef] [PubMed]
66. Kerstan, M.; Makos, I.; Nolte, S.; Tünnermann, A.; Ackermann, R. Two-Beam Femtosecond Coherent Anti-Stokes Raman Scattering for Thermometry on CO_2. *Appl. Phys. Lett.* **2017**, *110*, 021116. [CrossRef]
67. Santagata, R.; Scherman, M.; Toubeix, M.; Nafa, M.; Tretout, B.; Bresson, A. Ultrafast Background-Free Ro-Vibrational fs/ps-CARS Thermometry Using an Yb:YAG Crystal-Fiber Amplified Probe. *Opt. Express* **2019**, *27*, 32924–32937. [CrossRef] [PubMed]
68. Kearney, S.P.; Scoglietti, D.; Kliewer, C. Hybrid Femtosecond/Picosecond Rotational Coherent Anti-Stokes Raman Scattering Temperature and Concentration Measurements Using Two Different Picosecond-Duration Probes. *Opt. Express* **2013**, *21*, 12327–12339. [CrossRef]
69. Peng, J.H.; Pestov, D.; Scully, M.O.; Sokolov, A.V. Simple Setup for Hybrid Coherent Raman Microspectroscopy. *J. Raman Spectrosc.* **2009**, *40*, 795–799. [CrossRef]
70. Chen, D.; Liu, Z.; Liu, S.; Niu, H. Ultra-Broadband Three-Color Coherent Anti-Stokes Raman Scattering Spectroscopy. *Opt. Commun.* **2022**, *519*, 128367. [CrossRef]
71. Pestov, D.; Wang, X.; Ariunbold, G.O.; Murawski, R.K.; Sautenkov, V.A.; Dogariu, A.; Sokolov, A.V.; Scully, M.O. Single-Shot Detection of Bacterial Endospores via Coherent Raman Spectroscopy. *Proc. Natl. Acad. Sci. USA* **2008**, *105*, 422–427. [CrossRef]
72. Peng, J.; Sokolov, A.V. Epi-detected hybrid coherent Raman micro-spectroscopy. *J. Mod. Opt.* **2009**, *56*, 1964–1969. [CrossRef]
73. Cao, T.; Huang, L.; Peng, J. Simple configurations for hybrid coherent Raman microspectroscopy. In Proceedings of the 2016 Progress in Electromagnetic Research Symposium (PIERS), Shanghai, China, 8–11 August 2016; IEEE: Piscataway, NJ, USA, 2016; p. 7734949.
74. Miller, J.D.; Roy, S.; Gord, J.R.; Meyer, T.R. Communication: Time-Domain Measurement of High-Pressure N_2 and O_2 Self-Broadened Linewidths Using Hybrid Femtosecond/Picosecond Coherent Anti-Stokes Raman Scattering. *J. Chem. Phys.* **2011**, *135*, 201104. [CrossRef]
75. Kearney, S.P.; Scoglietti, D.J. Hybrid Femtosecond/Picosecond Rotational Coherent Anti-Stokes Raman Scattering at Flame Temperatures Using a Second-Harmonic Bandwidth-Compressed Probe. *Opt. Lett.* **2013**, *38*, 833–835. [CrossRef]
76. Kearney, S.P.; Guildenbecher, D.R. Temperature Measurements in Metalized Propellant Combustion Using Hybrid fs/ps Coherent Anti-Stokes Raman Scattering. *Appl. Opt.* **2016**, *55*, 4958–4966. [CrossRef] [PubMed]
77. Kearney, S.P. Hybrid fs/ps rotational CARS temperature and oxygen measurements in the product gases of canonical flat flames. *Combust. Flame* **2015**, *162*, 1748–1758. [CrossRef]
78. Kearney, S.P.; Danehy, P.M. Pressure measurements using hybrid femtosecond/picosecond rotational coherent anti-Stokes Raman scattering. *Opt. Lett.* **2015**, *40*, 4082–4085. [CrossRef]
79. Hellerer, T.; Enejder, A.M.; Zumbusch, A. Spectral Focusing: High Spectral Resolution Spectroscopy with Broad-Bandwidth Laser Pulses. *Appl. Phys. Lett.* **2004**, *85*, 25–27. [CrossRef]
80. Gershgoren, E.; Bartels, R.A.; Fourkas, J.T.; Tobey, R.; Murnane, M.M.; Kapteyn, H.C. Simplified setup for high resolution spectroscopy that uses ultrashort pulses. *Opt. Lett.* **2003**, *28*, 361–363. [CrossRef] [PubMed]
81. Pestov, D.; Wang, X.; Murawski, R.K.; Ariunbold, G.O.; Sautenkov, V.A.; Sokolov, A.V. Pulse shaping for mode-selective ultrafast coherent Raman spectroscopy of highly scattering solids. *JOSAB* **2008**, *25*, 768–772. [CrossRef]
82. Pegoraro, A.F.; Ridsdale, A.; Moffatt, D.J.; Jia, Y.; Pezacki, J.P.; Stolow, A. Optimally Chirped Multimodal CARS Microscopy Based on a Single Ti:Sapphire Oscillator. *Opt. Express* **2009**, *17*, 2984–2996. [CrossRef]
83. Langbein, W.; Rocha-Mendoza, I.; Borri, P. Coherent Anti-Stokes Raman Micro-Spectroscopy Using Spectral Focusing: Theory and Experiment. *J. Raman Spectrosc.* **2009**, *40*, 800–808. [CrossRef]

84. Mohseni, M.; Polzer, C.; Hellerer, T. Resolution of Spectral Focusing in Coherent Raman Imaging. *Opt. Express* **2018**, *26*, 10230–10241. [CrossRef]
85. Cole, R.A.; Slepkov, A.D. Interplay of Pulse Bandwidth and Spectral Resolution in Spectral-Focusing CARS Microscopy. *J. Opt. Soc. Am. B* **2018**, *35*, 842–850. [CrossRef]
86. Chen, K.; Wu, T.; Wei, H.; Li, Y. Dual-Soliton Stokes-Based Background-Free Coherent Anti-Stokes Raman Scattering Spectroscopy and Microscopy. *Opt. Lett.* **2016**, *41*, 2628–2631. [CrossRef] [PubMed]
87. Chen, B.C.; Sung, J.; Lim, S.H. Chemical Imaging with Frequency Modulation Coherent Anti-Stokes Raman Scattering Microscopy at the Vibrational Fingerprint Region. *J. Phys. Chem. B* **2010**, *114*, 16871–16880. [CrossRef] [PubMed]
88. Rocha-Mendoza, I.; Langbein, W.; Borri, P. Coherent Anti-Stokes Raman Microspectroscopy Using Spectral Focusing with Glass Dispersion. *Appl. Phys. Lett.* **2008**, *93*, 201103. [CrossRef]
89. Brückner, L.; Buckup, T.; Motzkus, M. Exploring the Potential of Tailored Spectral Focusing. *J. Opt. Soc. Am. B* **2016**, *33*, 1482–1491. [CrossRef]
90. Porquez, J.G.; Cole, R.A.; Tabarangao, J.T.; Slepkov, A.D. Spectrally-broad coherent anti-Stokes Raman scattering hyper-microscopy utilizing a Stokes supercontinuum pumped at 800 nm. *Biomed. Opt. Express* **2016**, *7*, 4335–4345. [CrossRef]
91. Langbein, W.; Rocha-Mendoza, I.; Borri, P. Single source coherent anti-stokes Raman microspectroscopy using spectral focusing. *Appl. Phys. Lett.* **2009**, *95*, 081109. [CrossRef]
92. Chen, K.; Wu, T.; Wei, H.; Zhou, T.; Li, Y. Quantitative chemical imaging with background-free multiplex coherent anti-Stokes Raman scattering by dual-soliton Stokes pulses. *Biomed. Opt. Express* **2016**, *7*, 3927–3939. [CrossRef]
93. Tomita, K.; Kojima, Y.; Kannari, F. Selective coherent anti-Stokes Raman scattering microscopy employing dual-wavelength nanofocused ultrafast plasmon pulses. *Nano Lett.* **2016**, *18*, 1366–1372. [CrossRef]
94. Herdzik, K.P.; Bourdakos, K.N.; Johnson, P.B.; Lister, A.P.; Pitera, A.P.; Guo, C.Y.; Mahajan, S. Multimodal spectral focusing CARS and SFG microscopy with a tailored coherent continuum from a microstructured fiber. *Appl. Phys. B* **2020**, *126*, 84. [CrossRef]
95. Pope, I.; Langbein, W.; Watson, P.; Borri, P. Simultaneous hyperspectral differential-CARS, TPF and SHG microscopy with a single 5 fs Ti:Sa laser. *Opt. Express* **2013**, *21*, 7096–7106. [CrossRef]
96. Kearney, S.P. Bandwidth optimization of femtosecond pure-rotational coherent anti-Stokes Raman scattering by pump/Stokes spectral focusing. *Appl. Opt.* **2014**, *53*, 6579–6585. [CrossRef] [PubMed]
97. Konorov, S.O.; Blades, M.W.; Turner, R.F. Lorentzian Amplitude and Phase Pulse Shaping for Nonresonant Background Suppression and Enhanced Spectral Resolution in Coherent Anti-Stokes Raman Scattering Spectroscopy and Microscopy. *Appl. Spectrosc.* **2010**, *64*, 767–774. [CrossRef] [PubMed]
98. Yang, L.X.; Iyer, R.R.; Sorrells, J.E.; Renteria, C.A.; Boppart, S.A. Temporally Optimized and Spectrally Shaped Hyperspectral Coherent Anti-Stokes Raman Scattering Microscopy. *Opt. Express* **2024**, *32*, 11474–11490. [CrossRef]
99. Mecker, N.T.; Courtney, T.L.; Patterson, B.D.; Escofet-Martin, D.; Peterson, B.; Kliewer, C.J.; Linne, M. Optimising hybrid rotational femtosecond/picosecond coherent anti-Stokes Raman spectroscopy (HR-CARS) in nitrogen at high pressures and temperatures. *JOSAB* **2020**, *37*, 1035–1046. [CrossRef]
100. Sorrells, J.E.; Yang, L.; Iyer, R.R.; Chaney, E.J.; Renteria, C.A.; Boppart, S.A. Programmable hyperspectral coherent anti-Stokes Raman scattering microscopy. *Opt. Lett.* **2024**, *49*, 2513. [CrossRef]
101. Xu, X.G.; Konorov, S.O.; Hepburn, J.W.; Milner, V. Noise autocorrelation spectroscopy with coherent Raman scattering. *Nat. Phys.* **2007**, *4*, 125–129. [CrossRef]
102. Li, B.; Warren, S.W.; Fisher, M.C. Phase-Cycling Coherent Anti-Stokes Raman Scattering Using Shaped Femtosecond Laser Pulses. *Opt. Express* **2010**, *18*, 25825–25833. [CrossRef]
103. Bernhard, V.V.; Motzkus, M.M. Time-Resolved Two-Color Single-Beam CARS Employing Supercontinuum and Femtosecond Pulse Shaping. *Opt. Commun.* **2006**, *264*, 488–493.
104. Bernhard, V.V.; Motzkus, M.M. Time-Resolving Molecular Vibration for Microanalytics: Single Laser Beam Nonlinear Raman Spectroscopy in Simulation and Experiment. *Phys. Chem. Chem. Phys.* **2008**, *10*, 681–691.
105. Isobe, K.; Suda, A.; Tanaka, M.; Hashimoto, H.; Kannari, F.; Kawano, H.; Mizuno, H.; Miyawaki, A.; Midorikawa, K. Single-Pulse Coherent Anti-Stokes Raman Scattering Microscopy Employing an Octave Spanning Pulse. *Opt. Express* **2009**, *17*, 11259–11266. [CrossRef]
106. Postma, S.; van Rhijn, A.C.W.; Korterik, J.P.; Gross, P.; Herek, J.L.; Offerhaus, H.L. Application of Spectral Phase Shaping to High Resolution CARS Spectroscopy. *Opt. Express* **2008**, *16*, 7985–7996. [CrossRef] [PubMed]
107. Konorov, S.O.; Xu, X.G.; Turner, R.F.B.; Blades, M.W.; Hepburn, J.W.; Milner, V. Pulse Optimization for Raman Spectroscopy with Cross-Correlation Frequency Resolved Optical Gating. *Opt. Express* **2007**, *15*, 7564–7572. [CrossRef]
108. Polack, T.; Oron, D.; Silbergerg, Y. Control and measurement of a non-resonant Raman wavepacket using a single ultrashort pulse. *Chem. Phys.* **2005**, *318*, 163–169. [CrossRef]

109. Roy, S.; Wrzesinski, P.J.; Pestov, D.; Dantus, M.; Gord, J.R. Single-Beam Coherent Anti-Stokes Raman Scattering (CARS) Spectroscopy of Gas-Phase CO2 via Phase and Polarization Shaping of a Broadband Continuum. *J. Raman Spectrosc.* **2009**, *41*, 1194–1199. [CrossRef]
110. Roy, S.; Wrzesinski, P.; Pestov, D.; Gunaratne, T.; Dantus, M.; Gord, J.R. Single-Beam Coherent Anti-Stokes Raman Scattering Spectroscopy of N2 Using a Shaped 7 fs Laser Pulse. *Appl. Phys. Lett.* **2009**, *95*, 074102. [CrossRef]
111. Li, H.; Harris, D.A.; Xu, B.; Wrzesinski, P.J.; Lozovoy, V.V.; Dantus, M. Standoff and Arms-Length Detection of Chemicals with Single-Beam Coherent Anti-Stokes Raman Scattering. *Appl. Opt.* **2009**, *48*, B17–B22. [CrossRef] [PubMed]
112. Katz, O.; Natan, A.; Silberberg, Y.; Rosenwaks, S. Standoff Detection of Trace Amounts of Solids by Nonlinear Raman Spectroscopy Using Shaped Femtosecond Pulses. *Appl. Phys. Lett.* **2008**, *92*, 171116. [CrossRef]
113. Oron, D.; Dudovich, N.; Silberberg, Y. All-Optical Processing in Coherent Nonlinear Spectroscopy. *Phys. Rev. A* **2004**, *70*, 023415. [CrossRef]
114. Bernhard, V.V.; Buckup, T.; Motzkus, M. Highly Sensitive Single-Beam Heterodyne Coherent Anti-Stokes Raman Scattering. *Opt. Lett.* **2006**, *31*, 2495–2497.
115. Tournois, P. Acousto-optic programmable dispersive filter for adaptive compensation of group delay time dispersion in laser systems. *Opt. Comm.* **1997**, *140*, 245–249. [CrossRef]
116. Oksenhendler, T.; Forget, N. Pulse-Shaping Techniques: Theory and Experimental Implementations for Femtosecond Pulses. In *Advances in Solid State Lasers Development and Applications*; Grishin, M., Ed.; IntechOpen: London, UK, 2010.
117. Forget, N.; Canova, L.; Chen, X.; Jullien, A.; Lopez-Martens, R. Closed-Loop Carrier-Envelope Phase Stabilization with an Acousto-Optic Programmable Dispersive Filter. *Opt. Lett.* **2009**, *34*, 3647–3649. [CrossRef] [PubMed]
118. Canova, L.; Chen, X.; Trisorio, A.; Jullien, A.; Assion, A.; Tempea, G.; Forget, N.; Oksenhendler, T.; Lopez-Martens, R. Carrier-Envelope Phase Stabilization and Control Using a Transmission Grating Compressor and an AOPDF. *Opt. Lett.* **2009**, *34*, 1333–1335. [CrossRef]
119. Ohno, K.; Tanabe, T.; Kannari, F. Adaptive Pulse Shaping of Phase and Amplitude of an Amplified Femtosecond Pulse Laser by Direct Reference to Frequency-Resolved Optical Gating Traces. *J. Opt. Soc. Am. B* **2002**, *19*, 2781–2790. [CrossRef]
120. Schubert, M.; Eisele, M.; Crozatier, V.; Forget, N.; Kaplan, D.; Huber, R. Rapid-Scan Acousto-Optical Delay Line with 34 kHz Scan Rate and 15 as Precision. *Opt. Lett.* **2013**, *38*, 2907–2909. [CrossRef]
121. Cousin, S.L.; Forget, N.; Grün, A.; Bates, P.K.; Austin, D.R.; Biegert, J. Few-Cycle Pulse Characterization with an Acousto-Optic Pulse Shaper. *Opt. Lett.* **2011**, *36*, 2803–2805. [CrossRef] [PubMed]
122. Zhi, M.; Wang, K.; Hua, X.; Sokolov, A.V. Pulse-Shaper-Assisted Phase Control of a Coherent Broadband Spectrum of Raman Sidebands. *Opt. Lett.* **2011**, *36*, 4032–4034. [CrossRef]
123. Zhi, M.; Wang, K.; Hua, X.; Strycker, B.D.; Sokolov, A.V. Shaper-Assisted Phase Optimization of a Broad "Holey" Spectrum. *Opt. Express* **2011**, *19*, 23400–23408. [CrossRef] [PubMed]
124. Li, Z.; Leng, Y.; Li, R. Further Development of the Short-Pulse Petawatt Laser: Trends, Technologies, and Bottlenecks. *Laser Photonics Rev.* **2023**, *17*, 2100705. [CrossRef]
125. Shivkumar, S.; Ranann, D.; Metais, S.; Suresh, S.; Forget, N.; Bartels, R.; Oron, D.; Rigneault, H. Selective Detection in Impulsive Low-Frequency Raman Imaging Using Shaped Probe Pulses. *Phys. Rev. Appl.* **2023**, *19*, 054075. [CrossRef]
126. Lindley, M.; Gala de Pablo, J.; Kinegawa, R.; Hiramatsu, K.; Goda, K. Highly Sensitive Fourier-Transform Coherent Anti-Stokes Raman Scattering Spectroscopy via Genetic Algorithm Pulse Shaping. *Opt. Lett.* **2021**, *46*, 4320–4323. [CrossRef]
127. Zhang, S.; Zhang, L.; Zhang, X.; Ding, L.; Chen, G.; Sun, Z.; Wang, Z. Selective Excitation of CARS by Adaptive Pulse Shaping Based on Genetic Algorithm. *Chem. Phys. Lett.* **2007**, *433*, 416–421. [CrossRef]

molecules

MDPI

Article

Ultrasensitive Chemical Analysis on Gold Nano Popcorn Substrate Using Digital Surface-Enhanced Raman Scattering

Soyeon Lee and Jaebum Choo *

Department of Chemistry, Chung-Ang University, Seoul 06974, Republic of Korea; dearosie@naver.com
* Correspondence: jbchoo@cau.ac.kr

Abstract: This study presents a digital surface-enhanced Raman scattering (SERS) method to enhance the sensitivity of SERS detection for low-concentration analytes. Conventional SERS analysis using average Raman intensity faces limitations in distinguishing low concentrations due to the substrate's sparse distribution of target molecules. To overcome this challenge, we used a binary code-based data analysis approach. Gold nano popcorn substrates were utilized for high-sensitivity detection, with malachite green isothiocyanate (MGITC) as the target molecule. Raman mapping data were analyzed using both the conventional average Raman intensity method and the proposed digital SERS approach. In the digital SERS method, a threshold value was established based on the mean and standard deviation of Raman signals in the absence of target molecules. Pixels with Raman intensities exceeding this threshold were assigned a value of "1", while those below were assigned "0". Quantification was then performed based on these digital counts corresponding to MGITC concentrations. Our results demonstrate that the digital SERS method significantly improved the ability to distinguish and quantify analytes in low-concentration ranges that were indiscernible using the conventional approach. This analytical technique shows promise for ultrasensitive chemical analysis and expands the capabilities of SERS-based detection methods, potentially revolutionizing the field of trace analyte detection.

Keywords: digital SERS; Au nano popcorn substrate; surface-enhanced Raman scattering; digital analysis; ultrasensitive chemical analysis

1. Introduction

Surface-enhanced Raman scattering (SERS) has emerged as a powerful analytical technique capable of detecting molecules at ultra-low concentrations, even down to single-molecule levels [1–3]. This remarkable sensitivity is achieved by amplifying Raman scattering signals through localized surface plasmon resonance (LSPR) occurring on metal surfaces or nanostructures [4–7]. The widespread applications of SERS in fields such as disease diagnosis, chemical analysis, and nanotechnology underscore its significance in modern scientific research [8–12]. Despite its potential, SERS faces challenges in signal sensitivity and reproducibility, particularly at very low analyte concentrations. These limitations stem from the uneven distribution of "hot spots"—areas of intense electromagnetic field enhancement that account for a disproportionate amount of the total Raman signal [13,14]. As analyte concentration decreases, the contribution of SERS signals to the overall signal diminishes, leading to increased fluctuations and reduced detection sensitivity.

To address these challenges, researchers have primarily focused on developing plasmonic nanostructures with enhanced amplification effects [15–20]. However, this approach alone may not fully resolve the issues of reproducibility and sensitivity at extremely low concentrations. In response to these limitations, a new data processing method called digital SERS was proposed [21,22]. The digital SERS method divides the analysis area into micro-compartments and digitizes each compartment for data processing. Recently, a Au nano popcorn substrate with highly uniform nanogaps has been developed by growing gold particles on Au nano dimple substrates [23]. This study aims to implement reproducible high-sensitivity SERS detection by introducing the digital SERS technique to these Au nano popcorn structures. This approach is particularly effective for very low analyte concentrations, where traditional methods struggle to provide reliable results.

To demonstrate the efficacy of the digital SERS method, we used malachite green isothiocyanate (MGITC), widely used as a Raman reporter in SERS detection, as a model analyte. Raman mapping images were obtained by loading different concentrations of MGITC onto the Au nano popcorn substrates and continuously measuring Raman signals over specific areas. We then analyzed the Raman mapping data using both conventional methods and our proposed digital SERS approach. In the conventional SERS analysis, characteristic Raman peak intensities for each pixel were averaged to calculate quantitative values. For the digital SERS method, a threshold value based on the measured Raman peak intensities for the digital SERS method was established. Signals above this threshold were assigned a value of "1", while those below were assigned "0". These binary values were then decoded into black and white colors for analysis [24,25]. The primary objective of this research is to demonstrate the potential of digital SERS detection technology in overcoming the limitations of reproducibility and sensitivity inherent in conventional SERS. By introducing this novel data processing approach, we aim to expand the capabilities of SERS-based detection methods and pave the way for more reliable and sensitive analytical techniques in various scientific disciplines.

2. Results and Discussion

The Au nano popcorn substrate used in the experiment was fabricated based on previous papers reported by our research group [26–29]. The Au nano popcorn substrate has uniform nanogaps on its surface, creating numerous hot spots where Raman signals are greatly amplified due to LSPR when analytes are positioned on the structure's surface. The enhancement factor (EF) of the Au nano popcorn substrate, calculated by comparing the Raman signal of MGITC measured on a glass substrate, was evaluated to be 7.18×10^7. Therefore, the Au nano popcorn substrate enables high-sensitivity detection of target molecules. The overall experimental process is illustrated in Figure 1a. A 3×3 μm^2 piece of Au nano popcorn substrate was placed in a 1.6 mL microtube, and 200 μL of MGITC solution diluted with ethanol was added. After reacting for 1 h at 500 rpm, the remaining solution was removed using a blower. After washing, the nano popcorn substrate was fixed on a slide glass, and Raman signals were measured. Figure 1b shows a conceptual diagram of the Raman mapping image measurement converted to digital code after acquiring the Raman signal after loading MGITC. Figure 1b displays the Raman mapping image obtained by monitoring the intensity change in MGITC's characteristic Raman peak at 1614 cm^{-1}. In the conventional data analysis, the average value of Raman peak intensities is correlated with the concentration. Herein, Raman peak intensities measured at all pixels are averaged to analyze the results (Figure 1c). In contrast, digital SERS sets a threshold value and converts the Raman intensity of each pixel into a digital code for analysis. If a pixel's Raman intensity exceeds the threshold value, it is considered a "1" or positive event and marked in black. If it does not exceed the threshold value, it is regarded as a "0" or negative event and marked in white (Figure 1d). A "1" indicates that a SERS intensity that is distinguishable from the background signal was obtained, while a "0" means it cannot be distinguished from the background signal.

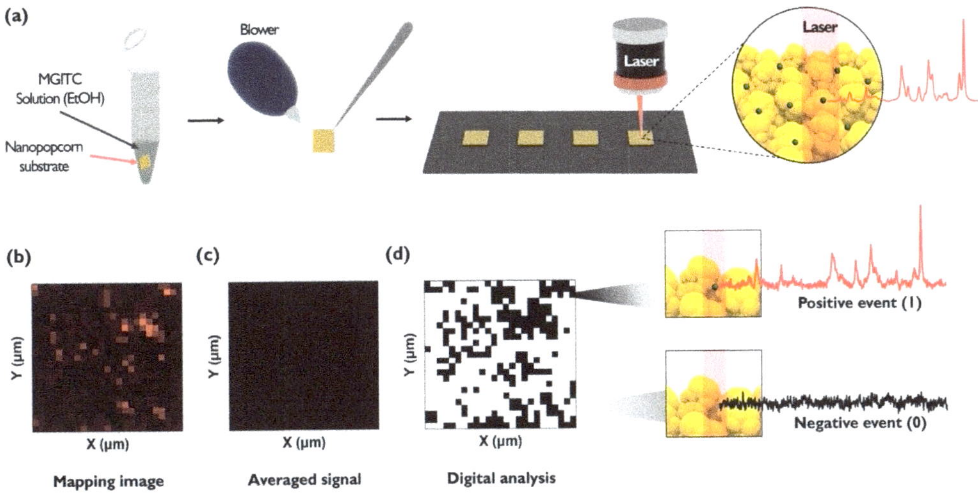

Figure 1. Illustration of digital SERS. (**a**) Experimental procedure to obtain Raman mapping image of MGITC on Au nano popcorn substrates. (**b**) SERS mapping image obtained from raw Raman mapping data. (**c**) Raman image using average intensity for all pixels of Au nano popcorn substrate. (**d**) Digital decoding SERS image.

An experiment was conducted to evaluate the reproducibility of Au nano popcorn substrates by assessing the fluctuation in Raman signal intensity between different substrates. As shown in Figure 2, five distinct Au nano popcorn substrates were used for this purpose. For quantitative experiments at ultra-low analyte concentrations, it is crucial that the fluctuation between different substrates is minimal, and that the Raman intensity remains consistent. First, five Au nano popcorn substrates were cut to identical sizes. Second, concentration and other reaction conditions were kept constant for all substrates. Third, using a computer-controlled x-y transitional stage, a 48 μm × 48 μm area was scanned with a 2 μm step size, resulting in a 625-pixel image (Figure 2b). Finally, Raman mapping images were obtained by color decoding the Raman intensity at 1614 cm^{-1}, characteristic of the MGITC molecule. Figure 2a shows the five Raman mapping images obtained from this experimental process. A color scale bar on the right side of the image indicates different Raman intensities. The uniformity of pixel colors in these images signifies the homogeneity of the plasmonic substrate, which minimizes substrate-to-substrate intensity fluctuations and increases reproducibility. Figure 2c compares each substrate's average Raman peak intensities in a bar graph. The relative standard deviation (RSD) of the Raman peak intensity at 1614 cm^{-1} for five substrates was calculated to be 2.24%, indicating excellent reproducibility of the plasmonic substrates used in this study. Figure 2d plots all 625 Raman spectra for easy comparison. The RSD calculated from the characteristic Raman peak intensity of MGITC at 1614 cm^{-1} was 6.54%, demonstrating outstanding pixel-to-pixel reproducibility within the substrate. Therefore, it could be concluded that the nano popcorn substrate used in this study proved highly suitable for ultra-low concentration quantitative analysis using the digital SERS method due to its excellent reproducibility between substrates as well as between spots in a single substrate.

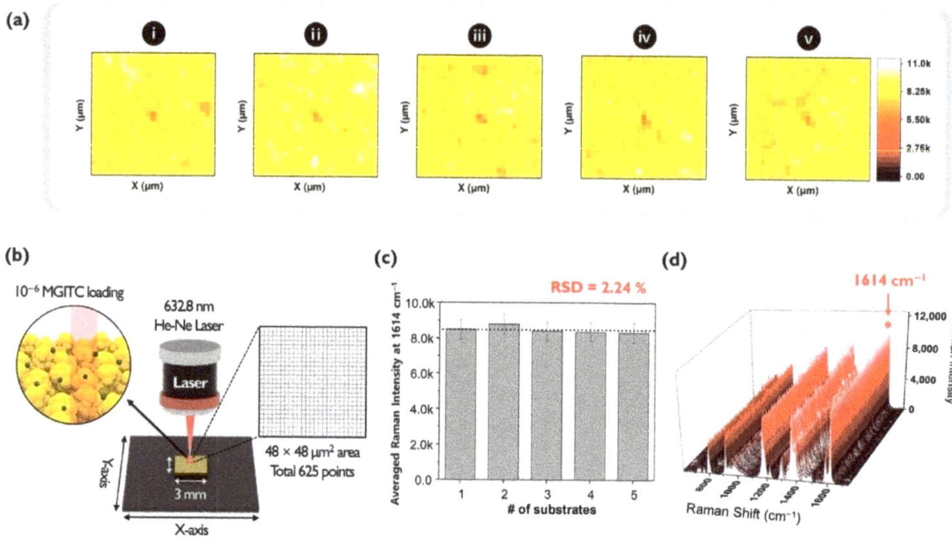

Figure 2. (**a**) Raman mapping images of MGITC for five different nano popcorn substrates (**i**~**v**). (**b**) Raman mapping method for Au nano popcorn substrate. The total area of each Au nano popcorn substrate was scanned over an area of 48 × 48 μm^2 range with 2 × 2 μm^2 mapping steps, totaling 625 pixels. The samples are exposed to 10% laser power with 0.1 s exposure time. (**c**) Relative standard deviation (RSD) to demonstrate substrate-to-substrate fluctuations. (**d**) Raman spectral data obtained from all mapping points in substrate to demonstrate spot-to-spot fluctuations.

After completing the reproducibility assessment, MGITC solutions were diluted to various concentrations and loaded onto the Au nano popcorn substrate to measure Raman peak intensities. Color decoding was performed using the characteristic Raman peak intensity at 1614 cm^{-1} of MGITC (Figure 3a–f). A 48 μm × 48 μm area was mapped with a 2 μm step size, resulting in Raman data for a total of 625 pixels. The reason for choosing a 2 μm × 2 μm pixel size for the mapping image is due to the approximate laser beam size. The Raman mapping measurements used a 20× magnification/NA 0.4 objective and a 633 nm He-Ne laser, which produces a focused laser beam size of 1.93 μm. The step size (2 μm) was set similarly to the laser beam size to achieve high spatial resolution. When measuring the Raman mapping images using this method, it was observed that the overall Raman mapping image became darker as the concentration of the MGITC solution decreased. Figure 3g shows the average Raman spectra of 625 pixels for each concentration. As the concentration of MGITC decreases, the intensity of the characteristic Raman peak at 1614 cm^{-1} also decreases. Since it is difficult to discern the low concentration range (below 10^{-9} M) in this figure, a histogram was created using only the Raman peak intensities at 1614 cm^{-1} for the low concentration range (10^{-8} M to blank) (Figure 3h). This graph shows that even in the low concentration range, the average Raman peak intensity gradually decreases as the concentration decreases, but it is not easy to distinguish the Raman peak intensity at concentrations below 10^{-10} M using this conventional SERS mapping method.

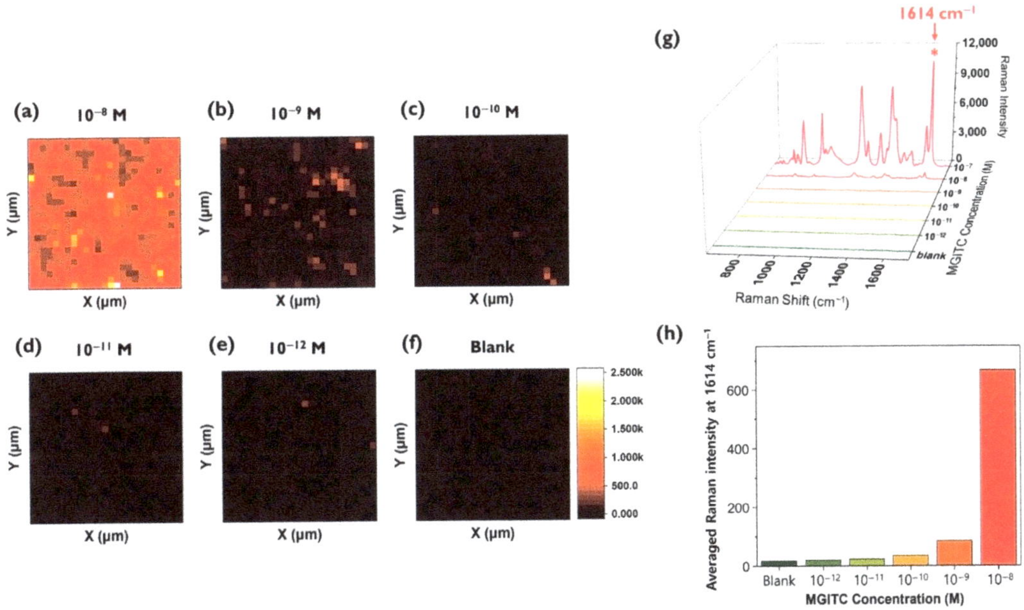

Figure 3. (**a**–**f**) Raman mapping images of MGITC in the 0 to 10^{-8} M range using the characteristic Raman peak intensity at 1614 cm^{-1}. A 48 μm × 48 μm area was mapped with a 2 μm step size, resulting in Raman data for a total of 625 pixels. (**g**) Average Raman spectra of 625 pixels for each concentration. (**h**) Histogram of the average Raman peak intensity in the 0 to 10^{-8} M range.

The Raman mapping data measured in Figure 3 were analyzed using the digital SERS technique (Figure 4a–g). The average Raman peak intensity and standard deviation of a blank sample without the analyte MGITC determined the threshold value. The threshold was set to minimize the influence of background noise signals so that over 99% of the 625 pixels in the blank fell within this range [21,22]. This method is identical to determining the limit of detection (LOD) in conventional bioassays [27,28]. While traditional LOD determination methods use the average and standard deviation of intensities for each concentration to plot a graph and then determine the LOD using the blank sample's average intensity + 3× standard deviation formula, digital SERS distinguishes between positive and negative for each pixel in the mapping image based on the threshold value, without calculating the average Raman peak intensity. As seen in Figure 3a–g, average Raman intensity values from SERS data can accurately analyze Raman data for high concentrations, but there are limitations in distinguishing low-concentration samples. Because many analyte molecules exist on the nanostructured surface, it is possible to distinguish high-concentration samples. In the case of low-concentration samples, enhanced Raman signals appear only in extremely limited areas at low concentrations because the analyte molecules are not uniformly distributed on the nanostructured surface and rarely exist at hotspots [14]. However, when analyzing very low analyte concentrations using conventional SERS, the SERS intensity from hotspots contributes little to the overall signal intensity, making it difficult to distinguish from background noise signals and reducing sensitivity in the quantitative analysis due to increased Raman signal fluctuation.

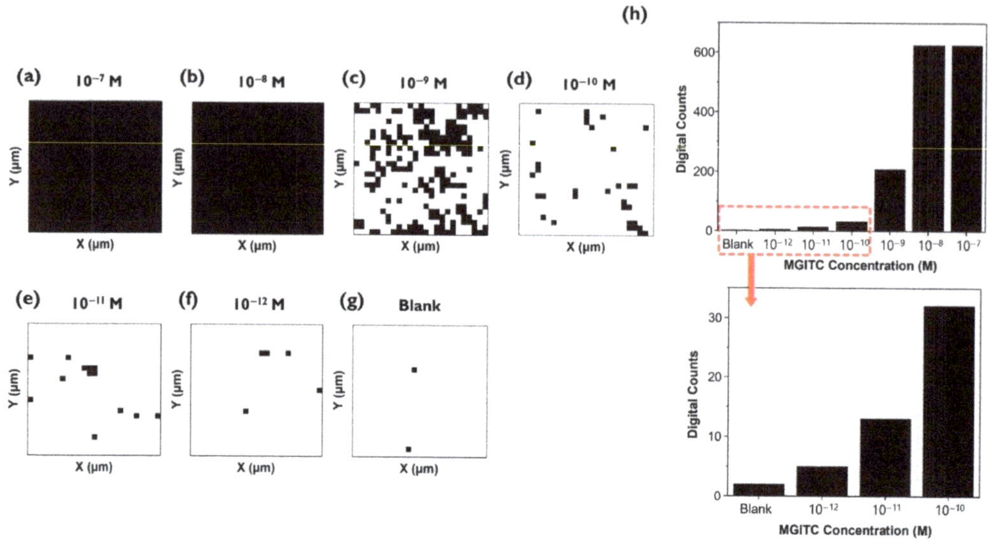

Figure 4. (**a**–**g**) Digital SERS mapping image of MGITC on Au nano popcorn substrate in the 0 to 10^{-7} M range. (**h**) Histogram of the number of digital counts ("1") in the 0 to 10^{-7} M range (**top**). Enlarged histogram for the low concentration range, 0 to 10^{-10} M (**bottom**).

To solve this problem, digital SERS technology can be introduced, which defines cases exceeding the threshold value for Raman peak intensity of each pixel as "1", and those that do not as "0", then counts the number of "1"s. This method makes accurate quantitative analysis possible even when the distribution of molecules on the Au nano popcorn substrate is not uniform in the low concentration range. As seen in Figure 4a, at high concentrations, MGITC molecules are distributed across the entire substrate, showing a signal of 1 in all pixels (Figure 4a,b). In the low concentration range, which conventional methods could not distinguish, the number of black pixels gradually decreases as the concentration decreases (Figure 4c–e). Figure 4h shows the count of black pixels from Figure 4a (top). The low concentration range of 10^{-10} M to 10^{-12} M and the blank are enlarged as they are difficult to discern visually (bottom). This figure confirms the trend of decreasing digital counts as the concentration decreases.

Raman spectra were measured three times for the same concentrations of MGITC sample under identical measurement conditions (633 nm laser, power 10% = 0.8 mW, exposure time 1 s, objective lens 20×, 48 μm × 48 μm, 2 μm step size). Three Raman mapping images, each consisting of 625 pixels, were obtained and analyzed using the conventional SERS mapping and digital SERS methods to obtain corresponding calibration curves. Figure 5a shows a histogram for the average SERS peak intensity at 1614 cm^{-1} for 625 points measured three times for each concentration. The *y*-axis represents the average Raman peak intensity, and the error bars indicate the standard deviations for each concentration. Using a 4-parameter fitting equation, the calibration curve was determined and yielded an R^2 value of 0.998, and the calculated LOD was 1.19 nM. Figure 5b shows the results of analyzing the same Raman spectral data using digital decoding. As seen in Figure 4a, for relatively high concentrations such as 10^{-7} M and 10^{-8} M, Raman intensity values for all pixels are saturated, resulting in a positive event ("1") for all pixels, making it impossible to determine quantitative relationships. Therefore, the histogram shows the change in digital counts according to the change in low MGITC concentrations below 10^{-9} M. The average value and standard deviation of digital counts were calculated from images obtained by mapping three times for each concentration. The threshold value was set as the LOD in the blank sample, which is the theoretical LOD of conventional SERS, as shown in Figure 3a. The average value of digital counts is represented as the y-value, and the standard deviation is represented as error bars. Similarly, using the 4-parameter fitting, the quantification curve yielded an R^2 value of 0.998, and the calculated LOD was 4.69 pM, about 300 times lower than the conventional SERS method. The R^2 values of both graphs ($\geq$0.99) confirm that the quantification curves were determined meaningfully and reliably. As expected, the digital SERS method allowed for accurate quantitative analysis at lower concentrations than the conventional SERS method. According to Figure 5b, the LOD for digital SERS was determined to be 4.69 pM, which is included in the concentration range of the quantitative curve. However, evaluating the usefulness of the digital SERS analysis method based on Raman spectral data for only four MGITC concentrations in the range below 10^{-9} appears to be of little significance. Therefore, additional Raman measurements for more concentrations were necessary for more reliable quantitative analysis.

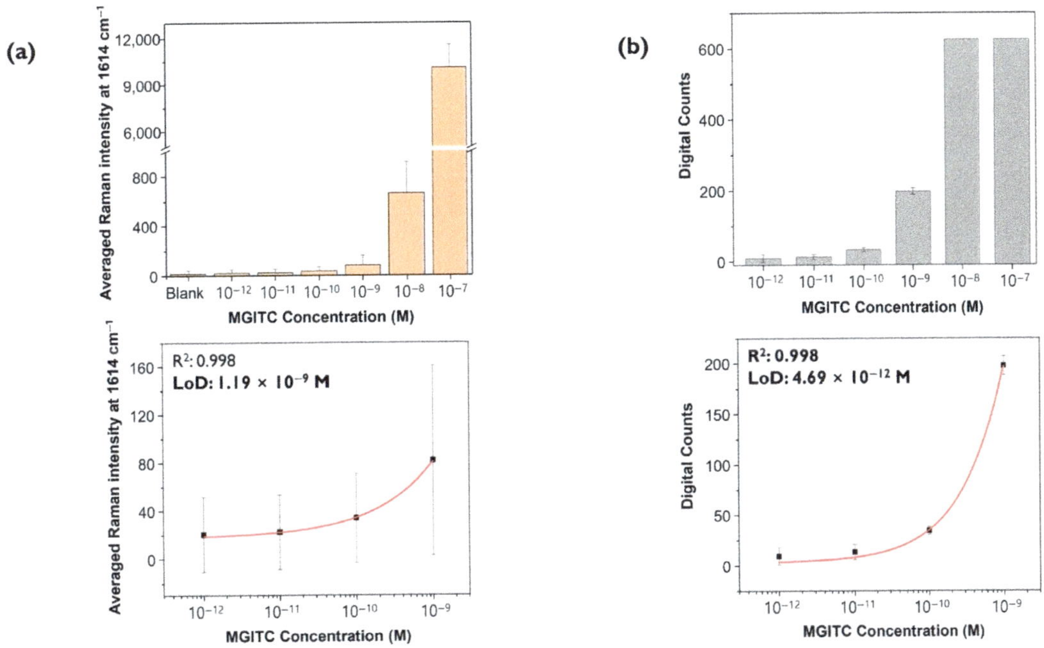

Figure 5. (**a**) Histogram of the average Raman peak intensity at 1614 cm^{-1} (**top**) and corresponding calibration curve (**bottom**). (**b**) Histogram of the number of digital counts (**top**) and corresponding calibration curve (**bottom**). Error bars indicate the standard deviations from three measurements.

To demonstrate that the digital SERS method can analyze target analytes at low concentrations more accurately, experiments were conducted at more finely divided intervals at lower concentrations from 5×10^{-10} M to 5×10^{-13} M (Figure 6). Figure 6a shows conventional Raman mapping images for additional MGITC concentrations, and Figure 6b shows corresponding digital SERS images. The scale bar for the Raman mapping image was set the same as in Figure 3a. Figure 6a shows that it is difficult to distinguish between the color-decoding Raman mapping images for all concentration ranges. In contrast, the binarized images using digital SERS could be visually distinguished for all images as shown in Figure 6b. Figure 6c,d compares the calibration curves for conventional SERS and digital SERS. For conventional SERS, a clear decrease in Raman intensity with concentration was not observed, and there was a large standard deviation and loss of linearity. In contrast, for digital SERS, the number of pixels showing "1" changed significantly with concentration changes. Only for digital SERS, it was possible to determine a meaningful quantitative curve, with an R^2 value of 0.991 and a LOD of 6.43 pM (Figure 6d). The LOD for digital SERS in this experiment was similar to the previous experimental result determined for smaller concentrations (Figure 5b). A significant reduction in error bar fluctuation was also observed for digital SERS. These results confirmed that digital SERS can achieve higher sensitivity and lower LOD values than conventional methods.

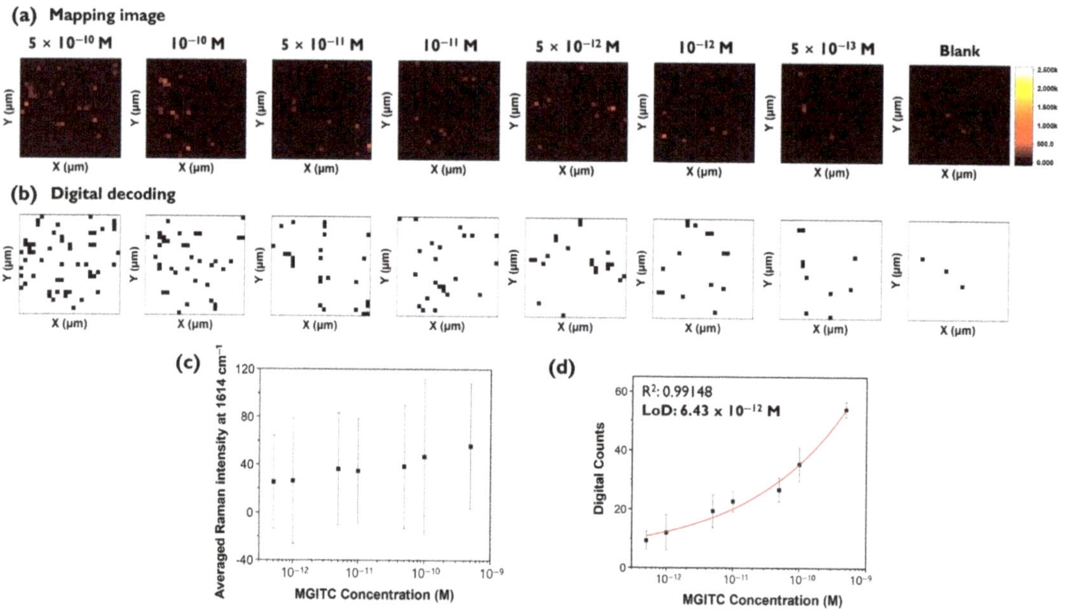

Figure 6. (**a**) Raman mapping and (**b**) digital decoding images of MGITC in the 0 to 5 × 10^{-10} M range. Corresponding calibration curves for (**c**) Raman mapping and (**d**) digital decoding images. Error bars indicate the standard deviations from three measurements.

3. Materials and Methods

Ethanol (99.5%) and tetrahydrofuran (THF) was purchased from Sigma-Aldrich (St. Louis, MO, USA). Malachite green isothiocyanate (MGITC) was purchased from Invitrogen Corporation (Carlsbad, CA, USA). Ultrapure water (0.055 μs/cmc) was obtained from a Laboratory Water System (Göttingen, Germany). A 125 μm thick polyethylene naphthalate (PEN) polymer substrate (Dupont, Wilmington, DE, USA) was used after removing the protective film. A polymer nano dimple pattern was fabricated on the PEN film by O_2 ion beam bombardment in a linear moving substrate. A linear O_2 ion beam with a width of 300 mm was generated using a linear ion source [30]. An O_2 flow rate was 70 sccm with the vacuum process chamber pressure of 0.9 mTorr. The PEN substrates were reciprocated at a linear moving speed of 10 mm s^{-1}, followed by 60 scans. The ion dose per scan was 2.3 ± 0.2 × 10^{15} cm^{-2} using a Faraday cup that reduced secondary electron emission by magnetic fields. The mean ion energy was 700 ± 70 eV measured with an ion energy analyzer [31]. A 100 nm thick Au layer was deposited directly on PEN nano dimples at a deposition rate of 2.0 Å s^{-1} using a thermal evaporation system (LAT, Osan, Republic of Korea). The base pressure of the chamber was 9.6 × 10^{-6} Torr. Then, the prepared Au/PEN nano-dimple substrates were processed with 97% perfluorodecanethiol (Sigma-Aldrich, St. Louis, MO, USA). A total of 10 μL of 97% perfluorodecanethiol solution was poured into a glass Petri dish, the Petri dish lid was put to the Au/PEN nano dimple substrate, and the lid was closed for 2 h [23]. Next, an 80 nm thick Au layer was deposited onto the PFDT-treated Au/PEN nano dimple substrate at a deposition rate of 0.3 $Ås^{-1}$ through a thermal evaporation process (LAT, Osan, Republic of Korea). The base pressure of the chamber was 9.6×10^{-6} Torr. The deposition rate was watched using a quartz crystal microbalance.

Before the experiment, the substrate was cut into 3×3 μm^2 size and washed with tetrahydrofuran and distilled water. Then, MGITC solution was diluted to the concentration using pure ethanol as buffer (10^{-7} M~5×10^{-13} M, blank). The cut substrate was added to 200 μL different concentrations of MGITC solution and reacted by shaking (500 rpm) for 1 h. After the reaction, the substrate was added to 200 μL of DW, washed by shaking for 5 min, and fixed on a slide glass to dry. To evaluate reproducibility between substrates, five Au nano popcorn substrates of 3×3 μm^2 size were prepared and reacted in 10^{-6} M of MGITC solution for 1 h.

The SERS spectra and Raman mapping images were obtained by an inVia Renishaw Raman microscope system (Renishaw, New Mills, UK). A He-Ne laser operating at 632.8 nm was used as the excitation source. Raman mapping images were obtained using a 20× objective lens and measured with an exposure time of 1 s and a laser power of 10%. To evaluate reproducibility, Raman mapping images were obtained using a 20× objective lens and measured with an exposure time of 0.1 s and laser power of 10%. The characteristic Raman peak of MGITC at 1614 cm^{-1} was scanned over an area of 48×48 μm^2 range with 2×2 μm^2 mapping steps, for a total of 625 pixels. The baseline correction of Raman spectra was carried out using WiRE V 5.3 software (Renishaw, Newmills, UK). Spectral analysis and Raman mapping image, digital decoding image generation were performed using Origin 2017 64Bit software (OriginLab Corporation, Northampton, MA, USA).

The data were processed using the mean and standard deviation of pixels. An area of 48×48 μm^2 range was measured three times at 2 μm step size, yielding a total of 1875 points of data. For data processing, we acquired the mean and standard deviation of the 1875 points to obtain a calibration curve. Based on the mean and standard deviation of the intensity of the blank (without analyte), the mean + 3 × standard deviation was set as the threshold value, which was set to "0" if it was less than the threshold value and "1" if it was greater than the threshold value. For each concentration, the number of pixels higher than the threshold value was counted and calculated for each concentration.

4. Conclusions

Recently, the SERS technique is gaining popularity as an analytical technique due to its high sensitivity, quantitative analysis, multiplexed detection, and rapid detection capabilities. SERS allows for analysis of trace amounts of substances, even down to a single molecular level. However, when the concentration of the analyte becomes extremely low, the signal-to-noise ratio decreases dramatically due to the influence of background noise, making it difficult to distinguish SERS signals generated from the plasmonic substrate's hotspots from noise. In this study, Raman reporter molecule MGITC was loaded at different concentrations onto Au nano popcorn substrates previously developed by our research group. The resulting Raman mapping images were analyzed using both conventional SERS and digital SERS methods. The conventional SERS method involves color decoding of Raman peak intensities for each pixel. However, at low concentrations, the signal-to-noise ratio decreases, making it difficult to distinguish color changes corresponding to concentration. Consequently, quantitative analysis using spectral data in conventional SERS results in increased error deviation, reducing accuracy, sensitivity, and reproducibility. The LOD for MGITC determined by conventional SERS was at the nM level, with concentrations below this becoming impossible to quantify due to excessive error fluctuations.

To address this issue, we introduced the digital SERS method. This approach involves setting a threshold value for Raman peak intensity, defining pixels as "1" if above the threshold and "0" if below, and transforming Raman mapping images into digital Raman images for analysis. This method allows for accurate quantitative analysis by counting the number of pixels showing "1", even when low-concentration MGITC molecules are unevenly distributed on the substrate surface. The introduction of the digital SERS method enabled accurate quantitative analysis for MGITC concentrations below 10^{-9} M by reducing error fluctuations. The LOD was significantly improved to the pM range, showing greater sensitivity compared to the conventional SERS method. Analysis of the same Raman mapping data yielded an LOD of 1.19 nM using conventional SERS and 4.69 pM using digital SERS. Since improving sensitivity through experimental methods alone is very challenging, it is expected that complementary use of the digital SERS analysis method can greatly enhance the sensitivity and reproducibility of analyte detection.

Author Contributions: S.L. and J.C.; methodology, software, validation, formal analysis, investigation, S.L.; resources, writing—original draft preparation, writing—review and editing, supervision, project administration, funding acquisition, J.C. All authors have read and agreed to the published version of the manuscript.

Funding: This research was supported by the Technology Innovation Program (Grant number RS-2024-00432382) funded by the Ministry of Trade, Industry, and Energy (MOTIE, Republic of Korea).

Institutional Review Board Statement: Not applicable.

Informed Consent Statement: Not applicable.

Data Availability Statement: The original contributions presented in this study are included in the article, further inquiries can be directed to the corresponding author.

Acknowledgments: This paper commemorates the contributions of Jaan Laane to the field of molecular spectroscopy.

Conflicts of Interest: The authors declare no conflicts of interest.

Abbreviations

The following abbreviations are used in this manuscript:

SERS	Surface-enhanced Raman scattering
MGITC	Malachite green isothiocyanate
LSPR	Localized surface plasmon resonance

References

1. Schlücker, S. Surface-enhanced Raman spectroscopy: Concepts and chemical applications. *Angew. Chem. Int. Ed.* **2014**, *53*, 4756–4795. [CrossRef] [PubMed]
2. Langer, J.; Jimenez de Aberasturi, D.; Aizpurua, J.; Alvarez-Puebla, R.A.; Auguié, B.; Baumberg, J.J.; Bazan, G.C.; Bell, S.E.; Boisen, A.; Brolo, A.G. Present and future of surface-enhanced Raman scattering. *ACS Nano* **2019**, *14*, 28–117. [CrossRef] [PubMed]
3. Blanco-Formoso, M.; Pazos-Perez, N.; Alvarez-Puebla, R.A. Fabrication and SERS properties of complex and organized nanoparticle plasmonic clusters stable in solution. *Nanoscale* **2020**, *12*, 14948–14956. [CrossRef] [PubMed]
4. Wu, K.; Li, T.; Schmidt, M.S.; Rindzevicius, T.; Boisen, A.; Ndoni, S. Gold nnoparticles sliding on recyclable nanohoodoos—Engineered for surface-enhanced Raman spectroscopy. *Adv. Funct. Mater.* **2018**, *28*, 1704818. [CrossRef]
5. Wang, Z.; Zong, S.; Wu, L.; Zhu, D.; Cui, Y. SERS-activated platforms for immunoassay: Probes, encoding methods, and applications. *Chem. Rev.* **2017**, *117*, 7910–7963. [CrossRef]

6. Chen, S.; Zhang, Y.; Shih, T.-M.; Yang, W.; Hu, S.; Hu, X.; Li, J.; Ren, B.; Mao, B.; Yang, Z. Plasmon-induced magnetic resonance enhanced Raman spectroscopy. *Nano Lett.* **2018**, *18*, 2209–2216. [CrossRef]
7. Baumberg, J.J.; Aizpurua, J.; Mikkelsen, M.H.; Smith, D.R. Extreme nanophotonics from ultrathin metallic gaps. *Nat. Mater.* **2019**, *18*, 668–678. [CrossRef]
8. Zheng, J.; He, L. Surface-enhanced Raman spectroscopy for the chemical analysis of food. *Compr. Rev. Food Sci. Food Saf.* **2014**, *13*, 317–328. [CrossRef]
9. Aoki, P.H.; Furini, L.N.; Alessio, P.; Aliaga, A.E.; Constantino, C.J. Surface-enhanced Raman scattering (SERS) applied to cancer diagnosis and detection of pesticides, explosives, and drugs. *Annu. Rev. Anal. Chem.* **2013**, *32*, 55–76. [CrossRef]
10. Liang, J.; Liu, H.; Huang, C.; Yao, C.; Fu, Q.; Li, X.; Cao, D.; Luo, Z.; Tang, Y. Aggregated silver nanoparticles based surface-enhanced Raman scattering enzyme-linked immunosorbent assay for ultrasensitive detection of protein biomarkers and small molecules. *Anal. Chem.* **2015**, *87*, 5790–5796. [CrossRef]
11. Yadav, S.; Sadique, M.A.; Ranjan, P.; Kumar, N.; Singhal, A.; Srivastava, A.K.; Khan, R. SERS based lateral flow immunoassay for point-of-care detection of SARS-CoV-2 in clinical samples. *ACS Appl. Bio Mater.* **2021**, *4*, 2974–2995. [CrossRef] [PubMed]
12. Chen, R.; Liu, B.; Ni, H.; Chang, N.; Luan, C.; Ge, Q.; Dong, J.; Zhao, X. Vertical flow assays based on core–shell SERS nanotags for multiplex prostate cancer biomarker detection. *Analyst* **2019**, *144*, 4051–4059. [CrossRef] [PubMed]
13. Fang, Y.; Seong, N.-H.; Dlott, D.D. Measurement of the distribution of site enhancements in surface-enhanced Raman scattering. *Science* **2008**, *321*, 388–392. [CrossRef] [PubMed]
14. Dos Santos, D.P.; Temperini, M.L.; Brolo, A.G. Intensity fluctuations in single-molecule surface-enhanced Raman scattering. *Acc. Chem. Res.* **2019**, *52*, 456–464. [CrossRef]
15. Zhang, D.; Huang, L.; Liu, B.; Ni, H.; Sun, L.; Su, E.; Chen, H.; Gu, Z.; Zhao, X. Quantitative and ultrasensitive detection of multiplex cardiac biomarkers in lateral flow assay with core-shell SERS nanotags. *Biosens. Bioelectron.* **2018**, *106*, 204–211. [CrossRef]
16. Liu, B.; Zheng, S.; Li, H.; Xu, J.; Tang, H.; Wang, Y.; Wang, Y.; Sun, F.; Zhao, X. Ultrasensitive and facile detection of multiple trace antibiotics with magnetic nanoparticles and core-shell nanostar SERS nanotags. *Talanta* **2022**, *237*, 122955. [CrossRef]
17. Bhamidipati, M.; Cho, H.-Y.; Lee, K.-B.; Fabris, L. SERS-based quantification of biomarker expression at the single cell level enabled by gold nanostars and truncated aptamers. *Bioconjug. Chem.* **2018**, *29*, 2970–2981. [CrossRef]
18. Ko, J.; Lee, C.; Choo, J. Highly sensitive SERS-based immunoassay of aflatoxin B1 using silica-encapsulated hollow gold nanoparticles. *J. Hazard. Mater.* **2015**, *285*, 11–17. [CrossRef]
19. Yang, J.-Y.; Park, S.-G.; Jung, S.; Byeon, E.-Y.; Kim, D.-g.; Jung, H.S.; Kim, H.J.; Lee, S. SERS substrates based on self-organized dimple nanostructures on polyethylene naphthalate films produced via oxygen ion beam sputtering. *Appl. Surf. Sci.* **2022**, *572*, 151452. [CrossRef]
20. Wu, Y.; Yu, Q.; Joung, Y.; Jeon, C.S.; Lee, S.; Pyun, S.H.; Joo, S.-W.; Chen, L.; Choo, J. Highly uniform self-assembly of gold nanoparticles by butanol-induced dehydration and its sers applications in SARS-CoV-2 detection. *Anal. Chem.* **2023**, *95*, 12710–12718. [CrossRef]
21. Zhang, Y.; Noji, H. Digital bioassays: Theory, applications, and perspectives. *Anal. Chem.* **2017**, *89*, 92–101. [CrossRef] [PubMed]
22. Kuo, C.-W.; Smith, A.M. Digital and absolute assays for low abundance molecular biomarkers. *Acc. Chem. Res.* **2023**, *56*, 1031–1042. [CrossRef]
23. Park, S.G.; Xiao, X.; Min, J.; Mun, C.; Jung, H.S.; Giannini, V.; Weissleder, R.; Maier, S.A.; Im, H.; Kim, D.H. Self-assembly of nanoparticle-spiked pillar arrays for plasmonic biosensing. *Adv. Funct. Mater.* **2019**, *29*, 1904257. [CrossRef]
24. de Albuquerque, C.D.L.; Sobral-Filho, R.G.; Poppi, R.J.; Brolo, A.G. Digital protocol for chemical analysis at ultralow concentrations by surface-enhanced Raman scattering. *Anal. Chem.* **2018**, *90*, 1248–1254. [CrossRef] [PubMed]
25. Noji, H.; Minagawa, Y.; Ueno, H. Enzyme-based digital bioassay technology–key strategies and future perspectives. *Lab Chip* **2022**, *22*, 3092–3109. [CrossRef]
26. Dang, H.; Joung, Y.; Jeong, C.; Jeon, C.S.; Pyun, S.H.; Park, S.-G.; Choo, J. Nanoplasmonic assay platforms for reproducible SERS detection of Alzheimer's disease biomarker. *Bull. Korean Chem. Soc.* **2023**, *44*, 441–448. [CrossRef]
27. Chen, H.; Park, S.-G.; Choi, N.; Moon, J.-I.; Dang, H.; Das, A.; Lee, S.; Kim, D.-G.; Chen, L.; Choo, J. SERS imaging-based aptasensor for ultrasensitive and reproducible detection of influenza virus A. *Biosens. Bioelectron.* **2020**, *167*, 112496. [CrossRef]
28. Chen, H.; Park, S.-G.; Choi, N.; Kwon, H.-J.; Kang, T.; Lee, M.-K.; Choo, J. Sensitive detection of SARS-CoV-2 using a SERS-based aptasensor. *ACS Sens.* **2021**, *6*, 2378–2385. [CrossRef]
29. Chen, H.; Park, S.-K.; Joung, Y.; Kang, T.; Lee, M.-K.; Choo, J. SERS-based dual-mode DNA aptasensors for rapid classification of SARS-CoV-2 and influenza A/H1N1 infection. *Sens. Actuators B* **2022**, *355*, 131324. [CrossRef]

30. Lee, S.; Byun, E.-Y.; Kim, J.-K.; Kim, D.-G. Ar and O_2 linear ion beam PET treatments using an anode layer ion source. *Curr. Appl. Phys.* **2014**, *14*, S180–S182. [CrossRef]
31. Böhm, C.; Perrin, J. Retarding-field analyzer for measurements of ion energy distributions and secondary electron emission coefficients in low-pressure radio frequency discharges. *Rev. Sci. Instrum.* **1993**, *64*, 31–44. [CrossRef]

molecules MDPI

Article

Non-Isothermal Melt Crystallization of a Biodegradable Polymer Studied by Two-Dimensional Infrared Correlation Spectroscopy

Isao Noda

Department of Materials Science and Engineering, University of Delaware, Newark, DE 19716, USA; noda@udel.edu; Tel.: +1-513-600-9255

Abstract: The non-isothermal melt crystallization process of poly[(R)-3-hydroxybutyrate-*co*-(R)-3-hydroxyhexanoateate] (PHBHx) was monitored using attenuated total reflection infrared (ATR IR) measurement. The resulting time- and temperature-dependent spectra were subjected to the two-dimensional correlation spectroscopy (2D-COS) analysis. The C=O stretching region of the PHBHx sample consisted of several distinct IR contributions attributable to the population of amorphous component, well-ordered type I lamellar crystal, and less ordered inter-lamellar type II crystal. The spectral intensity change in type I crystal occurs in the earlier stage of the crystallization at a higher temperature range compared to the overall intensity decrease in the amorphous component occurring throughout the crystallization process. The growth of the type II crystal started in a later stage at a lower temperature than the creation of the type I crystal. An early decrease in a small but distinct portion of the amorphous component may be related to a crystallization precursor species with some level of molecular order. Hetero-mode correlation analyses revealed that the crystalline band intensity changes in the C-H stretching and fingerprint regions all occur later than the population changes in crystalline species reflected by the carbonyl stretching bands. This observation suggests that the spectral intensity changes in the C-H stretching and fingerprint regions do not directly represent the population dynamics of the crystalline and amorphous species but probe instead the molecular state of the crystalline entities still undergoing the evolutionary changes.

Keywords: poly(hydroxyalkanoate); two-dimensional correlation infrared spectroscopy; melt crystallization

1. Introduction

Professor Jaan Laane has been a true intellectual leader conducting the research effort in the field of molecular spectroscopy, including Fourier transform infrared (FT-IR), fluorescence, and laser Raman spectroscopy. His work has greatly contributed to the understanding of molecular vibrations and structures. He has published a number of articles and edited books on the subject of molecular spectroscopy. This author had the privilege and pleasure of contributing chapters to his books [1,2], which included some discussions on the IR study of a bio-based biodegradable polymer. The current article is a continuation of the molecular spectroscopy work on a similar bioplastics system to further explore the in-depth analysis of the material.

Polyhydroxyalkanoates or PHAs are totally bio-based and biodegradable aliphatic semicrystalline polyesters that have been explored as possible alternative materials to replace conventional petroleum-based plastics [3–7]. Among many types of PHAs, the interest in poly[(R)-3-hydroxybutyrate-*co*-(R)-3-hydroxyhexanoateate] or PHBHx has especially increased. Because of the molecular architecture of this copolymer, PHBHx possesses a unique set of practically useful properties, like ductility, toughness, and superior processing compatibility [8,9]. The material is nowadays produced by a large scale industrial manufacturing process using the fermentation of renewable resources, like vegetable oils. Because PHBHx has a set of attractive physical properties similar to those of conventional plastics made from petroleum, the additional unique property of biodegradability has attracted keen interest as a potentially useful sustainable alternative material. Melt processing is the preferred processing method for thermoplastics, like PHBHx, where the material is heated to become soft and fluid and then cooled to solidify in the shape of the final products. For a semicrystalline polyester, understanding the detailed crystallization process becomes a critical requirement for controlling the effective solidification and subsequent properties of the products.

As Laane has repeatedly demonstrated in his professional life, spectroscopy is an excellent tool to track the important but sometimes complex dynamics of various physical and chemical processes, including polymer crystallization, at the molecular level in order to provide useful insights into mechanistic understanding. Among many commonly available spectroscopic techniques, attenuated total reflectance (ATR) infrared (IR) spectroscopy is well suited for the in situ monitoring of thermal transition processes, like the crystallization of a polymer from the melt. While the crystallization of a PHAs melt has often been studied under the idealized isothermal condition [10,11], a more realistic melt crystallization process for industrial polymers is typically carried out under non-isothermal conditions with both varying temperatures and time. Although such time- and temperature-dependent ATR IR measurements can provide high-quality spectroscopic data that are rich in the latent information content, analysing and interpreting the very complex and highly convoluted results often become a major challenge.

Two-dimensional correlation spectroscopy (2D-COS) is a versatile data analysis tool employed by many researchers [12–14]. Evolution patterns of spectral intensities induced by the application of external perturbation, such as thermal processing, are systematically analyzed using a form of cross correlation approach. It was noted that certain parts of a spectrum can change simultaneously in a coordinated fashion, while others may vary differently in out-of-phase sequences. Such characteristic variation patterns are conveniently displayed in the form of planar spectral mapping schemes defined by two independent wavenumber axes. By spreading the overlapped bands along the second dimension, apparent spectral resolution is often enhanced. By following a simple set of interpretation rules for correlation peaks appearing in the 2D-COS maps, one can deduce the streamlined mechanistic picture of the evolving system represented by the spectra. In this work we apply the 2D-COS analysis to the ATR IR spectra of a bioplastic PHBHx sample undergoing a non-isothermal melt crystallization process to elucidate the intricate formation steps of crystalline components, which eventually control the final performance properties of the products.

2. Background

A very brief overview is provided here for the 2D correlation spectroscopy (2D-COS) to be applied toward the time- and temperature-dependent ATR IR spectra obtained during the non-isothermal melt crystallization of a PHBHx sample. A more detailed description of this general technique is available elsewhere [1,2,12–14]. Consider a set of m IR spectra $A(\widetilde{\nu}, t_i)$, with $i = 1, 2, \cdots m$, which are measured as functions of wavenumber $\widetilde{\nu}$ and another well-defined physical variable t_i, such as time or temperature; we define the dynamic spectrum $\widetilde{A}(\widetilde{\nu},t_i)$ as

$$\widetilde{A}(\widetilde{\nu},t_i) = A(\widetilde{\nu},t_i) - \overline{A}(\widetilde{\nu}) \tag{1}$$

The reference spectrum $\overline{A}(\nu)$ is typically set to be the average spectrum taken during the observation interval between t_1 and t_m.

$$\overline{A}(\widetilde{\nu}) = \frac{1}{m}\sum_{i=1}^{m} A(\widetilde{\nu},t_i) \tag{2}$$

The synchronous and asynchronous 2D correlation spectra, $\Phi(\widetilde{\nu}_1, \widetilde{\nu}_2)$ and $\Psi(\widetilde{\nu}_1, \widetilde{\nu}_2)$, for $\widetilde{A}(\widetilde{\nu},t_i)$ are given by

$$\Phi(\widetilde{\nu}_1,\widetilde{\nu}_2) = \frac{1}{m-1}\sum_{i=1}^{m} \widetilde{A}(\widetilde{\nu}_1,t_i)\cdot\widetilde{A}(\widetilde{\nu}_2,t_i) \tag{3}$$

$$\Psi(\widetilde{\nu}_1,\widetilde{\nu}_2) = \frac{1}{m-1}\sum_{j=1}^{m} \widetilde{A}(\widetilde{\nu}_1,t_j)\cdot\sum_{i=1}^{m} N_{ij}\widetilde{A}(\widetilde{\nu}_2,t_i) \tag{4}$$

with the elements of the Hilbert–Noda transformation matrix defined as

$$N_{ij} = \begin{cases} \frac{1}{\pi(j-i)} & if\ i \neq j \\ 0 & otherwise \end{cases} \tag{5}$$

Synchronous spectrum $\Phi(\widetilde{\nu}_1, \widetilde{\nu}_2)$ corresponds to the covariance of spectral intensity variations. It characterizes the simultaneous or coordinated changes in band intensities at two IR wavenumbers $\widetilde{\nu}_1$ and $\widetilde{\nu}_2$. The magnitude of the so-called autopeak appearing at the main diagonal position, i.e., $\widetilde{\nu}_1 = \widetilde{\nu}_2$, represents the extent or variance of the spectral intensity at a given coordinate. Peaks appearing in the off-diagonal positions of a 2D-COS map are called cross peaks. If the sign of a synchronous cross peak at a coordinate position $(\widetilde{\nu}_1,\widetilde{\nu}_2)$ is positive, i.e., $\Phi(\widetilde{\nu}_1, \widetilde{\nu}_2) > 0$, the measured IR spectral intensities at $\widetilde{\nu}_1$ and $\widetilde{\nu}_2$ are either increasing or decreasing together in the same direction. The intensities are changing in opposite directions for a negative cross peak.

An asynchronous spectrum $\Psi(\widetilde{\nu}_1, \widetilde{\nu}_2)$, in turn, characterizes the out-of-phase or sequential changes in spectral intensities, which are presumably arising from different sources. Signals arising from the same source cannot vary independent of each other. Thus, the appearance of asynchronous cross peaks implies the presence of multiple species which are responding differently to the given perturbation, like temperature change. This feature also becomes useful in differentiating highly overlapped spectral bands if they are responding differently to the perturbation. Enhanced spectral resolution based on the asynchronous correlation is one of the unique advantages of 2D-COS analysis.

If the signs of both synchronous and asynchronous cross peaks at a given coordinate $(\widetilde{\nu}_1, \widetilde{\nu}_2)$ are the same, i.e., $\Phi(\widetilde{\nu}_1, \widetilde{\nu}_2)\cdot\Psi(\widetilde{\nu}_1, \widetilde{\nu}_2) > 0$, spectral intensity changes measured at $\widetilde{\nu}_1$ predominantly occur earlier than that at $\widetilde{\nu}_2$ along the variable t. On the other hand, if the two peak signs are different, this relationship is reversed. Thus, the 2D-COS cross peaks immediately provide useful knowledge about the sequential order of spectral intensity changes. Such information obviously is beneficial in establishing a streamlined mechanistic insight into convoluted multistep processes.

3. Results and Discussion

3.1. ATR IR Spectra

Figure 1 shows the overlaid spectral plots of three pertinent regions of the ATR IR spectra of PHBHx (12 mol% 3HHx) undergoing the non-isothermal crystallization process from the amorphous melt at 140 °C to the semicrystalline state at 40 °C. The effect of the crystal growth at the expense of the amorphous melt component is well reflected by the changes in the spectral intensities of IR bands. For example, spectral variations in the C-H stretching region (Figure 1a) are dominated by the increasing contributions of IR intensities associated with the crystalline component around 2975, 2968, 2932 and 2875 cm^{-1} and decreasing contributions from the amorphous component around 2984 and 2942 cm^{-1}. Moieties contributing to the molecular vibrations for these bands are obviously placed in different local environments characteristic to ordered crystal lattice and random amorphous coils.

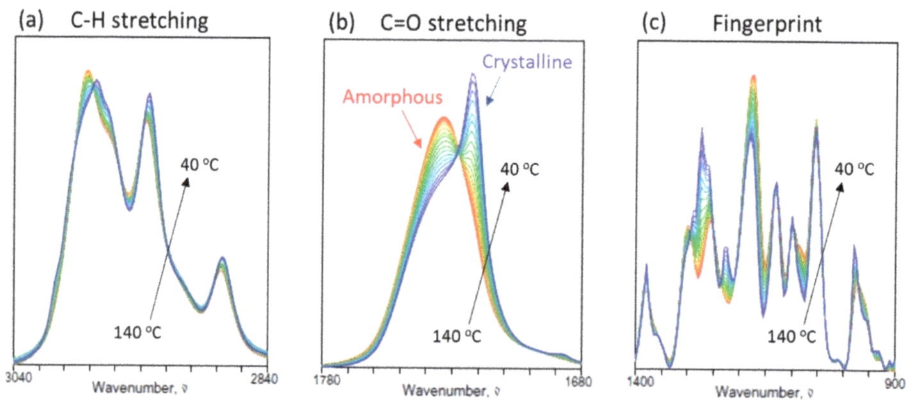

Figure 1. Temperature-dependent ATR IR spectra of PHBHx (12 mol% 3HHx) undergoing the non-isothermal melt crystallization process between 140 and 40 °C: (**a**) C-H stretching region, (**b**) C=O stret5ching region, and (**c**) fingerprint region.

The spectral region corresponding to the carbonyl stretching vibrations of the polyester (Figure 1b) provides the information most directly useful for tracking the crystallization of PHBHx. The completely molten PHBHx sample exhibits a single IR peak centered around 1736 cm^{-1} corresponding to the fully amorphous state of the polymer. Upon cooling the sample, an additional spectral feature appears at around 1720 cm^{-1}, while the peak intensity at around 1736 cm^{-1} gradually decreases. The temperature of the appearance of the 1720 cm^{-1}-band coincides with the onset of crystallization for this sample. It is, therefore, reasonable to assign the increasing feature around 1720 cm^{-1} to be associated with the contribution from the growing crystalline component population of the polymer. There remain enough amorphous features in the sample at 40 °C to indicate that this polymeric material is in the semicrystalline state at the final stage. It should be pointed out that the lower-wavenumber feature at around 1720 cm^{-1} actually originates from the enhanced hydrogen bonding interaction of carbonyl oxygen instead of the specific helical structure of polymer segments in the crystal lattice [15].

The fingerprint region of the IR spectrum (Figure 1c), primarily reflecting the contributions from C-C-O and C-O-C vibrations of PHBHx, also exhibits complex and rich information associated with the crystallization process. These bands are sensitive to the conformational state of the polymer backbone and the interactions of the neighboring segments. Similarly to the C=O stretching region, peaks associated with the growth of crystalline component increase their intensities at around 1389, 1276, 1263, and 1228 cm^{-1}, while those for the amorphous component at around 1302, 1259, and 1181 cm^{-1} decrease. Unlike the carbonyl stretching region, however, these bands in the fingerprint region coexist in a highly congested and overlapped manner to make the description of crystallization population dynamics much less straightforward.

3.2. 2D-COS Analysis of the C=O Stretching Region

Figure 2 shows the 2D-COS spectra constructed from the set of time- and temperature-dependent ATR IR spectra of PHBHx undergoing melt crystallization in the C=O stretching region (Figure 1b). The overlaid 1D spectra are provided at the top and left side of each 2D map for reference purposes, and negative cross peaks are indicated by the blue shading. A similar system was actually analyzed using 2D-COS in the past [1,2] but without the intensity correction by normalization, which has brought out the details of the crystalline and amorphous contributions more clearly this time. The synchronous spectrum (Figure 2a) shows the autopeaks along the main diagonal positions at around 1736 and 1722 cm^{-1}, which, respectively, match well with the major band positions of the amorphous and crystalline contributions. In parallel, there appear a pair of cross peaks at the corresponding spectral coordinates. The negative signs of the cross peaks indicate that the spectral intensity of the band around 1736 cm^{-1} is decreasing, while the other band around 1722 cm^{-1} is increasing. This observation agrees with the fact that the population of the amorphous component of PHBHx is decreasing while the crystalline contribution is increasing during the melt crystallization process.

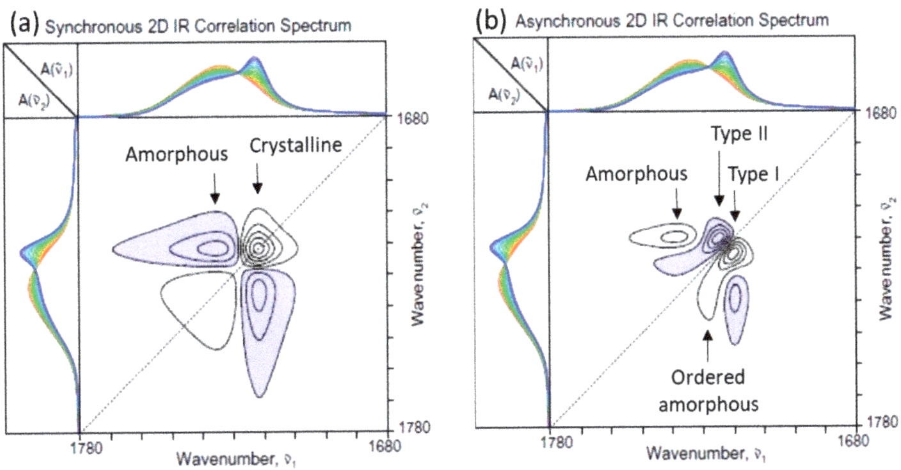

Figure 2. (**a**) Synchronous and (**b**) asynchronous 2D-COS spectra of PHBHx undergoing the melt crystallization process constructed from the set of 1D ATR IR spectra in the C=O stretching region (Figure 1b). One-dimensional spectra are provided at the top and left side of each 2D map, and negative cross peaks are indicated by the blue shading.

The asynchronous 2D-COS spectrum (Figure 2b) provides another interesting story. First of all, the spectrum itself exhibits a number of asynchronous cross peaks, indicating that the spectral intensity changes in this system during the melt crystallization are not simultaneously proceeding in a simple coordinated manner. The presence of asynchronous cross peaks around the coordinates of (1736, 1720) and (1720, 1736) indicates that the decrease in the amorphous component reflected by the declining behavior of the 1736 cm^{-1} band does not occur simultaneously with the increase in the major crystalline contribution from the band around 1720 cm^{-1}. A comparison of the peak signs between the synchronous and asynchronous cross peaks reveals that the intensity increase around 1720 cm^{-1} occurs predominantly in earlier stages at a higher temperature while the decrease around 1736 cm^{-1} continues to the later stage of the lower temperature range.

The reason for the asynchronicity between the crystalline 1720 cm^{-1} band and the amorphous 1736 cm^{-1} band may be understood by the appearance of an additional cross peak pair around (1720, 1724) and (1724, 1720). It appears that there are two distinct populations of crystalline contributions, such that the amorphous component is further consumed to create the second crystalline component at 1724 cm^{-1} even after the growth of the 1720 cm^{-1} component has slowed down. The presence of two types (I and II) of crystalline components during the melt crystallization of PHBHx has been observed in the past [1,2,15–19]. The type I crystal found at around 1720 cm^{-1}, which grows first, is believed to have a well-ordered lattice structure, corresponding most likely to the volume-filling lamellae of the crystalline spherulites. In turn, the type II crystal, which grows later, is speculated to have a less ordered structure associated with the secondary crystals formed within the confined inter-lamellar spacing. While the two types of PHBHx crystals exhibiting distinct growth behaviors are vividly revealed in the asynchronous 2D-COS plot (Figure 2b), it is actually rather difficult to detect the difference by a cursory observation of the set of 1D spectra, like the one shown in Figure 1b.

The asynchronous cross peak at (1720, 1724), exhibiting the foothill-like extension toward the region at around (1730, 1736), presents an interesting observation. This feature becomes visible only if the threshold of the contour line level is set sufficiently low, such that the existence of the additional peak was not always noticed in the past [1,2]. A previous Raman study of the isothermal crystallization of a similar PHBHx sample also observed a similar feature at around 1734 cm^{-1} [20]. While this band was tentatively assigned to the nucleating crystals with imperfect structures, this assignment may not be applicable here. The IR intensity at 1730 cm^{-1} definitely decreases during the crystallization, so it is reasonable to assign this band to be a part of the amorphous contribution. However, the decline of this band intensity occurs earlier than that of the rest of the amorphous region, and its behavior is actually well synchronized with the increase in the intensity of the type I crystal at around 1720 cm^{-1}. It is tempting to speculate that this band at 1730 cm^{-1} is related to a sort of precursor species for the formation of the well-ordered lamellar type I crystal. Because of the lower-wavenumber position of this band compared to the rest of the amorphous band, it is possible that a somewhat more ordered non-crystalline species may be present in the melt before the crystal formation.

3.3. *2D-COS in Other Spectral Regions*

Figure 3 shows the 2D-COS spectra constructed from the ATR IR spectra in the C-H stretching region (Figure 1a). The spectral intensity changes in this region show the time- and temperature-dependent dynamics of the increasing crystalline and decreasing amorphous contributions. Major increasing bands, connected by the positive synchronous cross peaks, are assignable to the symmetric methyl stretching mode at around 2875 cm^{-1} and antisymmetric methylene stretching at around 2932 cm^{-1}. They are clearly associated with the crystalline component. In turn, the decreasing band around 2942 cm^{-1} must be arising from the contribution of the amorphous component. An asymmetric methyl stretching contribution of crystal is observed at around 2975 cm^{-1}, as well as the amorphous band at around 2984 cm^{-1}. In addition, a pair of crystalline bands at around 2996 and 3007 cm^{-1} are observed. They are believed to arise from an unusual localized interaction between one of the methyl hydrogens and carbonyl oxygen, as observed in [21–24]. While C–H hydrogens do not usually participate in the hydrogen bonding interactions, the close proximity of the methyl hydrogen and carbonyl oxygen in the crystalline lattice seems to bring in this unique interaction.

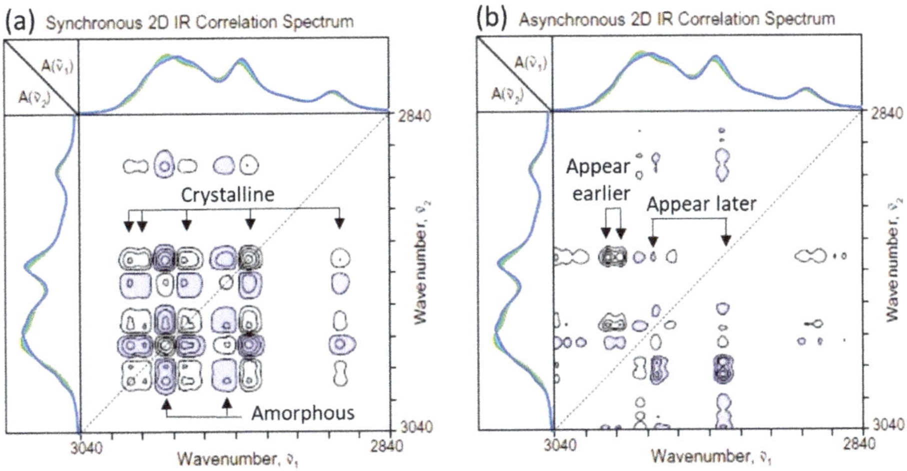

Figure 3. (**a**) Synchronous and (**b**) asynchronous 2D-COS spectra of PHBHx undergoing the melt crystallization process constructed from the set of ATR IR spectra in the C-H stretching region (Figure 1a).

The fact that the asynchronous spectrum (Figure 3b) exhibits a number of cross peaks indicates that the spectral intensity variations associated with various components in this region once again do not all occur uniformly. For example, the intensities of the two high-wavenumber crystalline bands at 2996 and 3007 cm^{-1} seem to increase well ahead of all the other bands. The increases in the rest of the bands associated with the crystalline contributions occurring later at a lower temperature and decreases in the bands at around 2984 and 2942 cm^{-1} for the amorphous component. The increases in the intensities at around 2975 and 2932 cm^{-1} occur later than that of the band at around 2968 cm^{-1}. Unlike the carbonyl stretching region, however, it is rather difficult to assign highly overlapped bands with increasing intensities in this region to any specific crystalline species.

Figure 4 shows the 2D-COS spectra constructed from the spectra in the fingerprint region (Figure 1c). Once again, bands associated with both increasing and decreasing components of PHBHx during the crystallization are observed. Even though many band profiles tend to overlap, the signs of cross peaks in the synchronous spectrum (Figure 4a) makes it possible for the initial classification of the pertinent bands. The intensities of two bands at around 1170 and 1080 cm^{-1} arising from the ester C-O and C-O-C stretching vibrations are decreasing. They are presumably assignable to the amorphous component, while other increasing bands are assignable to the crystalline contributions. The most prominent band is the increasing C-O-C stretching band at around 1276 cm^{-1}, which splits into two in the corresponding asynchronous spectrum (Figure 4b). The asynchronous spectrum also shows that the amorphous ester C-O stretching band at around 1170 cm^{-1} is accompanied by an overlapping adjacent crystalline band at around 1181 cm^{-1}. The enhanced resolution of the overlapped bands, achieved by spreading the peaks along the second dimension, is one of the key advantages of 2D-COS analysis.

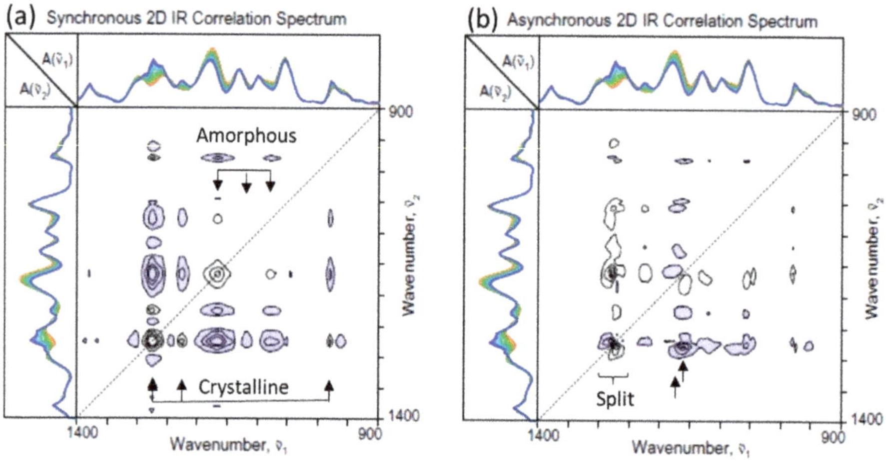

Figure 4. (**a**) Synchronous and (**b**) asynchronous 2D-COS spectra of PHBHx undergoing the melt crystallization process constructed from the set of ATR IR spectra in the fingerprint region (Figure 1c).

3.4. Hetero-Mode 2D-COS

While the resolution of overlapped bands arising from different contributing species obviously is a beneficial result by itself, proper classification and assignment of such individual spectral responses to specific contributing components of the system are even more desired. Hetero-mode correlation among different vibrational modes of the sample becomes a useful tool to achieve such a task. Hetero-mode correlation is the simplest form of the so-called 2D hetero-spectral correlation analyses. In hetero-spectral correlation, two sets of spectral datasets observed under the same perturbation conditions are compared. For example, IR and Raman spectra may be compared to each other to take advantage of the different sensitivities of individual probes to particular molecular responses. If one type of the dataset is better understood or better interpreted than the other, one can use the former to assist the analysis of the latter. Thus, hetero-spectral correlation is sometimes called the Rosetta Stone analysis, based on the historical discovery of a tablet written in previously less understood Egyptian hydrographs in parallel with the interpretable Greek text.

Figure 5 shows the hetero-mode correlation between the C=O stretching and C-H stretching regions of the ATR IR spectra. We have already established that the carbonyl stretching mode of this PHBHx sample consists of several distinct contributions, such as the amorphous band at around 1736 cm^{-1}, well-ordered type I lamellar crystals at around 1720 cm^{-1}, and less ordered type II inter-lamellar crystals at around 1724 cm^{-1}. The hetero-mode synchronous spectrum (Figure 5a) clearly reveals the specific correlations among crystalline and amorphous bands in the two different spectral regions. The vertical $\tilde{\nu}_2$ coordinates of all the maxima and minima of the synchronous peaks are observed at around 1736 and 1723 cm^{-1}. This result may seemingly suggest that predominant crystalline changes in the C-H stretching region are associated more so with the growth of less ordered type II crystals. The asynchronous spectrum (Figure 5b) also supports this tentative conclusion, since the peak maxima of the crystalline bands in the C-H stretching region are all found at around $\tilde{\nu}_2$ =1720 cm^{-1}, indicating that they are behaving somewhat differently from the formation of well-ordered type I crystals. On the other hand, the closer peak position of 1723 cm^{-1} instead of 1724 cm^{-1} also may suggest that signals are arising from both type I and type II crystals.

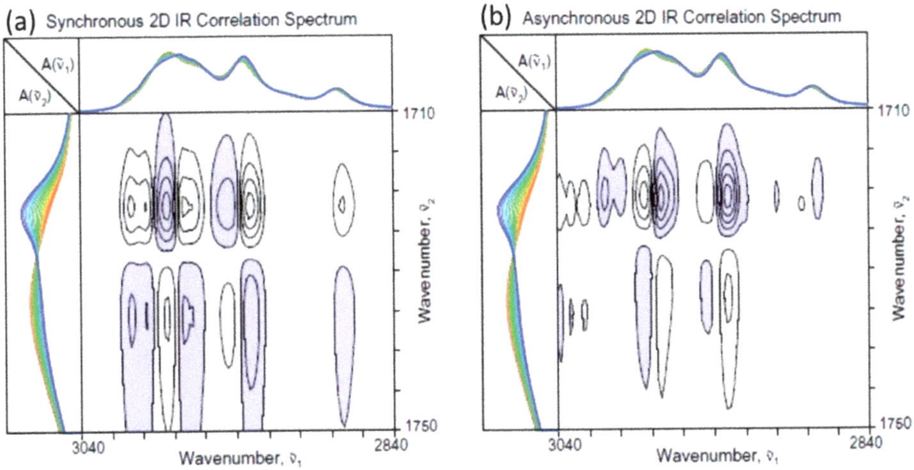

Figure 5. (**a**) Synchronous and (**b**) asynchronous 2D hetero-mode correlation spectra of PHBHx undergoing the melt crystallization process constructed by comparing the ATR IR spectra in the C-H stretching region (Figure 1a) and C=O stretching region (Figure 1b).

The sequential order analysis based on the cross peak signs of the hetero-mode synchronous and asynchronous spectra reveals a fascinating discovery. It was revealed that intensity changes in bands in the C-H stretching region all seem to happen later at a lower temperature range than those of the carbonyl stretching region. In other words, intensity variations in the C-H stretching region bands occur after the type I and II crystals are already formed. The band intensity changes, therefore, may not directly represent the formation of different species of crystals but rather reflect the subtle evolutionary rearrangement of the crystalline lattice structure. Gradual reorganization of the lamellar structure of PHBHx crystals toward higher perfection was indeed suggested in the past, based on a low-frequency Raman study of the isothermal crystallization process of PHBHx [20]. If this conjecture is correct, we may now have a potentially useful IR probe in the C-H stretching region for tracking the state of crystalline lattice structure evolution.

Figure 6 shows the hetero-mode correlation between the fingerprint and C=O stretching region. Once again, the asynchronous spectrum (Figure 6b) suggests that the crystalline signals in the fingerprint region are behaving somewhat differently from the well-ordered type I crystal at around 1720 cm^{-1} of the C=O stretching region. Compared to the C-H stretching region, however, the $\tilde{\nu}_2$ coordinates of the synchronous cross peaks for the apparent crystalline contributions in the fingerprint region are even closer to 1722 cm^{-1}. The result suggests that the IR signals increasing during the crystallization process in the fingerprint region are not necessarily originating from a distinct population of the crystalline species.

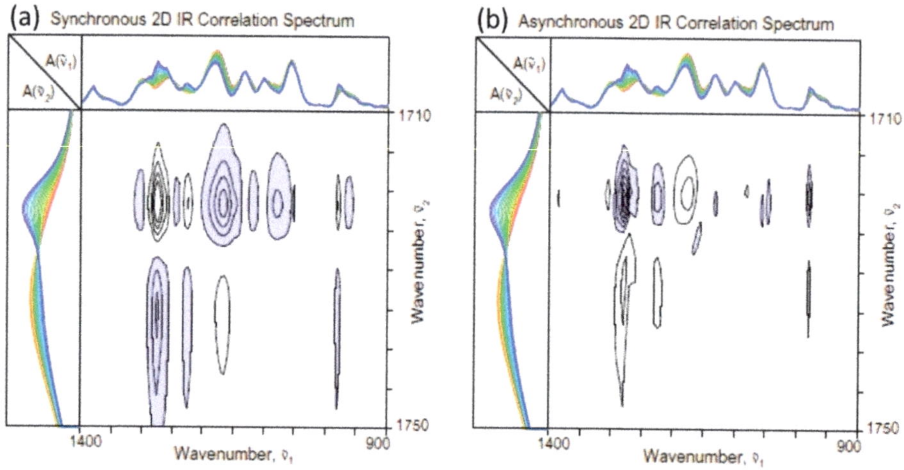

Figure 6. (**a**) Synchronous and (**b**) asynchronous 2D hetero-mode correlation spectra of PHBHx undergoing the melt crystallization process constructed by comparing the ATR IR spectra in the fingerprint region (Figure 1c) and C=O stretching region (Figure 1b).

The sequential order analysis based on the cross peak signs once again reveals that the band intensity changes in the fingerprint region are all occurring later than those in the carbonyl stretching region in a manner similar to the C-H stretching bands. The variations in band intensities in the fingerprint region are observed after the type I and II crystals are already formed. The nature of the spectral changes in this region, however, may be somewhat different. While the C-H stretching mode vibrations are more likely influenced by the perfection of crystalline lattice packing, the fingerprint region probably has higher susceptibility toward the conformational order of polymer chains and inter-chain interactions.

4. Experimental

4.1. Material

The sample of poly([R]-3-hydroxybutyrate-*co*-[R]-3-hydroxyhexanoate) (PHBHx) was prepared in the research laboratory of the Procter & Gamble Company in Cincinnati, OH, USA, by the fermentation of palm kernel oil using a strain of Gram-negative bacterium *Aeromonas caviae*. The polymer was extracted from the cell body using chloroform as the solvent and purified by repeated precipitation and dissolution using ethanol and hexane as the precipitating non-solvent. The content of comonomer 3-hydroxyhexanoate in the comonomer was about 12 mol%, with the rest being 3-hydroxybutyrate. The weight average molecular weight was about 894,000.

4.2. Method

The PHBHx sample was heated to 170 °C, which is well above the melt temperature of this copolymer at about 120 °C, and kept for one minute. The molten polymer sample was placed in the temperature-controlled attenuated total reflection (ATR) cell, initially kept at the starting temperature of 140 °C, and mounted on a Bio-Rad Model 165 FT-IR spectrometer (Hercules, CA, USA). The ATR IR spectra of the PHBHx sample were continuously collected while the temperature of the ATR cell was decreased at a constant cooling rate of 3.2 °C/min from 140 °C to 40 °C. The whole process took less than 30 min to complete. The collected spectra were normalized by the area under the curve between 1780 and 1760 cm^{-1} to minimize the superfluous interferences, such as sample contact effect. The correction factor for the normalization was only within 5% of the original data, but the treatment makes it easier to reveal the detailed dynamics of the spectral contribution of the crystalline component.

5. Conclusions

The non-isothermal crystallization of PHBHx (12% 3HHx) by cooling the melt from 140 °C to 40 °C at the rate of 3.2 °C/min was monitored using ATR IR measurement. The resulting time- and temperature dependent spectra were subjected to a 2D-COS analysis. It was revealed that the C=O stretching region of the PHBHx sample consisted of several distinct IR contributions attributable to the population of an amorphous component detected at around 1736 cm^{-1}, well-ordered type I lamellar crystal at around 1720 cm^{-1}, and less ordered inter-lamellar type II crystal at around 1724 cm^{-1}. The spectral intensity change in the type I crystal occurred in the earlier stage of the crystallization at a higher temperature range compared to the overall intensity change in the amorphous component occurring throughout the crystallization process. The growth of the type II crystal started at a stage much later than the creation of the type I crystal. In addition to the above observations, some early decreasing in a portion of the amorphous component was also observed at around 1730 cm^{-1}. This amorphous component may be a precursor species to the crystallization with some level of molecular order.

Dynamics of the spectral intensity variations in the C-H stretching and fingerprint regions are much more complex, even though it is possible to classify bands to amorphous and crystalline contributions according to the directions of the intensity changes. Crystalline bands did not increase in their intensity in a uniform manner, and there are noticeable asynchronous behaviors of individual bands during the crystallization process. A hetero-mode correlation analysis of the band intensity changes in the C-H stretching and fingerprint regions with respect to the C=O stretching region revealed that they all occur later than the population changes in crystalline and amorphous components reflected by the carbonyl stretching bands. In other words, the spectral intensity changes in these regions are observed after the type I and II crystals are already formed. This discovery suggests that the spectral intensity changes in the C-H stretching and fingerprint regions do not directly represent the population dynamics of specific crystalline and amorphous species during the crystallization but probe instead the molecular state of the already created crystals still undergoing the evolutionary changes.

Funding: This research received no external funding.

Institutional Review Board Statement: Not applicable.

Informed Consent Statement: Not applicable.

Data Availability Statement: Data are contained within the article.

Conflicts of Interest: The author declares no conflict of interest.

References

1. Noda, I. Generalized two-dimensional correlation spectroscopy. In *Frontiers of Molecular Spectroscopy*; Laane, J., Ed.; Elsevier: Amsterdam, The Netherlands, 2009; Chapter 13, pp. 367–381. [CrossRef]
2. Noda, I. Advances in two-dimensional correlation spectroscopy (2DCOS). In *Frontiers and Advances in Molecular Spectroscopy*; Laane, J., Ed.; Elsevier: Amsterdam, The Netherlands, 2018; Chapter 2, pp. 45–75. [CrossRef]
3. Palmerio-Sámchez, T.; O'Flaherty, V.; Lens, P.N.L. Polyhydroxyalkanoate bio-production and its rise as biomaterial of the future. *J. Biotechnol.* **2022**, *348*, 10–25. [CrossRef] [PubMed]
4. Koller, M. Advances in polyhydroxyalkanoate (PHA) production, Volume 3. *Bioengineering* **2022**, *9*, 328. [CrossRef] [PubMed]
5. Dalton, B.; Bhagabati, P.; De Micco, J.; Padamati, R.B.; O'Connor, K. A review on biological synthesis of the biodegradable polyhydroxyalkanoates and the development of multiple applications. *Catalysts* **2022**, *12*, 319. [CrossRef]

6. Kumar, M.; Rathour, R.; Singh, R.; Sun, Y.; Pandey, A.; Gnansounou, E.; Lin, K.-Y.A.; Tsang, C.W.; Thakur, I.S. Bacteria polyhydroxyalkanoates: Opportunities, challenges, and prospects. *J. Clean. Prod.* **2020**, *263*, 121500. [CrossRef]
7. Samui, A.B.; Kanai, T.R. Polyhydroxyalkanoates based copolymers. *Int. J. Biol. Macromol.* **2019**, *140*, 522–537. [CrossRef] [PubMed]
8. Noda, I.; Green, P.R.; Satkowski, M.M.; Schechtman, L.A. Preparation and properties of a novel class of PHA copolymers. *Biomacromolecules* **2005**, *6*, 580–586. [CrossRef] [PubMed]
9. Noda, I.; Lindsey, S.B.; Carraway, D. Nodax^{TM} class PHA copolymers. In *Plastics from Bacteria: Natural Functions and Applications*; Chen, G.-Q., Ed.; Springer: Heidelberg, Germany, 2010; Chapter 10, pp. 237–255. [CrossRef]
10. Nguyen, T.V.; Nagata, T.; Noso, K.; Kaji, K.; Masunaga, H.; Hoshino, T.; Hikima, T.; Sakurai, S.; Yamamoto, K.; Miura, Y.; et al. Effect of the 3-hydroxyhexanoate content on melt-isothermal crystallization behavior of microbial poly(3-hydroxybutyrate-co-3-hydroxyhexanate). *Macromolecules* **2021**, *54*, 8738–8750. [CrossRef]
11. Adar, F.; Street, R.; Noda, I. Isothermal crystallization of polyhydoxyalkanoate (PHA) utilizing Raman spectroscopy to follow chain packing as well as molecular motion. *Spectrochim. Acta Part A Mol. Biomol. Spectrosc.* **2023**, *285*, 121861. [CrossRef] [PubMed]
12. Noda, I. Generalized two-dimensional correlation method applicable to infrared, Raman, and other types of spectroscopy. *Appl. Spectrosc.* **1993**, *47*, 1329–1336. [CrossRef]
13. Noda, I.; Ozaki, Y. *Two Dimensional Correlation Spectroscopy: Applications in Vibrational and Optical Spectroscopy*; Wiley: Chichester, UK, 2004. [CrossRef]
14. Noda, I.; Dowrey, A.E.; Marcott, C.; Story, G.M.; Ozaki, Y.J.A.S. Generalized two-dimensional correlation spectroscopy. *Appl. Spectrosc.* **2000**, *54*, 236A–248A. [CrossRef]
15. Sobieski, B.J.; Noda, I.; Rabolt, J.F.; Chase, D.B. Observation of intermolecular hydrogen bonding interactions in biosynthesized and biodegradable poly[(R)-3-hydroxybutyrate-*co*-(R)-3-hydroxyhexanoate] in chloroform and 1,1,1,3,3,3-hexafluoro-2-propanol (HFIP). *Appl. Spectrosc.* **2017**, *71*, 2339–2343. [CrossRef] [PubMed]
16. Padermshoke, A.; Katsumoto, Y.; Sato, H.; Ekgasit, S.; Noda, I.; Ozaki, Y. Surface melting and crystallization behavior of polyhydrxyalkanoates studied by attenuated total reflection infrared spectroscopy. *Polymer* **2004**, *45*, 6547–6554. [CrossRef]
17. Padermshoke, A.; Sato, H.; Katsumoto, Y.; Ekgasit, S.; Noda, I.; Ozaki, Y. Crystallization behavior of poly(3-hydrxybutyrate-*co*-3-hydroxyhexanoate) studied by 2D IR correlation spectroscopy. *Polymer* **2004**, *45*, 7159–7165. [CrossRef]
18. Padermshoke, A.; Sato, H.; Katsumoto, Y.; Ekgasit, S.; Noda, I.; Ozaki, Y. Thermally induced phase transition of poly(3-hydrxybutyrate-*co*-3-hydroxyhexanoate) investigated by two-dimensional correlation spectroscopy. *Vib. Spectrosc.* **2004**, *36*, 241–249. [CrossRef]
19. Padermshoke, A.; Katsumoto, Y.; Ekgasit, S.; Noda, I.; Ozaki, Y. Melting behavior of poly(3-hydrxyalkanoate) investigated by two-dimensional infrared correlation spectroscopy. *Spectrochim. Acta Part A* **2005**, *61*, 541–550. [CrossRef] [PubMed]
20. Noda, I.; Roy, A.; Carriere, J.; Sobieski, B.J.; Chase, D.B.; Rabolt, J.F. Two-dimensional Raman correlation spectroscopy study of poly[(R)-3-hydroxybutyrate-*co*-(R)-3-hydroxyhexanoate] copolymer. *Appl. Spectrosc.* **2017**, *71*, 1427–1431. [CrossRef] [PubMed]
21. Sato, H.; Murakami, R.; Padermshoke, A.; Hirose, F.; Senda, K.; Noda, I.; Ozaki, Y. Infrared spectroscopy study of CH$\cdots$O hydrogen bondings and thermal behavior of biodegradable poly(hydroxyalkanoate). *Macrolecules* **2004**, *37*, 7203–7213. [CrossRef]
22. Sato, H.; Mori, K.; Murakami, R.; Ando, Y.; Takahashi, I.; Zhang, J.; Terauchi, H.; Hirose, F.; Senda, K.; Tashiro, K.; et al. Crystal and lamella structure and CH$\cdots$O=C hydrogen bonding and thermal behavior of poly(3-hydroxyalkanoate) studied by x-ray diffraction and infrared spectroscopy. *Macrolecules* **2006**, *39*, 1525–1531. [CrossRef]
23. Wang, H.; Tashoro, K. Reinvestigation of crystal structure and intermolecular interactions of biodegradable poly(3-hydroxybutyrate) α-form and the prediction of its molecular property. *Macromolecules* **2016**, *49*, 581–594. [CrossRef]
24. Brela, M.Z.; Boczar, M.; Wójcik, M.J.; Sat, H.; Nakajima, T.; Ozaki, Y. The Born-Oppenheimer molecular simulations of infrared spectra of crystalline poly-(R)-3-hydroxybutyrate with analysis of weak C–H$\cdots$O=C hydrogen bonds. *Chem. Phys. Lett.* **2017**, *678*, 112–118. [CrossRef]

MDPI AG
Grosspeteranlage 5
4052 Basel
Switzerland
Tel.: +41 61 683 77 34

Molecules Editorial Office
E-mail: molecules@mdpi.com
www.mdpi.com/journal/molecules

www.ingramcontent.com/pod-product-compliance
Lightning Source LLC
Chambersburg PA
CBHW061450210926
51850CB00008B/1676
9783725846214